The Biology of *Mallomonas*

Developments in Hydrobiology 63

Series editor
H. J. Dumont

The Biology of *Mallomonas*

Morphology, Taxonomy and Ecology

PETER A. SIVER
Department of Botany, Connecticut College, New London, Connecticut, U.S.A.

KLUWER ACADEMIC PUBLISHERS
DORDRECHT / BOSTON / LONDON

Library of Congress Cataloging-in-Publication Data

Siver, Peter Allan.
The biology of Mallomonas : morphology, taxonomy, and ecology / Peter A. Siver.
p. cm. -- (Developments in hydrobiology ; 63)
Includes bibliographical references and index.
ISBN 0-7923-1166-3 (alk. paper)
1. Mallomonas. I. Title. II. Series.
QK569.M27S58 1991
589.4'87--dc20 91-8231

ISBN 0-7923-1166-3

Published by Kluwer Academic Publishers,
P.O. Box 17, 3300 AA Dordrecht, The Netherlands.

Kluwer Academic Publishers incorporates the publishing programmes of
D. Reidel, Martinus Nijhoff, Dr W. Junk and MTP Press.

Sold and distributed in the U.S.A. and Canada
by Kluwer Academic Publishers,
101 Philip Drive, Norwell, MA 02061, U.S.A.

In all other countries, sold and distributed
by Kluwer Academic Publishers Group,
P.O. Box 322, 3300 AH Dordrecht, The Netherlands.

Printed on acid-free paper

Previously published photomicrographs by the author in the Nordic Journal of Botany (volume 8: 99–107 and 319–323); the Journal of Phycology (volume 26: 728–740); the Canadian Journal of Botany (volume 66: 1391–1403; volume 68: 374–380); the Journal of Paleolimnology (volume 5: 11–17) and the Transactions of the American Microscopical Society (volume 107: 152–161) have been reproduced with permission.

Illustrations: John R. Glew

Printed in the Netherlands

To Scott W. Siver

Table of Contents

Preface

I began work on this project in September, 1983 when I first sampled Bigelow Pond in Connecticut. I referred to the initial project as the Connecticut Chrysophyte Survey, but by 1985 it had expanded to include sites in the Adirondack Mountain region of New York State. Since that first collection in Bigelow Pond over 500 more have been made, the results of which are included in this work. Each of these collections was studied with light and scanning electron microscopy. Eighty days and over 600 hours of SEM time were spent studying the samples at the University of Connecticut. The last two years have been spent compiling the data, preparing the photographic plates and the figures, digitizing thousands of cells, scales and bristles and writing the guts of the chapters. It will seem funny when it is finally completed!

There are a number of people who were instrumental in helping me with completing this task. I especially have to thank Regina who let me work many odd hours during the last seven years when there were 'other duties' to attend to! Regina also accompanied me on most of the field collecting trips; she never thought that the Saturday and Sunday field trips would end. My good friend Bill Quinnell helped with the printing of the pictures and deserves many thanks. Jim Romano and Alan Wachtell at the University of Connecticut allowed me to use the SEM facility. John Glew, my roommate during sabbatical leave at Queen's University, prepared the illustrations and Kate Duff did a lioness share of the typing of the manuscript. Much of the writing was done while I was on sabbatical leave at Queen's University, Kingston, Ontario, in J.P. Smol's laboratory. Many thanks to JPS and his crew for providing me with a stimulating atmosphere from which to work. Special thanks to Dick Nixon for his friendship, encouragement and support. Thanks to F.R. Trainor, B.F. Cumming, J.P. Smol and R. Andersen for reading the manuscript and providing many valuable suggestions.

I set out to accomplish three primary goals in completing this work. First, I wanted to provide a compendium of scanning electron micrographs of the more common taxa of *Mallomonas* which would complement works consisting primarily of transmission electron micrographs. Second, I tried to summarize morphological characteristics of taxa within the genus. This was done in a comparative way for scales, bristles and whole cells. Third, I wanted to characterize each taxon according to its ecological traits. I hope that all of you find the information contained within this book informative and useful in your work.

I want to thank my family, especially my mon and dad, friends and the many teachers who helped me reach this point in my career and profession. This includes such people as my second grade teacher, Mrs. Morrison, who in 1959 helped me identify butterflies; my high school biology teacher, Mr Lavenberg, who first showed me an algal cell; and of course my university mentors, Dr. George Schumacher, Dr. Alan Baker and Dr. F.R. Trainor. Who knows why a person selects a given profession? I do know that I never would have written this book without the guidance and nurturing of many people who sparked an interest in the study and understanding of nature. For me the seed to become a scientist was first planted as a kid collecting insects in the fields of New England and catching fish in the Adirondacks with my father and brothers.

This volume is dedicated to the memory of my younger brother, Scott William Siver, who died several years ago at the age of 29 from liver cancer. During his fight with cancer, Scott maintained a positive attitude and will to live. Despite the treatments and side effects and the pain and suffering, Scott had an incredible spirit that allowed him to rise in the morning and live another day. He even found the time to ask me about the pictures of the microscopic siliceous critters. I can not begin to describe to you how it felt to live with Scott through his illness, but I can tell you that it changed my life. As I sit in the grass in front of the old cabin my family once owned when Scott and I were kids, the spirit that Scott had to face his disease is within me. That spirit not only provided the necessary stimulus to complete this book, but it is within each page of the text. I want to thank Scott for providing me with the drive and the spirit.

June 1, 1990
In front of an old cabin
Fish Creek Ponds
The Adirondacks, NY

PETER A. SIVER

CHAPTER I

Introduction to the Genus *Mallomonas*

General Characteristics

The genus *Mallomonas* Perty 1851 includes unicellular organisms with one bilobed or two chloroplasts, an organized covering of siliceous components and two parallel and apically inserted flagella. The siliceous cell covering consists of individual plates known as scales that are imbricated and arranged in a very precise manner (Takahashi, 1978; Wee, 1982; Asmund and Kristiansen, 1986; Siver and Glew, 1990). A second siliceous component, referred to as a bristle, also forms part of the cell covering. Bristles are elongated structures that are tucked under the distal portion of a scale and radiate outwards from the cell. Because bristles can rotate, their position may change relative to the longitudinal axis of the cell; they are streamlined when the cell is actively swimming and positioned closer to 90° at rest. The scales are arranged in spiral rows around the cell and aligned in a precise fashion (see Chapter II).

Cells are often spherical, ovoid or ellipsoidal in shape; a somewhat streamlined morphology is most common. Most species have cells that range in size from 10 μm to 50 μm · 5 μm to 20 μm, although cells as large as 100 μm (e.g. *M. caudata*) and less than 10 μm (e.g. *M. mangofera* f. *mangofera*) have been reported (see Chapter IV). Generally, one of the flagella is longer, bears two rows of hairs and is easily seen with light microscopy (LM), while the other is smooth, reduced and most often not visible with LM (Belcher, 1969; Hibberd, 1976; Moestrup, 1982; Asmund and Kristiansen, 1986). In *M. splendens*, one of the basal bodies is non-functional or barren in that it lacks a flagellum (Beech and Wetherbee, 1990a; 1990b). The arrangement of microtubule roots associated with the flagellar basal bodies is well summarized by Andersen (1987). Essentially, the R_1 microtubule root, from which other cytoskeletal microtubules originate, has a unique position in *Mallomonas* and *Synura* whereby it is looped in a clockwise manner around the flagella basal bodies (Andersen, 1987; Beech and Wetherbee, 1990b). One set of microtubules originates from the R_1 root and extends along the periphery of the cell just below the plasmalemma, forming an internal cytoskeleton (Andersen, 1987; Beech and Wetherbee, 1990b). A second set of microtules also originates at the R_1 root, but extends downward into the central region of the cell. Beech and Wetherbee (1990b) pointed out that in *M. splendens* the internal cytoskeletal microtubules were in groups of six to eight. According to Beech and Wetherbee (1990b) the R_1 root of *M. splendens* is morphologically similar to that of *S. uvella* (Andersen, 1985), however, the former taxon lacks a second microtubular root. The photoreceptor appears to consist of a pair of flagellar swellings that are not associated with an eyespot or with cell or plastid membranes (Andersen, 1987). Such swellings are absent in *M. splendens*.

The most comprehensive description of the structure of the flagellar apparatus was completed by Beech and Wetherbee (1990b) for *M. splendens*. The two basal bodies of *M. splendens* are parallel, apically inserted in a shallow depression and are connected by two fibrous bands. Basal bodies in other species of Synurophyceae are known to be held together with three connecting bands (Andersen, 1985). The flagellar-bearing basal body in *M. splendens* is connected or anchored to the surface of the cell by a group of fibers. A similar set of connecting fibers is not present on the barren or non-functional basal body (Beech and Wetherbee, 1990b). A thick fibrous capsule, perhaps unique to *M. splendens*, encloses both basal bodies. A well defined and thick rhizoplast originates within the fibrous capsule that surrounds the basal bodies and extends to the nucleus. The rhizoplast becomes splayed into numerous branches that form an inverted cone over the surface of the nucleus (Beech and Wetherbee, 1990b). In *M. splendens*, the fibrous capsule and rhizoplast provide a physical link between the nucleus and the basal bodies and may serve to anchor the latter during flagellar motion.

As has been demonstrated in other algal groups, the flagellum and basal bodies in *Mallomonas* undergo a developmental sequence known as flagellar transformation (Beech and Wetherbee, 1990a). In the case of *M. splendens*, the emergent flagellum is retracted by the cell just prior to cell division and concurrent with the development of two new flagella that arise from newly formed basal bodies (Beech and Wetherbee, 1990a). Each daughter cell eventually receives one new flagellum; the basal body corresponding to the original emergent flagellum of the parent becomes non-functioning or barren. Thus, the non-functioning basal body represents the older of the two.

Most cells have a golden or yellow-brown color as a result of their complement of pigments. The primary accessory pigments are chlorophyll c_1 and fucoxanthin; they lack chlorophyll c_2 (Andersen and Mulkey, 1983; Andersen, 1987). The structure of the chloroplast is similar to other taxa with chlorophyll c in having thylakoids grouped in threes, girdle lamella, and additional membranes surrounding the chloroplast proper, known as chloroplast endoplasmic reticulum (CER).

A rather large nucleus is centrally positioned between the chloroplast and just below a single large Golgi complex (Hibberd, 1976; Beech and Wetherbee, 1990b). A large chrysolaminarin vacuole is often found in the posterior end of the cell. Mignot and Brugerolle (1982) reported that the outer membrane of the nucleus was connected to the CER in the related genus *Synura*. However, other researchers believe that such a confluence is, at best, weakly developed (Andersen, 1985; Hibberd, 1978; Andersen, 1987). Regardless of whether the outer membranes of the nucleus and chloroplast are connected or not, cellular organelles are in fixed positions relative to one another. For example, chloroplasts are held in place within the cell by microtubules originating from the R_1 root. During scale and bristle production, the Golgi apparatus, nucleus and chloroplast(s) are most likely maintained in a fixed position (see below for details). Andersen (1985; 1987) summarized the fixed positions of the two known complexes of organelles.

The production of scales and bristles occurs endogenously and is undoubtedly associated with the Golgi apparatus-nuclear-chloroplast subcellular complex (Andersen, 1987). The site of scale and bristle formation has been shown to be restricted to a region along the outer surface of one of the chloroplasts for *Mallomonas papillosa* (Belcher, 1969), *M. caudata* (Wujek and Kristiansen, 1978), *M. splendens* (Beech et al., 1990) and *Synura petersenii* (Schnepf and Deichgräber, 1969; Mignot and Brugerolle, 1982; Leadbeater, 1990). Each scale and bristle are formed within a vesicle known as a silica deposition vesicle (SDV) that probably originates from the Golgi apparatus. Each SDV becomes oppressed against the anterior and outer surface of the chloroplast endoplasmic reticulum (CER) and is gradually moved along the CER in a helical pathway from the anterior to the posterior of the cell (Mignot and Brugerolle, 1982; Beech et al., 1990). As the SDV is moved along the CER, it is molded into the shape of a mature scale or bristle followed by the deposition of silica to produce the finished product (Wujek and Kristiansen, 1978; Mignot and Brugerolle, 1982; Leadbeater, 1986). Scales and bristles presumably become mature in the posterior end of the cell (Mignot and Brugerolle, 1982). It is well documented that microtubules (Mignot and Brugerolle, 1982; Leadbeater, 1986; Andersen, 1987) and actin-like microfilaments (Brugerolle and Bricheux, 1984; Leadbeater, 1990) are involved in scale and bristle formation, probably serving to position in place and mold into shape the SDV. A glue-like organic substance may be involved in cementing the siliceous components into a completed cell covering (Leadbeater, 1986; 1990; Andersen, 1987). An organic substance is also involved in connecting a bristle to a scale (Beech et al., 1990). Although it is unknown how the scales and bristles ultimately become arranged into the cell covering, hypothetical mechanisms recently proposed by Siver and Glew (1990) for *Mallomonas* and Leadbeater (1990) for *Synura* will be discussed in Chapter II. The formation, deployment and positioning of new posterior bristles for *M. splendens* during cell division (Beech et al., 1990) will also be reviewed.

Knowledge of the processes involved in producing a population of cells is very scarce and a complete life cycle is not known for any species (Asmund and Kristiansen, 1986). All species form a siliceous resting stage known as a stomatocyst, cyst or statospore, often during the declining portion of the growth phase. The stomatocyst is produced within the cell either as a result of an asexual or sexual process (Skuja, 1950; Cronberg, 1986). Sudden changes in environmental conditions (e.g. temperature or nutrient levels) are believed to provide at least one triggering mechanism for the formation of stomatocysts (Cronberg, 1980; Sandgren, 1981; Smol, 1984; Cronberg, 1986). The factors involved in determining what percentage of cells within a population actually encyst are unknown. It seems likely that the percentage would be greater if asexual processes were involved.

Stomatocysts generally range in diameter from ca. 4 μm to 30 μm (Sandgren and Carney, 1983), are spherical or oval, and have a wide variation in structure and ornamentation. Cysts may be smooth, reticulated or sculptured with spines, ridges and depressions of various designs. The morphology of the stomatocyst, especially of the pore-collar complex and surface, is of considerable taxonomic significance (Hibberd, 1977; Skogstad, 1984; Sandgren and Carney, 1983; Sandgren, 1989) and many species probably form species-specific designs (Cronberg, 1986; Sandgren, 1989). However, some species are now known to be able to produce cysts with different surface ornamentation patterns (e.g. *M. caudata*; Cronberg, 1988), while other closely related taxa are known to produce cysts with similar morphologies (Siver, 1991). It is also important to note that the stomatocysts are known for less than half of the species of *Mallomonas* and of these very few detailed studies exist. The lack of knowledge of stomatocysts currently limits their use in ecological studies (e.g. paleolimnological reference work; Smol, 1991).

The siliceous wall of the cyst is formed within a silica deposition vesicle (SDV) which encloses a portion of the cytoplasm, including the nucleus, chloroplast, Golgi apparatus and storage products (Sandgren, 1980; Cronberg, 1986; Sandgren, 1989). The SDV is formed from the fusion of vesicles originating from the Golgi or from a whole distal Golgi cisterna. The hole or pore within the cyst wall, from which the cell will eventually excyst, is formed secondarily and plugged with an organic substance.

The development of the stomatocyst wall is a continuous process, believed to occur in two phases, whereby the inner wall is formed prior to the surface ornamentation and the collar (Sandgren, 1980a; Cronberg, 1980; Skogstad, 1984; Sandgren, 1989). The first phase results in the formation of a thin primary inner cyst wall that is complete except for the pore (Sandgren, 1989). The formation of the primary cyst wall proceeds rapidly within the SDV in a proximal to distal fashion. The wall is subsequently thickened and the collar and surface ornamentation added during the slower and more controlled secondary phase. Thus, immature stomatocysts that lack a full compliment of ornamentation can be found. In addition, the degree of ornamentation of the stomatocyst is known to be influenced by the physiological state of the organism during encystment (Sandgren, 1983). Clearly, much more research is needed in order to fully document the degree of variation in the structure of the stomatocyst that could be found for individual taxa.

Hypothetically, a cyst sinks to the bottom of the waterbody where it remains until conditions trigger it to germinate (excystment). Many of the details concerning the actual process of germination of stomatocysts are unknown. Likewise, it is unclear what percentage of the cells in a population produce stomatocysts, what percentage of the cysts in the sediments germinate and whether cysts actually germinate while in the sediments or are first redistributed into the water column. Regardless of the details of germination, a naked flagellated cell would be released via the pore, after dissolution of the organic plug, and eventually produce a siliceous armour (see proposed mechanism, Chapter II). Once the cell is mature it can presumably undergo vegetative cell division.

At a given instant in time, the relative proportions of the cells of a population that formed by direct germination of cysts as opposed to cell division are not known. It seems likely that a larger percentage of the individuals of a population in the early stages of development may be formed from the germination of stomatocysts; in later stages progressively larger percentages of cells develop through cell division.

The ecological as well as taxonomical implications of questions concerning the life cycle of species of *Mallomonas* as posed above are many and will require much future work. A simple and highly hypothetical example of how the dynamics of the cell cycle may influence the ecology of a species as interpreted by the scientist is illustrated below. If we assume that the germination of cysts takes place within the water column, then the number of cysts that germinate is dependent, in part, on water movements and, thus, indirectly on climatic conditions. A particularly stormy spring period may mix more stomatocysts into the photic zone and provide a larger 'seed' source for the subsequent development of a population. If the growth of the population does not become limited by factors such as nutrient supply or light levels (i.e. the population does not reach the carrying capacity) during the period of the year when other factors (e.g. temperature) are favorable, then the maximum concentration of cells observed in the lake during that year is directly related to the number of cysts that were able to germinate at the start of the season. Obviously, the larger the size of the resultant population for a given year, the larger will be the subsequent deposition of scales, bristles and newly produced cysts into the sediments. It is clear that any interpretation made concerning the ecology of the species based directly on biomass measurements of the live population or the resultant paleo-population must include the effect of turbulence on cyst germination. In summary, only with further work will the influences of the life cycle of an organism on its resultant ecology become better understood.

Cytokinesis in *Mallomonas* proceeds along the longitudinal axis of the cell, beginning from the anterior end, and requires a short period of time (e.g. minutes) to complete (Conrad, 1927; Harris, 1953; Wawrik, 1979; Beech and Wetherbee, 1990a; Beech et al., 1990). Initial results indicate that the rhizoplast of *Mallomonas* may serve as a spindle organizing center and that the nuclear envelope is not persistent during mitosis (Beech and Wetherbee, 1990b). Although rare, sexual reproduction and the formation of zygotes has been observed in several species (see Asmund and Kristiansen, 1986 for a review). Essentially, the cells (hologametes) make apical or caudal contact, fuse and produce a zygote.

Taxonomic Position

Historically, the taxa with two discernible heterodynamic flagella, which also shared other morphological characteristics, were placed into the genus *Mallomonopsis* Matvienko (Matvienko, 1941; Wujek and Timpano, 1984). The primary difference between the two genera was related to the number of emergent flagella; cells of *Mallomonas* had one flagellum whereas those of *Mallomonopsis* possessed two (Wee, 1982; Wujek and Timpano, 1984). With further study it was found that all species have two flagella (see Andersen, 1987) prompting Belcher (1969) to recommend merging *Mallomonopsis* with *Mallomonas*. Such a proposal has been subsequently followed in the taxonomic works of Momeu and Péterfi (1979) and Asmund and Kristiansen (1986). According to Asmund and Kristiansen (1986), species with a discernible second flagellum are placed within the Sections Mallomonopsis, Multisetigerae and Papillosae.

Until recently, the genus *Mallomonas* (and *Mallomonopsis*) had been placed with the colonial genus *Synura* and several other poorly understood genera (e.g. *Conradella*) in the family Mallomonadaceae Diesing of the class Chrysophyceae (*sensu* Hibberd, 1976; 1986). However, based on major differences in biochemical and ultrastructural characteristics, Andersen (1987) removed the Mallomonadaceae from the class Chrysophyceae and established the new class, the Synurophyceae.

Bourrelly (1957) based identification of species of *Mallomonas* on morphological features of both the cells and scales as observed with LM. Harris and Bradley (1957, 1960) further emphasized the morphologies of scales and bristles, especially using electron microscopy (EM), and divided the genus into four major series: the Tripartitae, Planae, Quadratae and Torquatae. Several species, referred to as 'Isolated species', could not be placed into one of the four primary series. In the subsequent works of Péterfi and Momeu (1976), Takahashi (1978), Momeu and Péterfi (1979), Wee (1982) and Asmund and Kristiansen (1986) taxonomy of the genus *Mallomonas* became progressively more dependent on the ultrastructure of the siliceous armour as viewed with EM. Today, species are no longer considered valid unless they are based on EM descriptions (Wee, 1982; Asmund and

Kristiansen, 1986).

Momeu and Péterfi (1979) raised the four series of *Mallomonas* proposed by Harris and Bradley (1957; 1960) to the rank of section in accordance with the International Code of Botanical Nomenclature. By 1986, 115 taxa of *Mallomonas* in 16 sections were recognized using EM (Asmund and Kristiansen, 1986). Since the Asmund and Kristiansen (1986) monograph, at least 19 additional taxa (Gretz et al., 1985; Dürrschmidt, 1986; Croome and Tyler, 1986; Nicholls, 1987a; Nicholls, 1987b; Nicholls, 1988a; Nicholls, 1988b; Siver, 1988a; Siver and Skogstad, 1988; Siver et al., 1990) and one section (Siver, 1988a) have been described.

Although the current taxonomic system based on the ultrastructure of scales and bristles is artificial (Asmund and Kristiansen, 1986), it represents the only reliable way to distinguish between different taxa. Such a system has resulted in some difficulties and confusion concerning the older, pre-EM literature. The major problem is that many of the older descriptions can not be linked to those based on EM. It is most likely true that many taxa described with LM have been redescribed with different names using EM. Species that have not been described with EM are no longer considered valid by most workers in the field, however, several recent works (e.g. Starmach, 1985) still include both sets of names. For example, only about half of the taxa of *Mallomonas* listed in Starmach (1980) are known using EM (Asmund and Kristiansen, 1986). The older names based on cell morphology are accordingly still used by scientists not familiar with the current taxonomy.

Most species descriptions of *Mallomonas* are based on transmission (TEM) vs. scanning (SEM) electron microscopy. Fewer are based on both TEM and SEM or only SEM (Asmund and Kristiansen, 1986). Such a distribution of descriptions is most likely the result of the widespread availability of TEM as opposed to SEM instruments during the 1960's and 1970's when many of the species were originally described.

Asmund and Kristiansen (1986) made the statement that 'Descriptions based on TEM are usually the most complete ones ...'. Although some structural features of scales could be missed using SEM, I believe that careful analyses of scales, bristles and whole cells using SEM reveals a greater amount of information than obtained from TEM analysis. Thus, I disagree that TEM analyses are more complete than those made with SEM. In particular, the surface detail, the three dimensional structure and the arrangement of components on the cell covering are clearly more evident with SEM. I further believe that descriptions based on SEM, especially when it is obvious that the species is a new one, are as valid as those based on only TEM.

Descriptions of scales with TEM often make reference to an 'internal structure'. Often such interpretations are not true internal structures, but rather the result of the surface features and differences in thickness between regions on the scale. The 'internal structure' may simply be an artifact of the TEM image, that is, a result of a reduction in dimension from three-space to two-space. For example, within the description of the series Allorgeanae of the section Quadratae, Asmund and Kristiansen (1986) discuss '... an internal layer of honeycomb structure ...'. On TEM micrographs of scales of *M. lychenensis* such a honeycomb pattern is clearly evident, however, the large surface papillae covering the shield are virtually undetected. It seems possible that the more electron dense honeycomb pattern associated with the shield as seen with TEM is, in part, the result of the large papillae. If the papillae are hollow, as they presumably are, then the reticulation of the honeycomb pattern may simply represent the spaces between the papillae and the sides of the papillae, whereas the more electron transparent areas correspond to the papillae themselves. Another example of a 'false impression' often viewed in TEM images is the idea that some scales have a ridge along the posterior border of the dome; in reality the image of a ridge is often the result of the slope of the dome creating a more electron dense region in two-space. In summary, I believe that it is easier to visualize the TEM image of a scale from a SEM micrograph than the reverse; that is, an image in three-space can be reduced to two-space, but, an image in two-space cannot, with certainty, be made into three-space. As such, it is often impossible to determine if a TEM image of a scale is being viewed from the top or bottom, unless some other distinctive feature is known (e.g. the structure of the dome).

Viewing both TEM and SEM images will provide the maximum amount of information on the true structure of the cell covering. Such a practice will be especially useful for beginning taxonomists. To date, most of the taxonomic literature for *Mallomonas* emphasizes TEM. One purpose of this work is to provide a compendium of SEM photographs to accompany the existing literature based on TEM.

CHAPTER II

Structure of the Cell Covering

Scale Morphology

Scales from all species consist of a base plate and a posterior upturned rim (Figs. 2.1-2.2). The base plate is most often perforated with minute pores, referred to as base plate pores, although in a few species the pores are inconspicuous and perhaps lacking (e.g. Sectio Papillosae). Base plate pores may be spaced evenly over the scale, lie concentrated in certain regions (e.g. the posterior of the shield) or be lacking altogether from specific areas (e.g. flanges). For most species the dome is devoid of such pores. The posterior margin of the scale is upturned and bent (recurved) back over the scale forming a rim (Figs. 2.1-2.2). Posterior rims may be narrow or quite broad and usually encircle about one-half of the perimeter of the scale. The posterior rim usually extends an equal distance around both sides of the scale, however, for scales of some species (e.g. *M. retrorsa*) and often for scales positioned close to the ends of the cell, the rim may extend further along one margin (usually the right margin) resulting in an asymmetric design. The degree to which the rim overlaps the scale is herein referred to as the canopy of the rim (Fig. 2.2). The end of the scale with the rim is referred to as the proximal end, whereas the front end is designated as the distal end.

Scales of some taxa (e.g. *M. caudata*) consist simply of a perforated base plate with a posterior rim, or the same with a secondary layer of silica deposited in a specific pattern (e.g. *M. bronchartiana*) over the scale. However, scales of the majority of species have a prominent V-shaped ridge of silica, referred to as the V-rib, positioned such that the base of the V lies in the proximal region and the sides (or arms) of the V-rib extend towards the distal end of the scale (Figs. 2.1-2.2). The arms

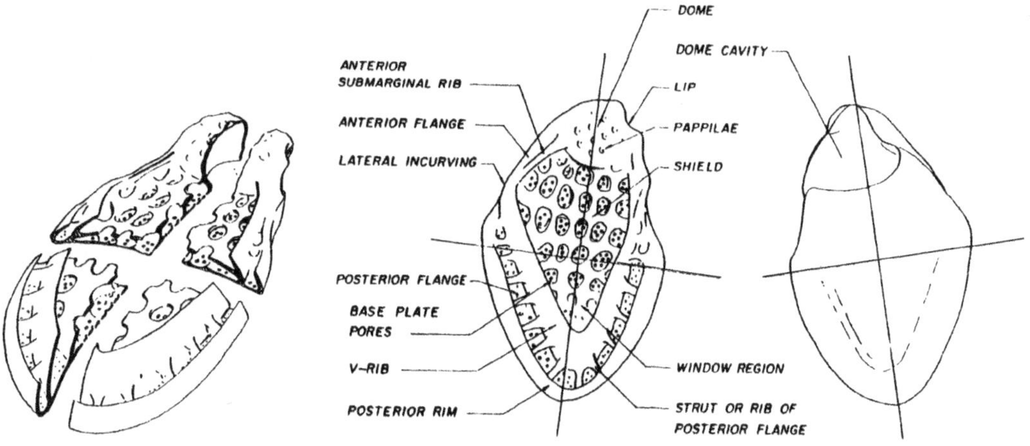

Fig. 2.1. Planar and three-dimensional drawings of a typical scale possessing a dome, posterior rim, V-rib and anterior submarginal ribs. Top and undersurface views are shown for the two-dimensional drawing. The three-dimensional figure represents the same scale divided along perpendicular axes. Note the dome cavity, the canopies formed by the posterior rim and the V-rib and the secondary ribbing covering the shield.

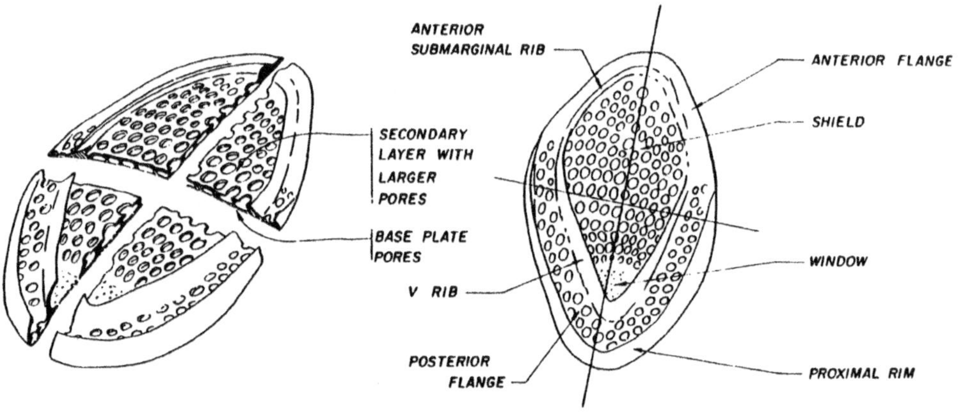

Fig. 2.2. Planar and three-dimensional drawings of a typical domeless scale possessing a posterior rim, V-rib and anterior submarginal ribs. The three-dimensional figure represents the same scale divided along perpendicular axes. Note the fusion of the anterior submarginal ribs in the distal end of the scale, the canopies formed by the posterior rim and the V-rib and the window region.

of the V-rib may terminate close to or at the perimeter of the scale (e.g. *M. cratis*), usually midway along the longitudinal axis, or extend to the dome (see below) (e.g. apical scales of *M. tonsurata*). Scales of many taxa possess two additional ribs, known as anterior submarginal ribs, that originate near the distal ends of the V-rib arms, extend more or less parallel to the margins of the scale and terminate in the distal region (Figs. 2.1-2.2). Most often, species whose scales are ornamented with a V-rib also possess anterior submarginal ribs and the junction of the two sets of ribs is of taxonomic significance. One of two situations is commonly found: 1) The proximal end of each anterior submarginal rib fuses close to or at the distal ends of the arms of the V-rib forming a distinct angle (e.g. *M. acaroides* var. *acaroides*, Fig. 3.52B); or 2) the distal ends of the arms of the V-rib curve and appear continuous with the anterior submarginal ribs (e.g. *M. elongata*, Fig. 3.34A). On domeless scales the distal ends of the anterior submarginal ribs usually meet and fuse or become extended into spines on posterior scales (e.g. *M. crassisquama*, Fig. 3.46D). On domed scales they either terminate along the lateral sides of the dome or continue partially or totally around the front end of the dome (e.g. anterior domed scale of *M. galeiformis*, Fig. 3.36D).

In a manner similar to that described for the posterior rim, the V-rib, especially near its base, extends over a portion of the scale forming a second canopy (Fig. 2.1); the extent of the canopy varies greatly between taxa. For many species the arms of the V-rib are commonly longer than the anterior submarginal ribs. Some species lack V-rib structures *per se*, but have analogous ribs (e.g. *M. punctifera* and *M. canina*); these are discussed in greater detail in Chapter III.

Scales that are associated with a single bristle (see below) commonly have a structure known as a dome (Figs. 2.1 and 2.3). A dome is a raised portion of the distal end of the base plate under which the proximal end of the bristle, or foot, is tucked. The shaft of the bristle emerges from an inverted U-shaped opening (referred to throughout this book as the U-shaped opening) located slightly to the right of center along the distal border of the dome (Fig. 3.26C); this latter feature imparts a slight asymmetry to the scale. The foot of the bristle fits under the dome in a ball and socket fashion such that the emergent shaft can rotate through a wide angle with the longitudinal axis of the cell. On some scales the rim of the inverted U-shaped opening, referred to as the lip, may be directed forward (Fig. 2.1). On apical scales, the left portion of the dome may also protrude forward forming a small wing-like structure. The range in height, surface area, shape and ornamentation of the dome varies greatly between taxa.

The V-rib and anterior submarginal ribs, collectively referred to as the submarginal rib, serve to divide the scale into distinct regions, that, in turn, are often differentially ornamented. The region bounded by the submarginal rib is referred to as the shield and that outside of the submarginal rib the flange (Figs. 2.1-2.2). The flange is further

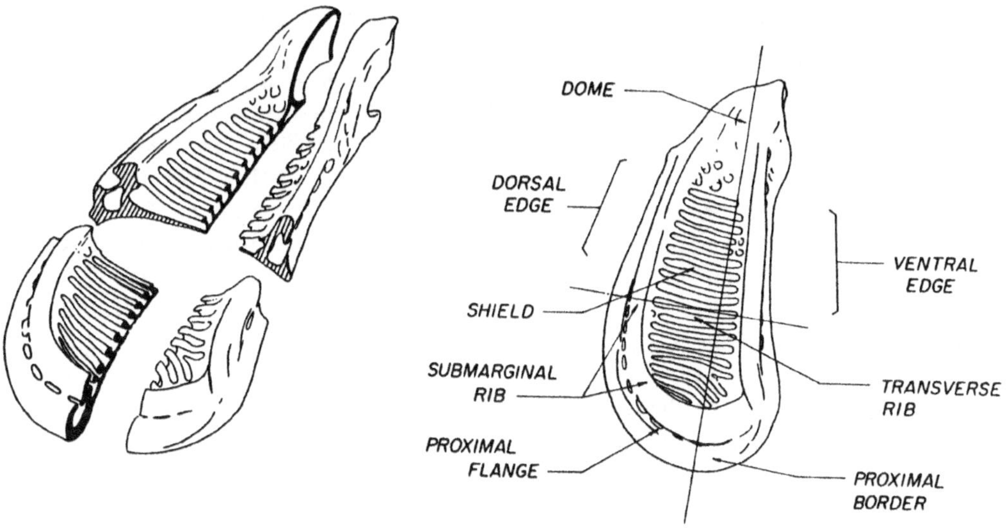

Fig. 2.3. Planar and three-dimensional drawings of a domed collar scale from the Sectio Torquatae. The three-dimensional figure represents the same scale divided along perpendicular axes. Note the asymmetrical design of the scale, the dorsal and ventral margins and the submarginal rib. These scales possess a single bristle which is tucked under and exits from the right hand margin of the dome.

subdivided into posterior and anterior flanges. The posterior flange is the area of the scale between the V-rib and posterior rim. The anterior flanges are the regions between the anterior submarginal ribs and the margin of the scale. The ornamentation of the shield and flange areas is most often of critical taxonomic importance.

The size, morphology and symmetry of scales varies greatly between taxa, within a given species, and even on an individual cell. Scales may range in length from less than 1 μm (e.g. rear scales of *M. dickii*, Fig. 3.57E) to over 10 μm (body scales of *M. pseudocoronata*, Fig. 3.51C), in surface area from less than 1 μm^2 (e.g. rear scales of *M. dickii*, Fig. 3.57E) to over 50 μm^2 (body scales of *M. pseudocoronata*, Fig. 3.51C), and in form factor from 0.1 (e.g. rear scales of *M. akrokomos*, Fig. 3.24D) to 1.0 (e.g. body scales of *M. caudata*, Fig. 3.7D) (see Chapter III for details). The planar shapes of scales are most often circular, elliptical, ovate, obovate or rhombic in nature. With the exception of collar scales, most exhibit essentially bilateral symmetry. However, except for domeless body scales, most are at least slightly asymmetric due to the shape of the dome, posterior rim and/or V-rib. Scales on the extreme ends of a cell usually exhibit the most asymmetry. Some scales possess lateral incurvings just above the distal ends of the posterior rim (Fig. 2.1). Most scales are not perfectly flat, but rather curved or bowed such that the inner concave surface conforms to the shape of the cell. In general, the larger the scale in relation to the size of the cell, and the thicker the scale, the greater is the degree of curvature (e.g. *M. lychenensis*, Fig. 3.11).

Many organisms have scales with features known as secondary structures (Takahashi, 1978; Wee, 1982; Asmund and Kristiansen, 1986). Secondary structure is herein defined as any siliceous material deposited over the base plate on the outer part of the scale. As such, the V-rib and anterior submarginal ribs, but not the dome *per se*, are secondary in nature. On scales of some taxa, such as *M. elongata*, the V-rib and anterior submarginal ribs are the only secondary structures. However, most species have additional secondary ornamentation, such as ribs and papillae, on the shield, flanges and/or dome (Figs. 2.1-2.3).

Secondary ribbing may be regularly (e.g. *M. asmundiae*, Fig. 3.26D) or irregularly (e.g. *M. heterospina*, Fig. 3.21) arranged, the resultant pattern related to the arrangement of pores on the base plate (Asmund and Kristiansen, 1986). Scales with an ordered arrangement of ribs most often have a regular arrangement of pores. The more random the pores, the more irregularly arranged are the ribs (Asmund and Kristiansen, 1986). Ribs, as well as other secondary structures, are positioned on the base plate between pores. The degree of ribbing on the shield and flanges may be minimal (e.g. *M. acaroides* var. *acaroides*, Fig. 3.52B) or so extensive as to form a reticulum of pores or meshes over the base plate (e.g. *M. crassisquama*, Fig. 3.46B). Scales usually lack secondary material at the base of the V-rib, a region referred to as the window (Figs. 2.1-2.2).

Papillae are often found on the shield, dome, sometimes the anterior flanges, but not on the posterior flange. Papillae located on the shield are normally regularly spaced, vary in density from low (e.g. *M. ouradion*, Fig. 3.2C) to high (e.g. *M. bronchartiana*, Fig. 3.12F), and are solid (e.g. *M. papillosa*, Fig. 3.4C) or hollow (e.g. *M. lychenensis*, Fig. 3.11B) in nature.

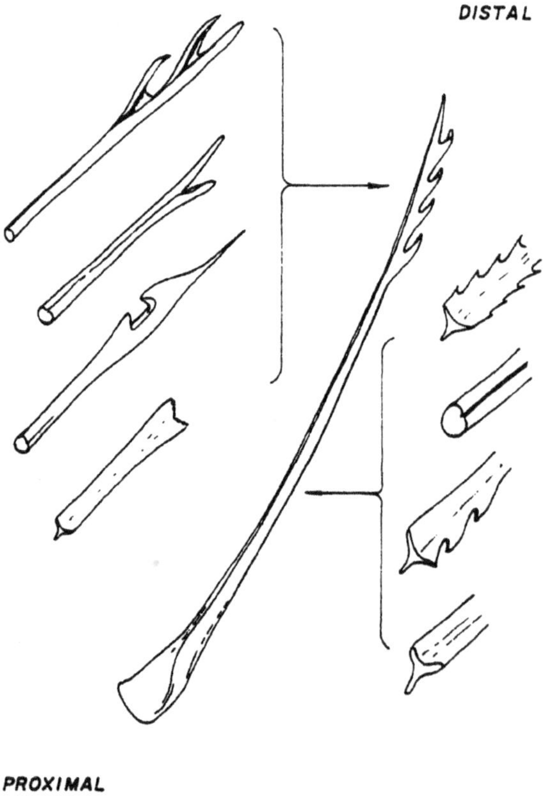

Fig. 2.4. Three-dimensional drawing of a typical bristle illustrating four different morphologies for both the distal tip and the shaft.

Bristle Morphology

Like scales, the morphologies of bristles are of great taxonomic importance, especially for taxa with similar scales. Bristles are divided into two regions, the foot and shaft (Fig. 2.4). The proximal portion of the bristle represents the foot, which is the end tucked under the distal end of the scale. The foot is most often flat and bent at a 30° to 90° angle relative to the shaft (Fig. 2.4). The shaft is usually curved, smooth or ribbed in nature and often serrated along the convex margin. For some species, the bristles appear to be rolled up (Wujek and Kristiansen, 1978; Mignot and Brugerolle, 1982) such that a slit occurs along the shaft at the point of convergence along the two folded margins. The slit may run the length of the shaft or be restricted to one end. Bristles with incompletely rolled-up shafts and toothed margins are referred to as craspedodont (Asmund and Kristiansen, 1986). Many species have shafts that are folded or ridged along their longitudinal axis (Fig. 2.4). One or two ridges is most common, although some taxa have bristles with three ridges (e.g. *M. punctifera*, Fig. 3.14A). Often a bristle with a slit or ridge appears smooth when viewed from the opposite side. The shafts are usually serrated along only one ridge, although some species are serrated along several ridges (e.g. *M. punctifera*, Fig. 3.14A). The teeth of the serration may be pointed (e.g. *M. corymbosa*, Fig. 3.35D), blunt (e.g. *M. caudata*, Fig.3.7E), recurved (e.g. *M. crassisquama*, Fig. 3.46F) or in turn bifurcate or trifurcate (e.g. *M. asmundiae*, Fig. 3.26B).

Like the shaft, the morphology of the bristle tip is also of taxonomic significance. Most often the distal tip is drawn out into a sharp point, although a few taxa have bristles with blunt tips (e.g. *M. hamata*, Fig. 3.18A). Often a second sharp tip, located in a subapical position, results in a forked or bifurcate appearance (e.g. *M. tonsurata*, Fig. 3.39). The subapical tip often represents the distal tip of one of the longitudinal ridges of the shaft that has folded over and terminated below the distal most tip (e.g. *M. tonsurata*, Fig. 3.40D; *M. corymbosa*, Fig. 3.42.B). Sometimes the subapical tooth is equivalent to the distal most tooth of the serration; bristles where the segment between the tip and the distal most tooth of the serration is the same length as the distal most tooth are referred to as pseudobifurcate (Asmund and Kristiansen, 1986). In other species the subapical tooth may be positioned on the side of the shaft opposite of the serration (e.g. *M. corymbosa*, Fig. 3.42B). Some bristles have a swollen or expanded distal end with a C-shaped or cleft-like opening; these bristles are referred to as helmet (e.g. *M. crassisquama*, Fig. 3.46E) or hooked (e.g. *M. heterospina*, Fig. 3.21C) bristles. Hooked bristles, found on species of the Sectio Heterospinae, are small, have extremely delicate shafts and greatly expanded C-shaped distal tips. A few taxa (e.g. *M. corcontica*, Fig. 3.27C) have a swollen and folded distal tip referred to as a lance bristle. Other rarer types of modifications, such as plume (e.g. *M. plumosa*; Croome and Tyler, 1983) and paddle (e.g. *M. retrorsa*; Siver, 1988a) bristles have been described.

Species may possess one or two morphologically distinct types of bristles that may be differentially distributed on a given cell (e.g. *M. hamata*, Fig. 3.17) or with respect to an environmental gradient (e.g. *M. crassisquama*). Siver and Skogstad (1988) found that cells of *M. crassisquama* had primarily serrated or helmet bristles when found in cold or warm waters, respectively. It is quite common for species to associate a smaller type of bristle with their apical scales and a longer, morphologically distinct, type of bristle with their body scales (e.g. *M. tonsurata*, Fig. 3.39C). However, a few species have a reversed arrangement; *M. hamata* has a morphologically smaller type of bristle distributed over its posterior region and a longer type of bristle associated with its apical and body scales. Still other taxa possess a random distribution of their bristle types.

The distribution of bristles on the cell is a characteristic of taxonomic significance. Most often bristles are either distributed over the entire cell (e.g. *M. asmundiae*, Fig. 3.25), restricted to the anterior one-half to two-thirds of the cell (e.g. *M. elongata*, Fig. 3.32) or are associated solely with apical scales (e.g. *M. dickii*, Fig. 3.56). Regardless of their distribution, bristles are free to rotate such that they commonly become streamlined when the cell is swimming and directed outward when the cell is at rest. Bristles can rotate because they are not physically attached to the scales. In most species the foot of the bristle is hinged within the cavity of the dome, similar to a ball-and-socket joint, and the shaft emerges from the inverted U-shaped opening. Scales with domes are associated with at most one bristle; scales that are associated with more than one bristle generally lack a dome (e.g. *M. caudata*). Bristles, especially for those scales lacking a well developed dome, may be held in place, in part, with a glue-like substance (Wujek and Kristiansen, 1978).

Arrangement of Scales on the Cell Surface

All species of *Mallomonas* have their scales arranged in overlapping spiral rows as to form a well ordered imbricated pattern (Fig. 2.5). The rows of scales are always spiralled to the right; that is, when traced from the posterior to the anterior of the cell each row of scales is skewed to the right around the cell. Each scale of the cell coat is overlapped by the scale positioned behind it within the same spiral row and by the scale(s) in the spiral row above it (Figs. 2.5-2.6). Thus, although scales within a given spiral row are overlapped in a bottom up manner, the spiral rows themselves are overlapped in a top down fashion (Siver and Glew,

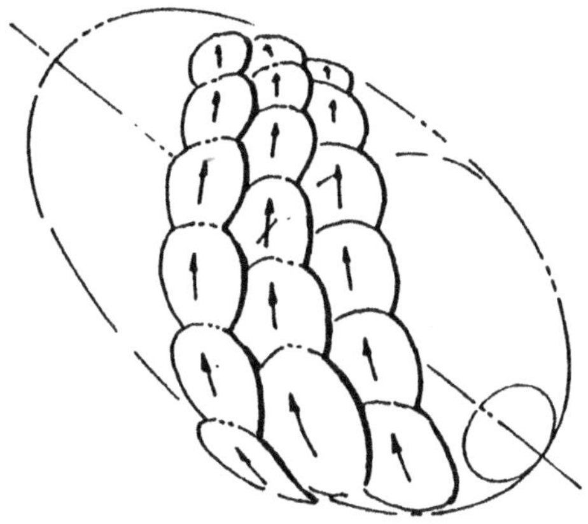

Fig. 2.5. A diagrammatic reconstruction of three spiral rows of scales on a cell of *Mallomonas* illustrating the pattern in which the scales are overlapped. Each scale is overlapped by the scales positioned behind it in the same spiral row and by those in the spiral row above it. This is an oblique posterior end view.

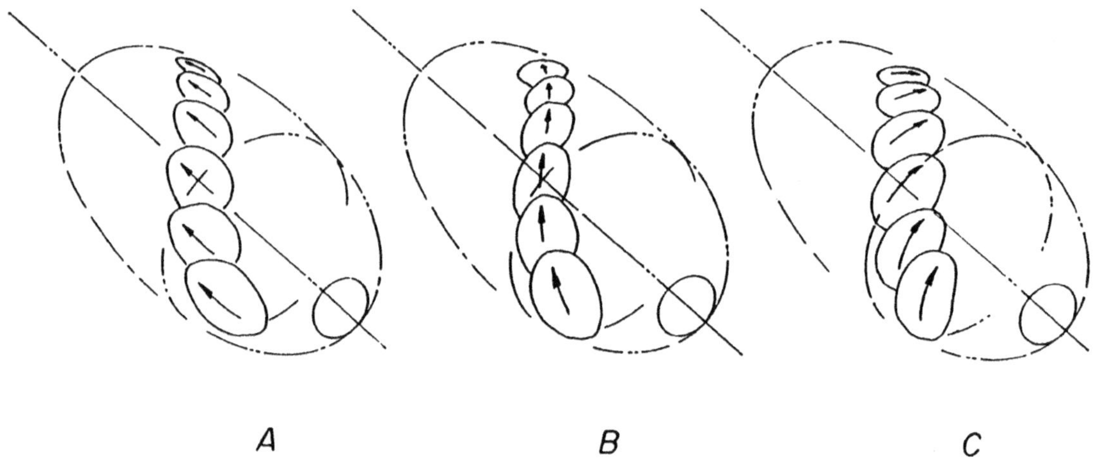

Oblique posterior end views

Fig. 2.6. The three basic ways in which scales are orientated on cells of *Mallomonas*. Scales are positioned with their longitudinal axes parallel (A), at an oblique angle (B), or perpendicular (C) to the longitudinal axis of the cell. Only one spiral row of scales is illustrated. The drawings represent oblique posterior end views.

1990). The only known exceptions to this pattern of scale overlap are *M. akrokomos* and *M. retrorsa*. In *M. akrokomos* the overlapping of scales within a given spiral row is identical to all other species of *Mallomonas*, however, adjacent spiral rows are overlapped in a posterior to anterior manner, which is opposite to that found in all other taxa. Such an overlapping of adjacent spiral rows is similar to that observed for *Synura* (Leadbeater, 1986; 1990; Siver and Glew, 1990). On the other hand, in *M. retrorsa* the pattern of overlap of adjacent spiral rows is similar to all other species of *Mallomonas*, but the overlapping of scales within a given row is opposite (Siver, 1988a; Siver and Glew, 1990). That is, each scale within a spiral row is overlapped by the scale positioned in front of it within the same spiral row. The number of spiral rows of scales and the tightness of the spirals varies between species, but is believed to remain relatively stable for a given taxon. Commonly, cells have from five to seven rows of scales.

Scales are aligned within a spiral row such that their longitudinal axes are parallel (Fig. 2.6A), perpendicular (Fig. 2.6C) or at an oblique angle (Fig. 2.6B) to the longitudinal axis of the cell. Scales are orientated such that their distal ends point towards the anterior of the cell (parallel arrangement), at 90° to the right (perpendicular arrangement) or at some angle, usually 40° to 60°, in between these two positions (oblique arrangement). Scales of *M. retrorsa* and most likely *M. fenestrata*, are arranged with their longitudinal axes parallel to the longitudinal axis of the cell; however, the distal ends of the scales are orientated in a backwards fashion, such that they face the posterior of the cell (Siver, 1988a). No species have been described with scales positioned such that their distal ends face to the left of the longitudinal axis of the cell.

It should be realized that the arrangement and alignment of scales within and between spiral rows is best observed using scanning electron microscopy (SEM). Care must be taken when interpreting a transmission electron micrograph (TEM) as the image may represent scales that are upside down.

For most species approximately 40 to 50% of the surface area of a given scale is overlapped by neighbouring scales. The posterior rim and posterior flange (if present) are always covered; the V-rib and sometimes the posterior region of the shield are also often overlapped. Most or all of the shield, anterior flanges and dome (if present) are exposed on intact cells (see Chapter IV). The distal end of a scale that overlaps the left side of the posterior flange of another scale terminates approximately midway along the longitudinal axis of the latter scale, such that the scales appear to also be positioned in a brickwork-like or escilon pattern. Thus, the domes of scales on a cell could also be viewed as being in a zig-zag alignment. This arrangement allows for the maximum distance between the points of emergence of the bristles and, thus, for the maximum angle of rotation of the bristles. Because of the brickwork-like pattern, scales from neighbouring spiral rows also become aligned in rings such that the number of scales within a ring is equal to the number of spiral rows of scales and the number of rings is equal to the number of scales within a spiral row.

Each spiral row of scales contains at least one of each type of scale found on a given cell. Thus, the anterior and posterior most scales of a spiral row would be represented by an apical and caudal scale, respectively. The number of spiral rows of scales equals the number of apical scales; Leadbeater (1990) pointed out that such may not be the case for cells of *Synura*. Since the apical scales form a precise pattern around the flagellar pore (Fig. 3.39D), they represent the apical most ring of scales. There is either an abrupt or gradual change in scale morphology along a given spiral row between apical and body scales and between body and posterior scales . Normally, the most abrupt change is found between the apical and body scales, especially for taxa within the Sectio Torquatae.

Proposed Mechanism for the Formation of the Cell Covering

The following discussion is a summary of the hypothetical mechanism recently proposed by Siver and Glew (1990) for the formation of the cell covering for *Mallomonas*. Ideas and observations recently proposed by Leadbeater (1990) and Beech et al. (1990) are also discussed.

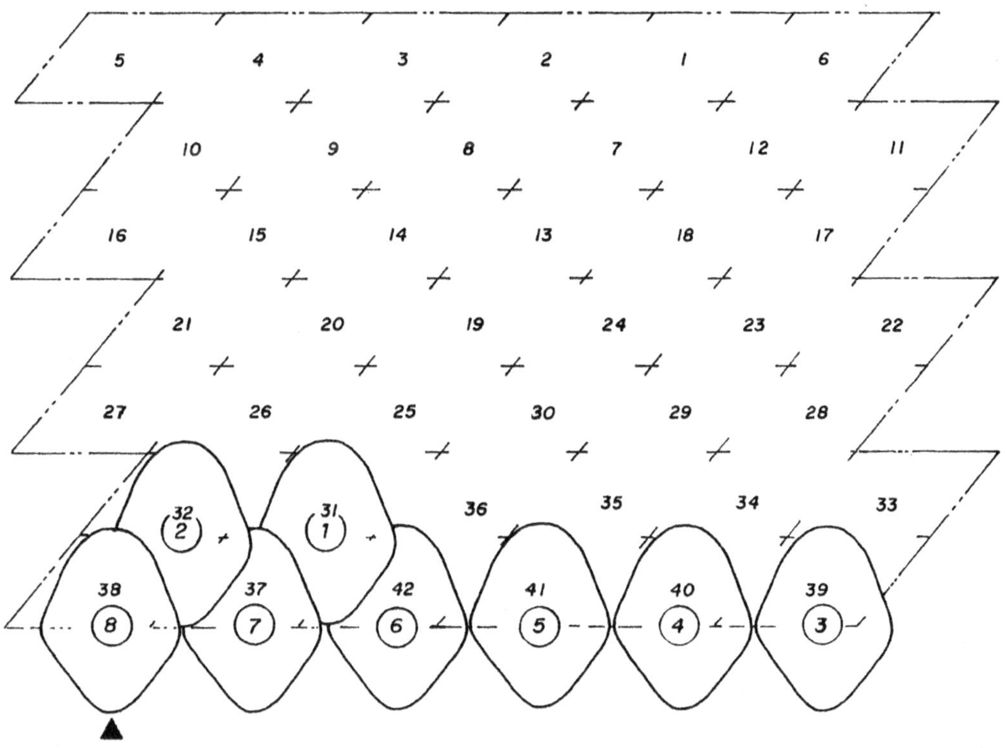

Fig. 2.7. A proposed pattern for the formation of a cell covering consisting of six spiral rows of 7 scales each for a total of 42 scales. The initial positions of the first eight scales upon release are represented by the larger numbers. The first six scales are ejected sequentially from the same site (marked by the solid triangle), each displacing the previous one laterally forming a ring. The seventh and eighth scales are then ejected such that they slide over scales one and two, respectively, and displace them towards the anterior of the cell. The smaller numbers correspond to the final positions of the 42 scales.

The mechanism proposed by Siver and Glew (1990) is based, in part, on three assumptions concerning the release of scales and bristles from the cell. First, scales and bristles are ejected from the cell at a fixed position close to where they become mature, presumably in the posterior region of the cell between the plasmalemmae and the chloroplast endoplasmic reticulum (CER). Second, the maturation and release of scales are a continuous 'assembly-line' process. Soon after a scale becomes mature it is ejected from the cell. Then another SDV, containing the next mature scale, moves into position at the ejection site. Third, as scales progress along the outer surface of the chloroplast and reach the point of ejection, they are all oriented in the same manner.

According to the hypothesis, if a cell has six spiral rows of scales, then the first six scales will be ejected in succession from the same point such that the most recently released scale displaces the previous one laterally, forming a ring around the cell (Fig. 2.7). The seventh scale is then ejected such that it overlaps the right posterior rim of scale number 1 and displaces the latter forward and to the right (Fig. 2.7). Ultimately, the first and seventh scales become aligned within the same spiral row. Scale number 8 is then ejected such that it overlaps and displaces scale number 2 and also comes to overlap the left posterior rim of scale number 7. Note that scale number 7 is positioned in the spiral row located below the one that contains scale number 8 (Fig. 2.7). If the cell had seven scales per spiral row, it would eventually eject 42 scales that would become positioned as illustrated in Figs. 2.7 and 2.8.

Practically all species of *Mallomonas*, especially those with the most precisely arranged cell armours, have scales with a V-rib. It was further proposed by Siver and Glew (1990) that the V-rib, and perhaps the posterior rim, acts as a guide in the proper spacing of the scales within the final cell covering and as a means by which a scale could dislodge the one in front of it. In the above example, scale number 7 is ejected and slid over scale number 1 until its left anterior border meets and abuts the right-hand side of the V-rib on scale number 1 (Fig. 2.7). The continued ejection of scale number 7 causes scale number 1 to move to an anterior position. In this manner the dome, anterior flanges and majority of the shield of each cell remain exposed on an intact cell.

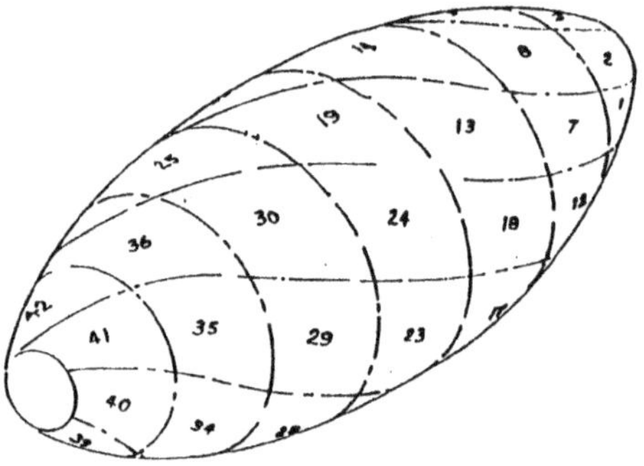

Fig. 2.8. A three-dimensional model of the scale pattern described in Figure 2.7. The posterior end is at the bottom left.

Individual scales and bristles are produced in separate silica deposition vesicles (Wujek and Kristiansen, 1978; Mignot and Brugerolle, 1982; Beech et al., 1990). Siver and Glew (1990) further proposed that for species where the bristle is associated with a dome, the ejection of the bristle occurs simultaneously with that of the scale, such that the proximal foot becomes positioned within the cavity of the dome upon release. Accordingly, bristles could be pushed out of the posterior end of the cell, the distal end first, with the angled proximal foot being fitted under the dome. If the proximal foot was not positioned upon release it could presumably be lost from the cell (see below). Thus, the scale and its associated bristle become a unit as they are released, regardless of their final position on the cell. Each scale-bristle unit would be subsequently moved into its final position upon the further ejection of scales. The inverted U-shaped opening from which the bristle emerges is situated to the right of center on the dome. Since the left distal margin of a scale is involved in pushing the scale positioned above it, the emergent bristle does not interfere with the sliding and movement of the scales.

Because the scales of all species of *Mallomonas* and *Synura* have a similar arrangement on the cell, it is believed that a universal mechanism for the formation of the cell covering exists (Siver and Glew, 1990). In 1986, Leadbeater proposed a partial hypothesis for *Synura petersenii*; a more complete model was presented in 1990. Besides the work of Leadbeater and the Siver and Glew (1990) model discussed above, no satisfactory proposal has been put forth to explain how the highly organized cell covering is initially produced by a naked cell. As noted above, Beech et al. (1990) described the formation, deployment and positioning of new posterior bristles for cells of *M. splendens*, already with cell armour, during cell division (see below). Siver and Glew (1990) believed that a mechanism should be able to explain how an excysted naked cell from a stomatocyst could produce a completed cell armour as well as how new spiral rows of scales could be positioned on recently divided cells. However, the mechanisms for both processes may in fact be different.

Siver and Glew (1990) suggested that the cell covering of a naked cell could be produced according to one of two basic hypotheses. For both hypotheses it was assumed that scales became mature in the posterior end of the cell, after they had been moved along the CER. According to the first hypothesis scales and bristles located in the posterior region would be moved around within the cell, positioned in the correct orientation and ejected at the point where they will be fitted into the finished cell covering. The second basic hypothesis is one where the scales are released from the cell and moved into position over the outside of the cell membrane.

The mechanism proposed by Leadbeater (1986) for the development of the cell coat in *Synura petersenii* esentially follows hypothesis number 1. Leadbeater (1986) suggested that scales were moved into proper position and alignment within the cell prior to their release. He further suggested that scales could be added to allow for expansion of the cell by spreading (unglueing) existing scales apart and inserting the new ones. According to Siver and Glew (1990) the insertion of single scales in this manner for *Mallomonas* does not seem probable because it would result in a disruption of the ordered arrangement of the scales in much the same way that a single brick could not be added to a completed brick wall. The addition of a scale to an existing ordered pattern would result in either too many scales at one point around a circumference of the cell or an increase in length of only one spiral row. Elongating one spiral row would require having to unglue and dislodge all the scales within the spiral in one direction.

In order to add scales according to the Leadbeater (1986) hypothesis and maintain an ordered and imbricated pattern, either an entirely new spiral row must be added or a similar type of scale must be added simultaneously to each spiral row. In either case, the cell would have to manufacture the correct number of scales along the outer surface of one chloroplast, move them around the inside of the cell, precisely align each scale, open the existing cell coat by ungluing the scales, simultaneously eject the new scales into place and reglue the new armour.

Siver and Glew (1990) listed a number of difficulties with hypothesis number 1. First, it would be mechanically difficult for a cell to move, position and align a group of scales and bristles into place on the inside of the cell without causing damage to organelles. For example, the length of body scales of *M. pseudocoronata* can reach almost half of the length of the cell. Recently, Leadbeater (1990) also suggested that for *Synura* a redistribution of mature scales within the cytoplasm was unlikely. Second, according to hypothesis number 1, multiple sites of ejection are needed. Third, the cell membrane would have to be severed along the entire longitudinal (if inserting a row) or median (if inserting a scale per row) axis. Fourth, a mechanism for ungluing and separating existing scales is needed.

The sliding mechanism proposed by Siver and Glew (1990) supports hypothesis number 2 where scales are moved into position along the surface of the cell simultaneously as they are released. The sliding mechanism overcomes several difficulties associated with hypothesis number 1. First, since scales and bristles, especially large ones, are moved around on the outside of the cell, there is substantially less chance of damaging organelles or their subcellular arrangements. There is some evidence that scales can be moved about over the outer cell surface during cytokinesis (Beech et al., 1990). Second, it suggests a single site for the release of all siliceous components. Third, the cell would require but a single template in order to properly align scales before their release (e.g. all scales positioned with their longitudinal axes at 45° to the longitudinal axis of the cell). It was further suggested that such a template may simply be the position in which the SDV is attached to the CER prior to release of the scale. Fourth, the ejection of scales from a posterior position would account for the observed fact that a scale within a spiral row is overlapped by the one positioned behind it. In order for such a pattern to develop according to hypothesis number 2 a posterior release of scales is necessary. It follows that scales would become mature close to their site of release. Fifth, the force needed to eject a scale becomes coupled with that needed to move scales into position on the cell covering. Sixth, the sliding mechanism is not in conflict with the idea that the cell would produce and release all of one type of scale (and associated bristle) before systematically switching to the production of the next type of scale found along the cell covering. For example, if a cell had six spiral rows of scales, then it would produce the six scales of the apical ring first, followed by a switch to the formation of body and ultimately posterior scales. Lastly, it offers an explanation as to why all species of both *Mallomonas* and *Synura* have scales with a posterior rim; the majority of species of *Mallomonas* also have a V-rib. On the other hand, scales from species of Paraphysomonadaceae that have a more unorganized coat of scales lack upturned rims.

Siver and Glew (1990) further proposed that a mature cell with a complete siliceous armour would not increase its number of scales, at least through the random addition of scales. Instead, they proposed that the cell could accommodate an increase in size through an expansion of the existing scale coat such that the degree of overlap between neighbouring scales was decreased. Only a slight expansion of the scales would be needed to significantly add to the volume of the cell. Such an expansion of the scales is often observed during formation of a stomatocyst. For the genus *Mallomonas*, the percentage of the surface area of a scale that remains exposed on an intact cell ranges from about 40% to 60%, and it is not uncommon for cells of a given population to have a difference range of up to 10% (see Chapter IV). Thus, room for expansion of the cell coat to accommodate growth of a cell exists within cells of a population.

Carter (1937) and Harris (1953) reported that groups of scales could become detached from the cell covering and subsequently replaced. Siver and Glew (1990) suggested that such a process would require that the cell have a 'map' of its scale covering and believed that it was, at best, a rare process.

Working with *Synura*, Leadbeater (1990) recently proposed that a spiral row of SDV's become aligned along the CER in the exact sequence in which they will become aligned in the scale coat. He further proposed that the developing scales become mature and are extruded in a posterior to anterior manner. Since the scales mature close to where they will be extruded, minimal movement of the scales within the cell is necessary, thus overcoming the primary difficulty with hypothesis number 1. This mechanism does not assume that all scales become mature in the posterior region of the cell as do the two hypotheses outlined above.

Cells of *M. splendens* are rather large and possess four apical and four posterior bristles (Asmund and Kristiansen, 1986; Beech and Wetherbee, 1990a). Several hours preceeding mitosis, cells of *M. splendens* produce and deploy four additional posterior bristles, resulting in cells that temporarily have eight posterior bristles and four anterior bristles (Beech et al., 1990). The corresponding new posterior base plate scales are produced and secreted after the bristles. Cells then undergo cytokinesis in an anterior to posterior manner, resulting in daughter cells each with four posterior and two anterior bristles (Beech et al., 1990). Each daughter cell produces two additional anterior bristles soon after cytokinesis.

The deployment and positioning of the new posterior bristles prior to mitosis by *M. splendens* is done in a most unusual way as described by Beech et al. (1990). The bristles are extruded from the cell with their basal ends first. As they are secreted, a cellular protuberance becomes associated with the proximal or basal ends. Once the bristles are fully extruded and their distal tips free from the cell, the bases of the bristles are pulled back to the cell. The process takes from 30 to 90 seconds to complete and is accompanied with a shortening of the protuberance. Anterior bristles are secreted with their distal tip first and do not require a reorientation like the posterior bristles.

In summary, much more information is needed before we will fully understand how species of *Mallomonas* build their siliceous cell coats.

CHAPTER III

Taxonomic Descriptions

Sectio Mallomonopsis (Matvienko) Asmund & Kristiansen

Series Matvienkoae Asmund & Kristiansen

Mallomonas matvienkoae f. *matvienkoae* (Matvienko) Asmund and Kristiansen

Synonym: *Mallomonopsis elliptica* Matvienko

Cells are obovoid to elliptical in outline, have two emergent flagella of unequal length, and are covered with bristles (Figs. 3.1A,B). The scales, which are arranged in spiral rows with their longitudinal axes positioned 60° to 90° to that of the cell, are not divided into distinct morphological types. Rather, all scales have the same basic morphology. Each scale is overlapped by the scale positioned behind it in the same spiral and by the scale(s) in the spiral row above it.

Scales are oval to elliptical in shape, have a base plate perforated with small evenly spaced pores and a posterior upturned rim encircling about one-half of the perimeter of the scale (Figs. 3.1C-E). Scales lack a dome and submarginal rib structure. A thick rib surrounds the distal half of the scale (Figs. 3.1D-E). Scales of this taxon are relatively easy to identify with LM because the distal one-half to two-thirds of the scale appears opaque (Harris, 1966; Asmund and Kristiansen, 1986). The opaque region is due to the development of a secondary layer or reticulum (Figs. 3.1D-E). The secondary layer consists of a series of ribs that originate on the base plate in the spaces between each pore and expand laterally a short distance above the base plate to form a more or less smooth and continuous covering with pores. The pores in the secondary covering are closely aligned with those of the base plate such that the rows of pores on the secondary layer are continuous with those of the base plate in the posterior region (Fig. 3.1E). Along the proximal border of the secondary layer of the scale illustrated in Fig. 3.1E a few base plate pores can be observed with only rudimentary ribs around them.

The construction of the secondary layer could result in a TEM image with three different degrees of electron density. Each pore of the secondary layer would represent the least electron dense region; the ribs originating from the base plate the most electron dense; and the expanded canopy of the ribs as an intermediate density surrounding each pore (e.g. Fig. 7d of Asmund and Kristiansen, 1986; Fig. 16 of Kristiansen, 1989). The same type of TEM image is often observed associated with the secondary covering on scales of some species of *Synura* (e.g. Figs. 171 and 174 of Wee, 1982).

A very distinctive feature of this taxon is the presence of a large slightly raised (or sometimes sunken) pore in the proximal region of the scale (Figs. 3.1C, E). One population collected from a small pond close to East Lake in Wuhan, P.R. of China had scales with two such pores (Fig. 3.1D). Kristiansen (1989) also observed a population where the scales had two large proximal pores; interestingly this population was also collected from a small pond (perhaps the same one) in Wuhan.

The shafts of the bristles are relatively straight and smooth except for a slit that runs more or less the length of the bristle (Figs. 3.1A-B). This slit becomes expanded (unrolled) to varying degrees near the distal tip. The distal tip is either blunt or bifurcate (Figs. 3.1A-B; 3.5A); the latter type being more often found on bristles where the slit has expanded significantly. The proximal foot is quite large, flat and usually bent at a 45° to 90° angle with the shaft. Bristles tend to be shorter near the apical end of the cell.

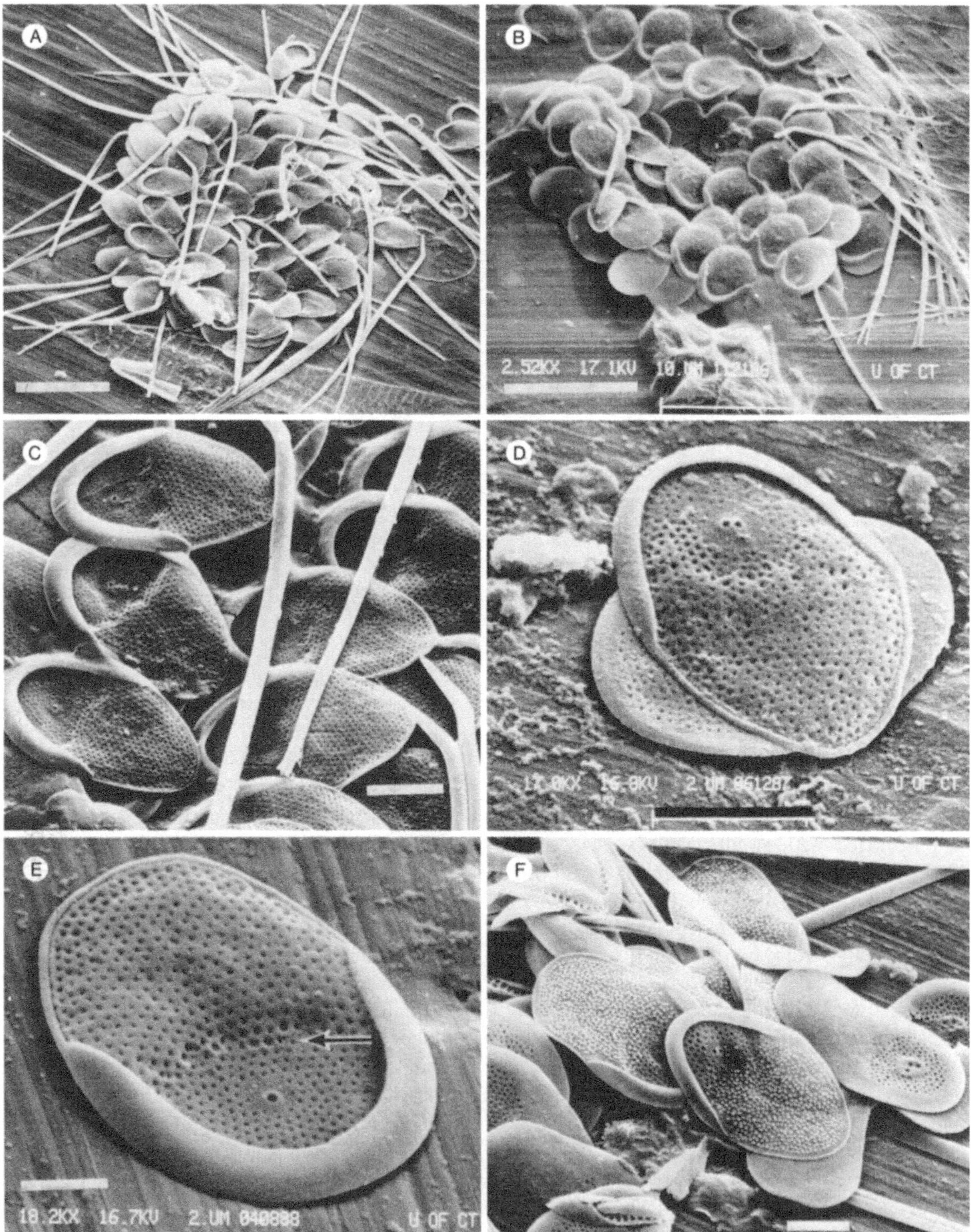
A
B
2.52KX 17.1KV
U OF CT
C
D
U OF CT
E
18.2KX 16.7KV 2.UM 040888
U OF CT
F

Mallomonas matvienkoae var. *myakkana* Siver var. nov.

Several populations of cells had scales with major differences from those of the type and are described here as a new variety, var. *myakkana*.

Diagnosis:
Squama in parte posteriore tres vel quinque poras majores exhibente, in parte anteriore strato secundario papillarum tecta, hic poris carente a var. *matvienkoae*.

Die 20 Martii anni 1986 in lacu Lower Mayakka areae Myakka River State Park regionis Sarasota civitatis americanae Florida sub numero CCS 251 a Scott Siver lecta, Fig. 3.1F typifica monstrata.

Scales of var. *myakkana* differ from the type in three ways: 1) they have a cluster of 3-5 large pores in the proximal region; 2) the secondary reticulum is replaced with a rather dense covering of papillae; and 3) the distal half of the base plate lacks pores. The structure of the bristles is similar to that of the type (Fig. 3.1F). I believe the differences from the type for this taxon are substantially greater than those separating f. *litteata* Dürrschmidt (Dürrschmidt, 1983b) from the type and therefore have given it variety status.

The type locality is Lower Lake Mayakka, for which the epithet refers, located within the Myakka River State Park, Sarasota, Florida, U.S.A. Specimens were collected on March 20, 1986 and March 16, 1987.

Mallomonas oviformis Nygaard

Synonyms: *Mallomonopsis elliptica* Matvienko var. *oviformis* (Nygaard) Harris

Mallomonopsis oviformis (Nygaard) Kristiansen

Cells are elliptical to ovoid with two emergent unequal flagella and are covered with bristles. Scales are positioned on the cell in the same manner as those in *M. matvienkoae*.

Scales are oval to elliptical, have a perforated base plate (Harris, 1966; Wee, 1982; Asmund and Kristiansen, 1986) and a proximal rim encircling one-half of the cell (Fig. 3.2A). The distal border of each scale consists of a thickened rib with a row of elliptical shaped papillae along its top surface; the papillae may give the distal edge a serrated appearance (Fig. 3.2A). Scales lack a dome and submarginal rib structures. Except in the zone adjacent to the proximal rim, the surface of the scale is covered with densely spaced papillae of varying sizes (Fig. 3.2A). The outer few rows of papillae that border the scale are regularly arranged, but the remainder form more irregularly arranged or wavy rows. A thin secondary layer, on which the papillae are situated, appears to cover the surface of the scale; because of this layer the pores of the base plate are not seen on SEM micrographs.

In TEM micrographs the central papillated region is observed to consist of short vermiform ribs, presumably formed by the fusion of papillae (Harris, 1966; Wee, 1982; Asmund and Kristiansen, 1986). The vermiform rib pattern is absent along the distal margin and the region next to the proximal rim (e.g. Fig. 66 of Takahashi, 1978; Fig. 14 of Harris, 1966). Also, in TEM micrographs the vermiform ribs usually mask the presence of the surface papillae resulting in the idea that the papillae are restricted to the distal margin (Wee, 1982). The vermiform pattern seen with TEM is

←

Fig. 3.1A-E. Mallomonas matvienkoae f. *matvienkoae*. Fig. A. Whole cell illustrating the arrangement of scales and bristles. Scale bar is 10μm. Fig. B. Portion of the cell coat. Note the bifurcate distal ends of the bristles. Scale bar is 10μm. Fig. C. Body scales. Close-up of the cell illustrated in Fig. A. Scale bar is 2μm. Fig. D. Body scales. Note the secondary layer covering the distal two-thirds of the shield, the thick rib surrounding the distal half of the scale and the presence of two large proximal pores. Scale bar is 2μm. Fig. E. Isolated body scale with one large proximal pore. Note that the pores in the secondary covering (distal portion of the shield) are aligned with those of the base plate seen in the proximal portion of the shield). A few base plate pores surrounded with only rudimentary ribs are seen (arrow). Scale bar is 1μm. Fig. F. *Mallomonas matvienkoae* var. *myakka*. Body scales showing the three main differences with f. *matvienkoae*: (1) a cluster of 3 to 5 large proximal pores; (2) a secondary layer of closely spaced papillae on the distal portion of the shield; and (3) lack of base plate pores on the distal half of the scale. Scale bar is 2μm.

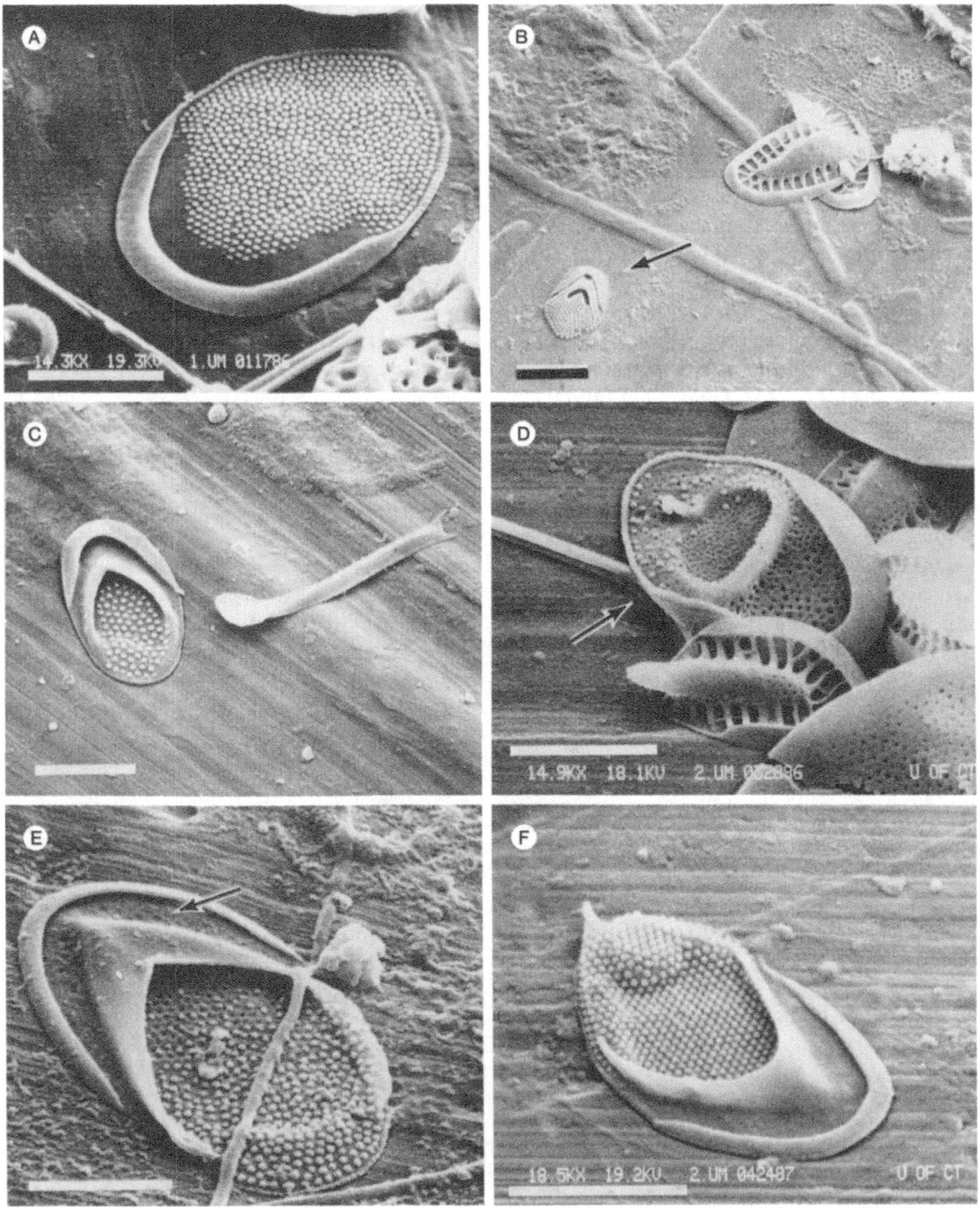
A
14.3KX 19.3KV 1.UM
B
C
D
14.9KX 18.1KV 2.UM
U OF CT
E
F
18.5KX 19.2KV 2.UM 042487
U OF CT

not observed in SEM images, but is undoubtedly partly the result of the closely spaced papillae. I believe that if the thin secondary layer of material was to differentially collapse between the papillae, it could result in the vermiform pattern (e.g. Fig. 13 of Harris, 1966).

In light microscope (LM) preparations scales have an appearance similar to that observed in *M. matvienkoae* with respect to the transmission of light (Harris, 1966). A U-shaped zone of greater light transmission separates the more opaque regions occupied by the proximal rim and the covering of papillae.

Bristles are most often short, slightly curved and with a bifurcate distal tip associated with an expansion of the longitudinal slit.

Series Ouradiotae Asmund & Kristiansen

Mallomonas ouradion Harris and Bradley

Synonym: *Mallomonopsis ouradion* Harris and Bradley (Harris)

Although only samples with isolated scales and bristles were observed in this study, cells have been reported as being ellipsoidal or ovoid and having scales each with one or more bristles (Harris and Bradley, 1958; Harris, 1966; Asmund and Kristiansen, 1986).

Scales are elliptical and symmetrical, with a posterior rim encircling less than half of the perimeter of the scale and a V-rib (Fig. 3.2C). Scales lack a dome and it is clear using SEM that pores do not penetrate the base plate (Fig. 3.2C). The shield is marked with evenly spaced papillae that become smaller and more widely spaced in the posterior region. The posterior flange is smooth and unornamented.

Bristles are short, thick, slightly curved with a wide 'spatula-like' proximal end and a bifurcated distal tip (Fig. 3.2C).

Isolated scales and bristles of *M. ouradion* could be mistaken for those of *M. calceolus* Bradley and *M. multisetigera* (Table 3.1). Except for the presence of a dome, scales of *M. calceolus* are virtually identical to those of *M. ouradion*. Scales of *M. multisetigera* are similar to those of *M. ouradion* in having papillae on the shield that become progressively smaller in size towards the posterior region and in often possessing more than one bristle (Fig. 3.2E). However, scales of *M. multisetigera* can be distinguished from those of *M. ouradion* in being more obovate (especially anterior scales) and having base plate pores, partially winged anterior submarginal ribs and a dome. Because the dome on scales of *M. multisetigera* is papillated in a similar fashion as the shield and is only slightly raised, it may appear inconspicuous. (See the discussion under *M. multisetigera*). Differences between scales of *M. ouradion* and the related taxa are summarized in Table 3.1.

Mallomonas parvula Dürrschmidt

Cells are small, spherical and covered with bristles (Dürrschmidt, 1982a). Most scales are obovate or rhomboid, small, lack base plate pores and a dome, have a broad posterior rim and a distinctive V-rib (Fig. 3.2B). Each arm of the V-rib is straight and consists of two parallel ridges giving the appearance of a 'double' V-rib. The shield is marked with a series of rather large papillae that cover all but a V-shaped window region in the angle

←

Fig. 3.2. Mallomonas oviformis. Body scale with a proximal rim and covered with papillae, except for the region adjacent to the rim. Scale bar is 2μm. Fig. B. *Mallomonas parvula.* Body scale (arrow). Note the pair of parallel ribs forming the 'double' V-rib, the V-shaped window devoid of papillae and the circular pit at the proximal border of the V-rib. Scale bar is 2μm. Fig. C. *Mallomonas ouradion.* Body scale and bristle. Note the V-rib and shield papillae on the scale, and 'spatula-like' proximal end and bifurcate distal tip of the bristle. Scale bar is 2μm. Fig. D. *Mallomonas bangladeshica.* Body scale (arrow). Note the centrally positioned transverse rib which forms the proximal border of a ring-like structure and the thickened rim surrounding the distal border. The diameters of the pores within the ring structure are smaller than those in the proximal region. Scale bar is 2μm. Fig. E. *Mallomonas multisetigera.* Apical scale. Note the extension of the left anterior submarginal rib into a small wing, the progressively smaller and fewer papillae on the proximal portion of the shield, the strongly hooded V-rib and the faint appearance of depressions on the posterior flange (arrow). Scale bar is 2μm. Fig. F. *Mallomonas paxillata.* Scale with papillae covering the shield, anterior flanges and dome, but lacking on the posterior flange. Note the extension of the left anterior submarginal rib into a sharp spike. Scale bar is 2μm.

Table 3.1. Summary of the major diagnostic features of scales and bristles as seen with SEM and used in this study to distinguish between several closely related taxa with papillated scales.

Taxon	Pores of base plate	Lateral incurvings	Anterior flanges	Posterior flanges
M. ouradion	Do not penetrate the base plate	-	Narrow, lacking ornamentation.	Smooth.
M. multisetigera	Present, more conspicuous in posterior of shield.	-	Narrow with papillae.	Smooth with base plate pores. Faint outline of subcircular depressions present.
M. calceolus	Do not penetrate the base plate.	-	Narrow, lacking ornamentation.	Smooth.
M. paxillata	Do not penetrate the base plate.	+, slight	Narrow, papillated.	Smooth.
M. papillosa	Do not penetrate the base plate. One large pore in posterior region of shield close to the angle of the V-rib.	-	Marked with series of parallel struts.	Smooth.
M. rasilis	Do not penetrate the base plate.	-	Narrow, marked with equally spaced papillae.	Smooth.

Table 3.1. continued

Dome features	Shield features	Anterior submarginal ribs	No. of bristles per scale	Bristles
–	Papillae smaller and more widely spaced in posterior region.	Inconspicuous.	1 to 5	Short, slightly curved, smooth shaft, bifurcated tip.
Small or broad and asymmetrically placed. Somewhat inconspicuous due to its shallow cavity. Ornamented with papillae.	Papillae smaller and more widely spaced in posterior region.	Ornamented with papillae; may be winged.	Anterior scales: 1–5 Body scales: 1	Short, slightly curved, shaft with crenulated edges, bifurcated tip.
Small, usually smooth.	Papillae are few and widely spaced.	Weakly developed.	1	Short, slightly curved, smooth shaft, bifurcated tip.
Small, covered with papillae in same density as on the shield.	Papillae dense and evenly spaced over entire shield.	Present, left one often extending forward past the dome forming a spine.	1	Short, slightly curved, smooth shaft, bifurcated tip.
Small, triangular, smooth of partially covered with papillae	Evenly spaced papillae, but density and area covered differs between varieties (see text).	Well develop. Terminate on sides of dome.	1	Short, slightly curved, serrated, lacks bifurcated tip.
Small, usually with out papillae.	Papillae dense and equally spaced over entire shield.	Lacking	1	Longer than others, slightly curved, unilaterally serrated, lacks bifurcated tip.

of the V-rib (Fig. 3.2B). The window is distinctive and easily recognized on both SEM and TEM micrographs. The posterior flange is covered by the broad hood of the posterior rim, except for a circular depression located at the proximal border of the V-rib.

Bristles are short, slightly curved, needle-like with a bifurcated distal tip (Dürrschmidt, 1982a).

Isolated scales are rather easy to identify due to the distinctive double V-rib and V-shaped window.

Series Peronoides Asmund & Kristiansen

Mallomonas bangladeshica (Takahashi & Hayakawa) Wujek & Timpano

Synonym: *Mallomonopsis peronoides* Harris var. *bangladeshica* Takahashi & Hayakawa

Cells are oblong to ellipsoidal and covered with bristles (Wujek and Timpano, 1984; Asmund and Kristiansen, 1986). Only isolated scales were observed in this study.

Scales are oval with a broad proximal rim, a perforated base plate and a centrally positioned transverse ridge (Fig. 3.2D). The diameter of the base plate pores increases in the posterior region of the scale. The proximal rim is continuous with a thickened rib that encircles the anterior half of the scale (Fig. 3.2D). Scales lack a true dome.

The shield is ornamented with an oval shaped ring of secondary material positioned on the anterior end (Fig. 3.2D); the thick transverse ridge forms the proximal border of the ring structure. A secondary layer is lacking on the inside of the ring, forming a subcircular depression which appears on TEM micrographs as an electron transparent region (e.g. Wujek and Timpano, 1984). The edge of the proximal transverse ridge, which may represent a modified V-rib, is often toothed (Fig. 3.2D). On the distal border of the subcircular depression is a raised umbrella-like or 'grapnel-like' (Wujek and Timpano, 1984) structure with 9-13 lobed arms. Wujek and Timpano (1984) reported that the grapnel-like structure on anterior and posterior scales may be less developed (e.g. Fig. 3.2D). The anterior end of the shield is usually marked with scattered papillae (Fig. 3.2D).

Bristles are thin, needle-like with an abrupt decrease in width near the distal tip (Wujek and Timpano, 1984).

Except for the presence of the raised grapnel-like structure, scales of *M. bangladeshica* are virtually identical to those of *Mallomonas peronoides* (Harris) Momeu and Péterfi. The grapnel-like ornament is also very similar to one produced on scales of *Mallomonas stellata* Cronberg.

Mallomonas cf. *stellata* Cronberg

Several isolated scales (Fig. 3.5C) were believed from a cell of *M. stellata*, a rare taxon reported only once from Zimbabwe. The morphology of the scales was very similar to that of *M. matvienkoae* f. *matvienkoae*, except for the presence of a sphere-like appendage on the distal end of the scale (Fig. 3.5C). The margin of the appendage was slightly undulated and may represent the finger-like projections on the star-like structure reported in the original description. Papillae were lacking.

Sectio Multisetigerae Asmund & Kristiansen 1986

Mallomonas multisetigera Dürrschmidt

Although whole cells were not observed, they have been described as being ellipsoidal, covered with bristles and possessing scales that are considerably overlapped and spirally arranged with their longitudinal axes perpendicular to the longitudinal axis of the cell (Dürrschmidt, 1982b). Each apical scale usually possesses more than one bristle.

Scales are of two types, body and apical scales. Body scales are oval, symmetrical with a small dome, while anterior scales are obovate with a broad and asymmetrically positioned dome (Dürrschmidt, 1982b). The shield and sometimes the posterior flange are perforated with minute pores that are most obvious in the posterior region of the shield (Fig. 3.2E). An additional series of large circular pores lining the posterior flange are seen on TEM micrographs (Takahashi, 1978; Nicholls, 1982; Dürrschmidt, 1982b; Asmund and Kristiansen, 1986). However, with SEM these structures appear as subcircular depressions that do not penetrate the base plate (Fig. 3.2E).

The V-rib is strongly hooded with arms that curve and become continuous with the anterior submarginal ribs. On apical scales, the left submarginal rib is often extended into a small wing-like structure (Fig. 3.2F). The shield, dome and anterior submarginal ribs are ornamented with papillae, the size and density of which decreases in the posterior region of the shield. Because the domes are rather shallow and ornamented in a similar fashion as the shield, they are rather inconspicuous when viewed with SEM. Except for the faint appearance of the subcircular depressions, the posterior flange is unornamented.

Bristles are short, slightly curved, have slightly crenulated edges and a bifurcated distal tip (Dürrschmidt, 1982b).

Although isolated bristles are not readily identified, with careful analyses scales of *M. multisetigera* can be distinguished from those of *M. ouradion*, *M. calceolus*, *M. paxillata* and other related taxa using the SEM (Table 3.1) or TEM.

Sectio Papillosae Asmund & Kristiansen 1986

Mallomonas conspersa Dürrschmidt

Cells are small and ovoid, range in length from 8 to 15 μm, and are covered with bristles (Dürrschmidt, 1986). Although whole intact cells were not observed with SEM, scales are spirally arranged (Dürrschmidt, 1986) and presumably overlapped in a similar manner as in other species of *Mallomonas*. Scales are small, tripartite in nature and of one of two types; apical domed scales with a forward projecting wing (Fig. 3.3D) or domed body scales lacking the wing (Figs. 3.3 C,E,F). All scales on a given cell are of approximately the same size (Dürrschmidt, 1986). There are about five to six apical scales surrounding the proximal end of the emergent flagellum.

Scales are elliptical to suboval in outline and possess a perforated base plate, dome, V-rib and a proximal rim encircling one-half of the perimeter. The V-rib is hooded, has a broadly rounded base and arms that are continuous with the anterior submarginal ribs (Figs. 3.3 C-E). On apical scales the left anterior submarginal rib is slightly winged and projects past the dome, forming a sharp apex (Fig. 3.3D). On body scales the anterior submarginal ribs are wingless, broad, ornamented with papillae and extend to the lateral sides of the dome (Fig. 3.3D); they are less prominent and more poorly delimited as compared to those on the apical scales. The dome is small, more or less symmetrically positioned and covered with papillae. Except for base plate pores the posterior flanges are smooth and lacking additional ornamentation (Fig. 3.3F). The shield is marked with a secondary reticulum of rectangular or polygonal shaped meshes each of which encloses three to four pores (Figs. 3.3C-F). The pattern on the shield is very similar to that described for *M. clavus* and *M. retifera* Dürrschmidt (Dürrschmidt, 1982a). Several scales were observed with a much thicker reticulum enclosing well defined circular, as opposed to polygonal meshes.

Bristles are short with a mean length of 6.5 μm, ribbed, slightly curved and terminate in a bifurcate tip, the branches of which are of unequal length (Fig. 3.3C).

M. conspersa is most appropriately placed within the Sectio Papillosae because of the presence of only domed scales, the similarity between all scales on a given cell, the small, tripartite scales with small and symmetrically positioned domes and the lack of secondary ornamentation on the posterior flange. In addition, some species within the Sectio Papillosae are similar to *M. conspersa* in bristle morphology (e.g. *M. calceolus*) and in possessing a forward projecting and winged anterior submarginal rib (e.g. *M. paxillata*). Dürrschmidt (1986) commented on the similarity of *M. conspersa* with *M. calceolus*. Lastly, species within the Sectio Papillosae have scales with shields marked with papillae (Asmund and Kristiansen, 1986). Scales of *M. conspersa* have papillae on the dome and anterior submarginal ribs as well as polygonal shield meshes which may represent the fusion of papillae.

I must comment on the apparent close relationship between *M. conspersa* and *M. retifera*. In my opinion both of these rare taxa are very similar and with further observation may be found to be the same species. The size and shape of both cells and scales, the presence of only tripartite scales and the morphology of bristles are the same for both taxa. The only significant difference between scales of *M. conspersa* and *M. retifera* is the presence of a transverse shield rib positioned just behind the dome on the latter taxon (compare Fig. 7.29 of Dürrschmidt, 1986 with Fig. 11 of Dürr-

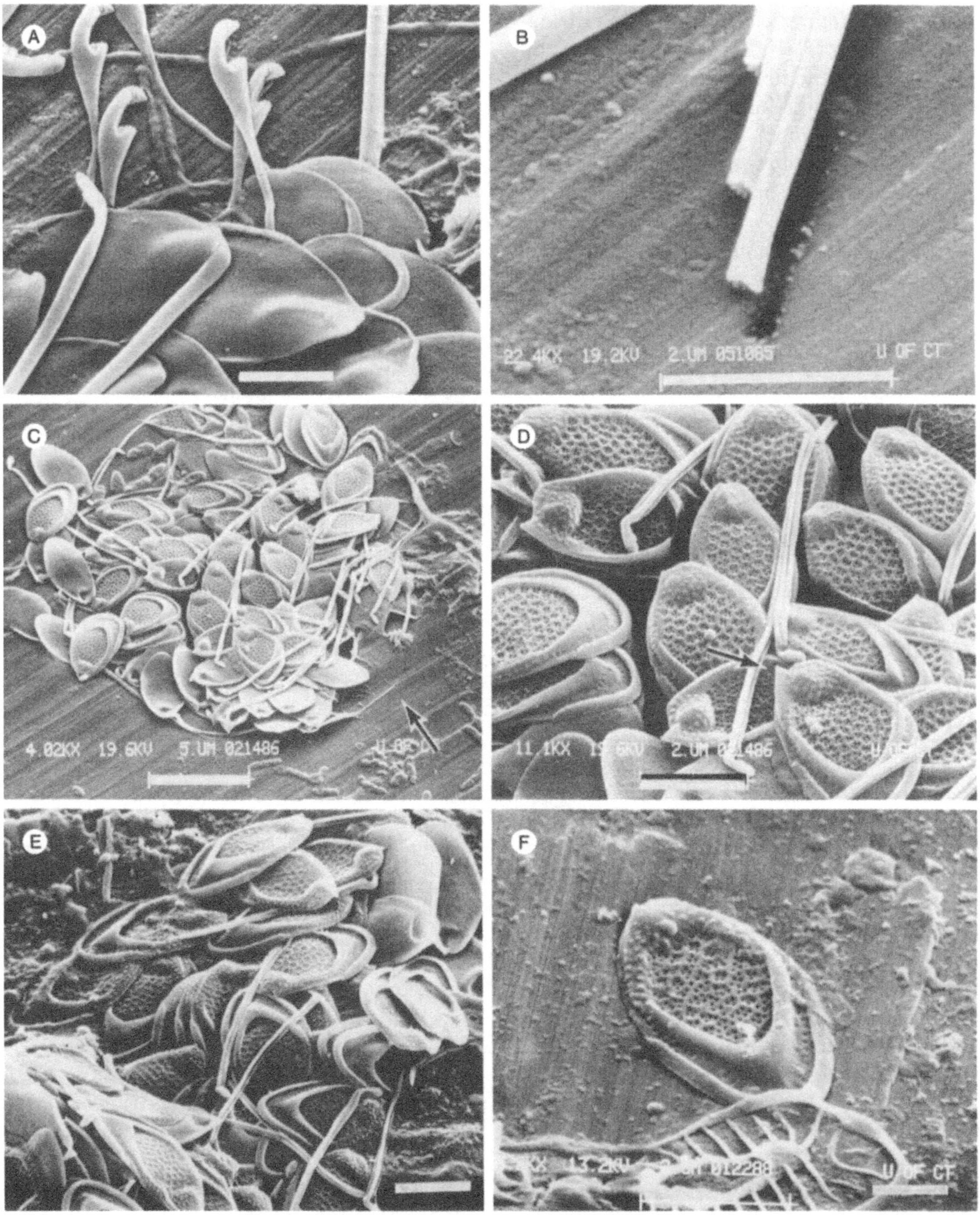

Fig. 3.3A-B. Mallomonas hamata. Fig. A. Body scales and short hooked bristles. Scale bar is 2μm. High magnification of needle bristles depicting the serrated distal tips. Scale bar is 2μm. Figs. C-F. *Mallomonas conspersa.* Fig. C. Siliceous remains of a whole cell. Note the ornamentation of the scales and the short bristles with bifurcate distal tips (arrow). Scale bar is 5μm. Fig. D. A close-up of the scales in Fig. C. Note the polygonal shaped meshes on the shield, each of which encloses three to four pores. The left anterior submarginal rib is extended forward into a small spike (arrow) on apical scales. Scale bar is 2μm. Fig. E. Body scales and bristles. Scale bar is 2μm. Fig. F. Body scale depicting the ornamentation of the shield and anterior submarginal ribs and the position of the V-rib. Scale bar is 1μm.

schmidt, 1982a); otherwise the ornamentation of the scales, especially the reticulation of the shield, is identical. Further work is clearly needed in order to determine whether *M. conspersa* and *M. retifera* are discrete taxa.

Mallomonas guttata Wujek

Synonym: *M. perforata* Cronberg and Hickel

Although whole cells were not observed they are reported as being small, 7 to 12 μm · 3 to 6 μm, ellipsoidal and covered with tripartite scales bearing rather short bristles (Wujek, 1984; Cronberg and Hickel, 1985). Scales are aligned in spiral rows and positioned with their longitudinal axis perpendicular to the longitudinal axes of the cell (Asmund and Kristiansen, 1986).

Only isolated scales were observed (Fig. 3.5B). Body scales are widely elliptical to oval with a broad proximal rim, a prominent V-rib with straight arms and a small, relatively indistinct dome. Scales lack anterior submarginal ribs. The shield and anterior flanges are covered with closely spaced, solid and large papillae. There are from 9 to 15 circular pits randomly spaced, but localized within the inner portion of the shield. The posterior flange is smooth and unornamented. There is a rib along each side of the dome. The inverted U-shaped opening from which the bristle emerges is also ribbed.

The morphology of the collar and spined scales is discussed by Wujek (1984) and Asmund and Kristiansen (1986). There seems to be some question as to whether the more broadly rounded, asymmetric scales with anterior submarginal ribs and a forward projecting wing (or spine) represent posterior (Wujek, 1984) or collar (Asmund and Kristiansen, 1986) scales.

Mallomonas papillosa Harris & Bradley

Cells are small, broadly ellipsoidal and covered with scales bearing short, curved bristles (Fig. 3.4A). All scales possess domes and are arranged in spiral rows with their longitudinal axes positioned at approximately a 90° angle to the longitudinal axis of the cell. The right side of the posterior flanges is overlapped by the scale positioned behind it in the same spiral row; the left side of the posterior flange is overlapped by the scales positioned in the spiral row above it.

Scales are oval and have a base plate with inconspicuous pores, a narrow posterior rim, an acutely angled, narrowly hooded V-rib, and a small dome (Figs. 3.4A-C). The edge of the hood near the angle of the V-rib is often toothed (Figs. 3.4A-C). On body scales the arms of the V-rib curve and become continuous with the anterior submarginal ribs; the latter terminate along the sides of the dome giving the scale a symmetrical appearance (Fig. 3.4C). The anterior flanges are marked with a series of 3 to 8 evenly spaced, parallel struts (Fig. 3.4C). Anterior scales may have fewer or no flange struts. The shield is covered with densely packed, regularly arranged, rows of papillae; each row commonly has from 12 to 15 papillae (Fig. 3.4C). Papillae may also be found on the dome, often on the right side (Fig. 3.4A), but are lacking on the anterior and posterior flanges. A single, sometimes several (Asmund and Kristiansen, 1986), circular pore with a raised bordered rim is often found in the posterior region of the shield near the V-rib (Fig. 3.4C); this pore is easily seen on the undersurface of the scale (Figs. 3.4A,F).

On apical scales one of the V-ribs extends along the lateral side of the dome, often protruding to form a small wing (Fig. 3.4A), or, more rarely, spine (Wee, 1982). As such, apical scales are asymmetrical and have only one submarginal rib.

Bristles are short, curved, unilaterally serrated along the convex margin with short teeth and terminate with a sharp point (Fig. 3.4A). Bristles emerge on the right side of the dome from a shallow U-shaped opening.

Two additional varieties have been formally described. Variety *ellipsoidea* Harris differs from the nominate variety in: a) possessing bristles with larger serrations; b) having papillae on the shield in a less dense pattern; and c) lacking papillae on the domes (Harris, 1967; Takahashi, 1978; Asmund and Kristiansen, 1986). Several populations had scales with fewer papillae (approximately 8 per row) and lacked papillae on the domes (Figs. 3.4E,F); these were identified as var. *ellipsoidea* even though the size of the teeth on the bristles was not significantly different from var. *papillosa* (Fig. 3.4E).

Variety *monilifer* Harris differs from var. *papillosa* by having scales where: a) a portion of the

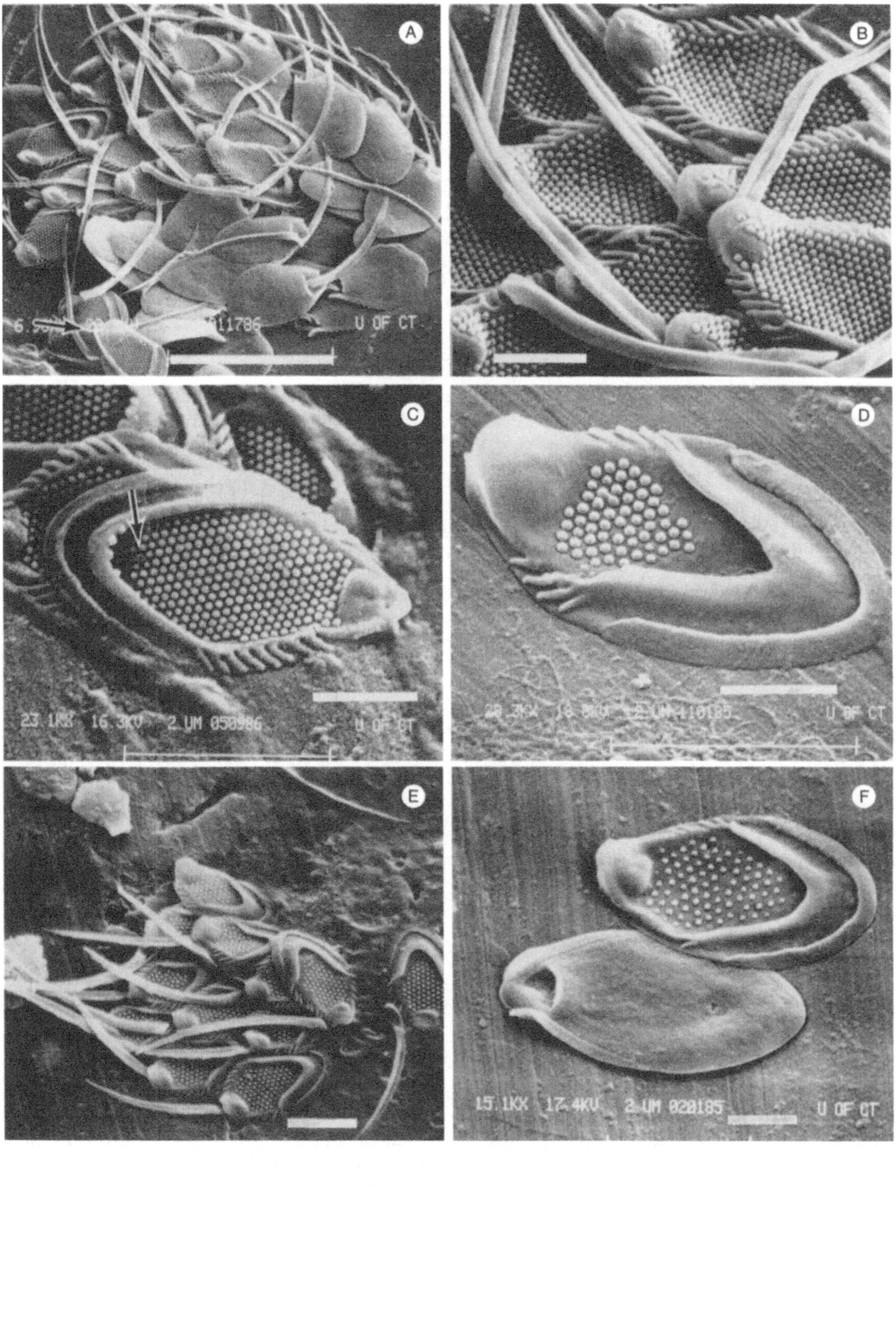
A
B
C
D
E
F
U OF CT
23.1KX 16.3KV 2.UM 050986
15.1KX 17.4KV 2.UM 020185
U OF CT

shield lacks papillae; b) a row of papillae line the posterior border of the dome; and c) the dome is devoid of papillae. Several isolated scales with shields only partially covered with papillae and presumably representing var. *monilifer* were observed (Fig. 3.4D); however, they lacked the row of papillae lining the posterior border of the dome.

The range in variability of the characteristics used to distinguish the varieties of *M. papillosa* is not fully known, making it often difficult to identify a given taxon. Clearly, further work is needed in order to delineate the characteristics of each taxon and to document the degree of morphological variability within each taxon.

Scales of *M. papillosa* could be confused with those of *M. rasilis* Dürrschmidt (Dürrschmidt, 1983c) and *M. paxillata* (see discussion for this species). Scales of the latter two taxa can be distinguished from the former because they lack the series of struts on the anterior flanges. In addition, scales of *M. rasilis* have an irregular series of ribs under the posterior rim; these struts are discernible only with TEM.

Mallomonas paxillata (Bradley) Péterfi and Momeu

Synonym: *Mallomonopsis paxillata* Bradley

Although whole cells were not observed, they have been reported as being ovoid and covered with bristles (Bradley, 1966; Asmund and Kristiansen, 1986).

Scales are oval with slight lateral incurvings and have a base plate that appears to lack pores, a hooded V-rib, a narrow proximal rim and a rather small dome (Fig. 3.2F). The arms of the V-rib are continuous with the anterior submarginal ribs. On some scales the left submarginal rib runs along the lateral side of the dome and projects forward forming a rather sharp spike (Fig. 3.2F). The dome, shield and anterior flanges are covered with equally spaced and densely packed papillae. In contrast, the posterior flange is smooth and unornamented (Fig. 3.2F).

Bristles are short, slightly curved, have a smooth shaft and a bifurcated distal end.

Scales of *M. paxillata* are separated from those of *M. papillosa* due to the lack of struts on the anterior flanges; *M. ouradion* because they have a dome, one bristle per scale and are more densely papillated; and *M. rasilis* because of the presence of anterior submarginal ribs and a denser papillated surface. Additional comparisons with these and other related taxa are summarized in Table 3.1.

Sectio Planae Momeu & Péterfi 1979

Series Fastigatae Momeu & Péterfi 1979

Mallomonas caudata Ivanov

Cells of *Mallomonas caudata*, the largest within the genus, are covered with distinctively shaped scales and bristles (Figs. 3.6; 3.7A-B). On most cells the posterior is tapered forming a 'tail' and yielding a spindle-like morphology (Fig. 3.6). Ovoid to cylindrical cells are also common. There is some evidence that cell shape varies within a given population (Takahashi, 1978; Weimann, 1933).

Scales are large, flat, oval to elliptical in outline, often slightly asymmetrical and have a simple morphology (Figs. 3.7C-D). Scales range in size from 5.1 to 8.7 μm in length and 3.0 to 6.6 μm in width; smaller scales with a circular outline can be found on the posterior tail. The proximal two-thirds of the base plate is perforated with small evenly spaced pores; pores are less dense near the distal end (Figs. 3.5F; 3.7D). A single, large,

←

Fig. 3.4. Mallomonas papillosa. Figs. A-C. *Mallomonas papillosa* var. *papillosa.* Fig. A. Remains of the siliceous armour of a cell. Note the evenly and closely spaced papillae, the small domes with a few papillae on the right side, the parallel struts on the anterior flanges and the teeth lining the angle of the V-rib on the scales. Bristles are short, slightly curved and unilaterally serrated. The arrow depicts an apical scale. Scale bar is 5μm. Fig. B. Close-up of Fig. A. showing the pattern of papillae on the scale. Scale bar is 1μm. Fig. C. Body scale. Arrow depicts the position of the posterior pore with a bordered rim. Scale bar is 1μm. Fig. D. *Mallomonas papillosa* var. *monilifera.* Body scale lacking papillae on the dome and a portion of the shield. Scale bar is 1μm. Figs. E-F. *Mallomonas papillosa* var. *ellipsoidea.* Fig. E. Body scales and bristles. Scale bar is 2μm. Fig. F. Two body scales. The papillae are less dense, more scattered and lacking on the dome. Note the isolated pore at the base of the V-rib observed on both the surface and undersurface of the scale.

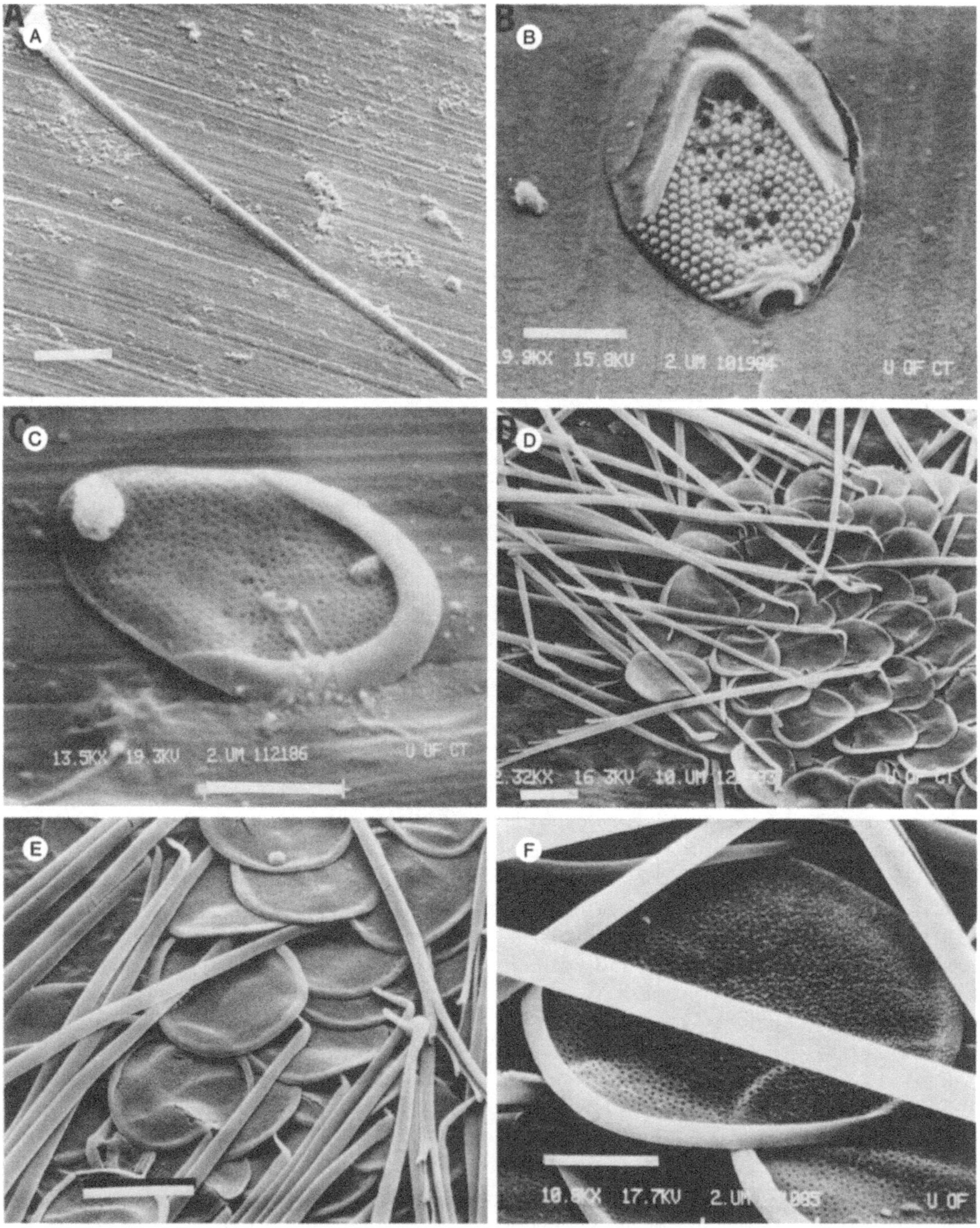

Fig. 3.5 A. Mallomonas matvienkoae f. *matvienkoae.* Bristle showing the smooth side of the shaft and bifurcate distal tip. Scale bar is 2μm. Fig. B. *Mallomonas guttata.* Body scale with large papillae on the shield and anterior flanges, prominent V-rib, small dome, a series of circular pits along the shield, and lacking anterior submarginal ribs. Scale bar is 1μm. Fig. C. *Mallomonas* cf. *stellata.* Body scale. Scale morphology is similar to that of *M. matvienkoae* f. *matvienkoae* except for the presence of a rounded appendage on the distal end. Scale bar is 2μm. Figs. D-F. *Mallomonas caudata.* Figs. D-E. Body scales and bristles. Note the somewhat uneven overlapping of the scales, the attachment of the bristles under the distal ends of the scales and the proximal foot of each bristle. Scale bars are 5μm. Fig. F. Body scale. Scale bar is 2μm.

elongated pore is often observed in the center of the posterior end of the scale (Figs. 3.5D-E; 3.7C-D). The posterior rim extends along one-half to two-thirds of the perimeter of the scale.

Cells are covered with distinctive bristles that radiate in all directions (Fig. 3.6). The bristles are shorter on the anterior end, but increase in mean length towards the middle and posterior of the cell. Each bristle consists of a folded siliceous membrane (Takahashi, 1978) that is fused along the lower two-thirds of the shaft and unfused along the distal one-third portion (Fig. 3.7E). The proximal fused portion of the shaft is relatively straight, smooth and hollow in cross section while the distal, unfused end is bent and serrated, with two to ten teeth along the convex margin (Fig. 3.7E). The tip of each tooth has many smaller teeth. The proximal foot is bent at a 45° to 90° angle with the shaft and tucked under the distal border of the scale; more than one bristle are often associated with a single scale.

As reported by Sandgren (1980; 1989) and Cronberg (1980; 1988), stomatocysts have an irregularly reticulated surface (Fig. 3.7F), although this feature may be quite variable (Cronberg, 1988; Sandgren, 1989)

Series Teilingianae Asmund and Kristiansen

Mallomonas cf. *bronchartiana* Compère

One sample had scales that are tentatively identified as *M. bronchartiana* (Fig. 3.12F). Body scales are large, oval, possess a wide proximal rim encircling one-half of the perimeter of the scale and are covered with closely-spaced papillae (Fig. 3.12F). Scales lack a dome and submarginal rib, although there is a V-shaped area along the proximal rim devoid of papillae. The area devoid of papillae is not visible on intact cells. These features compare well with those of *M. bronchartiana*.

Scales of *M. bronchartiana* are described as having struts under the proximal rim, an internal honeycomb pattern of ribs and an asymmetrically positioned proximal border. Although a few scales had a slightly asymmetrically placed rim, the presence of a honeycomb pattern and struts under the rim were not detected with SEM. Whole cells and bristles were not observed. Thus, further work will be necessary in order to make a definite identification.

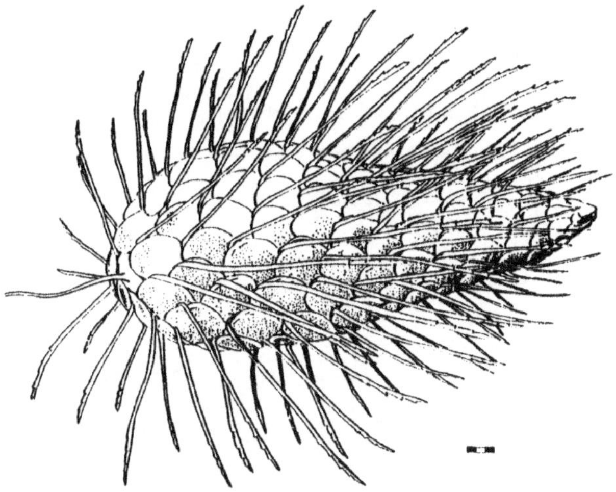

Fig. 3.6. Three-dimensional reconstruction of a whole cell of *Mallomonas caudata* depicting the distribution of scales and bristles. Scales lack domes and V-ribs and are positioned approximately 45° to the longitudinal axis of the cell. Cells are covered with long and distinctive bristles.

Sectio Retrorsa Siver

Mallomonas retrorsa Siver

Cells are lanceolate to spindle shaped, have a mean length of 33.9 μm, possess an apical ring of small forward-projecting paddle-like bristles that surround the flagellum (Fig. 3.9F), an elongated caudal tail (Fig. 3.9C) and three distinct types of scales (Figs. 3.8; 3.9A-B). Scales are differentiated into apical, body and caudal scales, all of which are aligned with their longitudinal axes parallel to the longitudinal axis of the cell (Siver, 1988a) (Figs. 3.9A-B). *M. retrorsa*, and probably *M. fenestrata* Cronberg and Hickel, are unique among species of *Mallomonas* because 1) the proximal portion of all scales (i.e. the end with the rim) faces the apical end of the cell and, as such, is orientated backwards, and 2) the bristles emerge from the posterior ends of the apical scales (Siver, 1988a).

M. retrorsa (and most likely *M. fenestrata*) differs from other species of *Mallomonas* regarding the pattern of overlap of its scales. As in all species, the scales are aligned in spiral rows. However, each scale is overlapped by the scale positioned in front of it within the same spiral row and by scale(s) in the spiral row above it (Fig. 3.9E). Thus, although the pattern of overlap of adjacent spiral rows is similar to other species of *Mallomonas*,

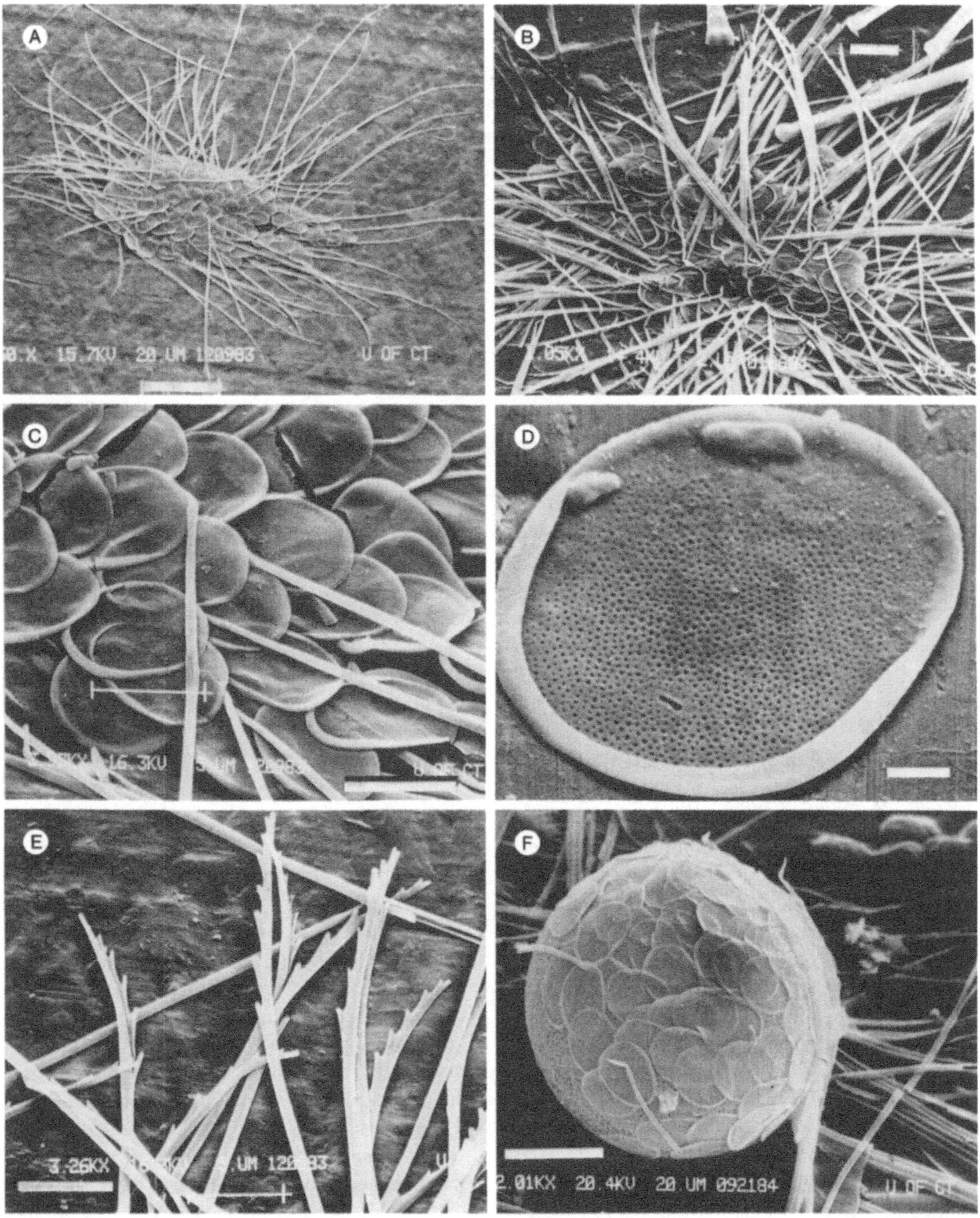

Fig. 3.7. Mallomonas caudata. Figs. A-B. Siliceous remains of two whole cells. Note the large number of scales and bristles. Scale bars are 20μm and 10μm, respectively. Fig. C. Close-up of body scales in the posterior region of the cell in Fig. A. Note that scales in this region are positioned such that they tend to face the posterior of the cell. Scale bar is 5μm. Fig. D. Body scale showing the distribution of base plate pores and the proximal rim. Note the presence of the large elongated pore in the proximal region. Scale bar is 1μm. Fig. E. The distal serrated portions of bristles. Scale bar is 5μm. Fig. F. Stomatocyst with an irregularly reticulated surface; scales are still attached. Scale bar is 10μm.

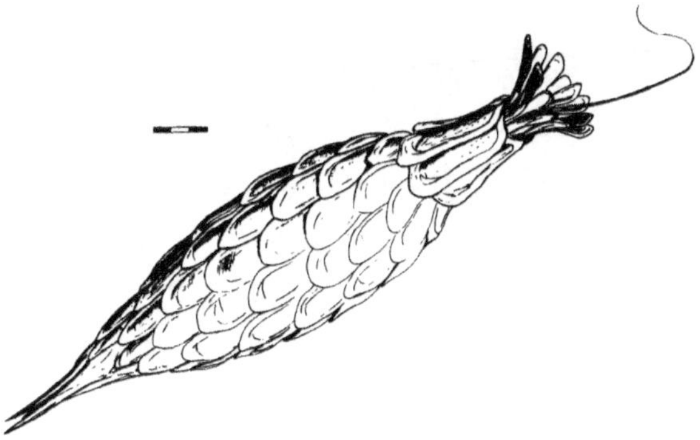

Fig. 3.8. Three-dimensional reconstruction of a whole cell of *Mallomonas retrorsa* depicting the distribution of scales and bristles. Scales are positioned parallel to the longitudinal axis of the cell and are arranged such that their distal ends face the posterior end of the cell. Note the distinctive ring of paddle-shaped bristles and the elongated caudal scales.

scales within a given row are overlapped in an opposite fashion.

Body scales are elliptical to circular in outline, domeless, have a perforated base plate and an asymmetrically positioned rim and V-rib (Figs. 3.9C-D). Thus, although the scale outline is not asymmetrical, the features of the scales are. The V-rib is broadly U-shaped with the right-handed arm extending almost to the distal border. Likewise, the right-handed side of the proximal rim also extends further along the margin of the scale than the left-handed part. The shield is covered with a thin secondary layer that gives a smooth to slightly undulate appearance (Fig. 3.9C). On some specimens the secondary layer forms a thumbprint-like pattern of irregularly shaped ribs. Since the secondary layer covers the base plate, the pores can be seen only with TEM or on the undersurface of the scale (Siver, 1988a).

The apical bristle-bearing scales are elongated, aligned into a distinct ring and are also asymmetrical in design (Figs. 3.8; 3.9F). The right side of the V-rib (submarginal rib) extends to the distal margin of the scale, while the left side is very short or more often absent (Fig. 3.9F). The shield of the caudal scales is much smaller, however, the arms of the V-rib are extended forming a long spine (Figs. 3.9C-D). The length of the spine, and hence the length of the caudal scales, gradually increases with proximity to the posterior most portion of the cell.

Bristles are restricted to the anterior end of the cell and are quite unique among species of *Mallomonas* (Figs. 3.8; 3.9F). Bristles are short, 3.2 to 4.5 μm in length, flat and paddle-shaped with a broadly-rounded, blunt distal end (Fig. 3.9F). Bristles gradually increase in diameter and become widest at the distal end. Each cell has from 6 to 12 bristles of more or less equal length that project forward forming an inverted cone-like structure around the base of the emergent flagellum (Siver, 1988a).

The relationship of *M. retrorsa* to *M. fenestrata* and *M. akrokomos*, the only other species known to have their body scales aligned parallel to the longitudinal axis of the cell, is discussed by Siver (1988a).

Sectio Quadratae Momeu & Péterfi 1979

Series Allorgeanae Asmund & Kristiansen 1986

Mallomonas lychenensis Conrad

Synonyms: *M. robusta* Matvienko
M. tessellata Nygaard
M. lychenensis var. *tessellata* (Nygaard) Bourrelly

Mallomonas allorgei (Deflandre) Conrad

Synonym: *Pseudomallomonas allorgei* Deflandre

Cells and scales of *M. lychenensis* and *M. allorgei* are often confused and difficult to identify. Based on the electron microscopy descriptions by Harris and Bradley (1960) and Asmund and Kristiansen (1986) the majority of the populations found in Connecticut localities are of *M. lychenensis*. However, cells within a population as well as individual cells often expressed features common to both taxa rendering a satisfactory identification difficult to make. In addition, I have received conflicting identifications from other taxonomists leading me to believe that the current distinctions between the two species are not sufficient and that the difference needs clarification. As a result I have chosen to describe the features of *M. lychenensis*, the primary ways that it differs from *M. allorgei* and the problems that can be encountered.

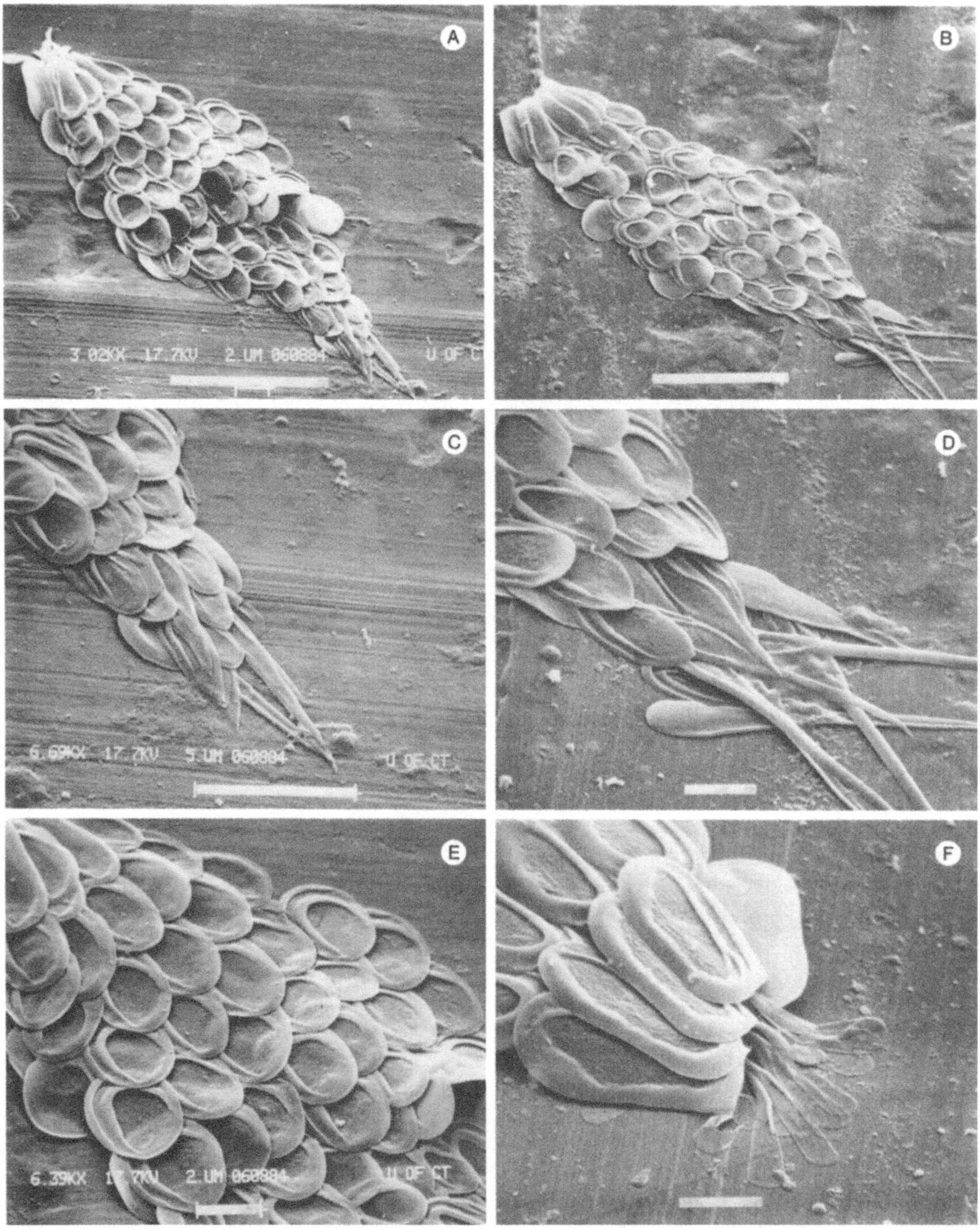

Fig. 3.9. Mallomonas retrorsa. Figs. A-B. Whole cells denoting the lanceolate shape, backwards arrangement of the scales and the elongated caudal tail. Scale bars are 10μm. Figs. C-D. Close-ups of the caudal tail regions of the cells illustrated in Figs. A and B. Note the extension of the arms of the V-rib forming the long distal spines. Scale bars are 5μm and 2μm, respectively. Fig. E. Close-up of the body scales of the cell depicted in Fig. A. Each scale is overlapped by the scale positioned in front of it within the same spiral row and the scales in the spiral row above it. Note the asymmetry of the scales; the right arm of both the V-rib and proximal rim extend further along the margin of the scale. Scale bar is 2μm. Fig. F. Ring of apical bristle-bearing scales with a tuft of paddle-shaped bristles. Scale bar is 2μm.

Cells are ovoid and have a small cell to scale ratio, bristles associated with apical scales only and short stout spines on posterior scales (Figs. 3.10; 3.11A,C,E; 3.12A). Scales are in spiral rows and are aligned such that their longitudinal axes are perpendicular to the longitudinal axis of the cell. Each scale is overlapped by the scale positioned behind it in the same spiral and by the scale(s) in the spiral above it (Figs. 3.11A,C,E).

Three basic types of scales are found: an apical ring of bristle-bearing, asymmetric and forward projecting scales (Figs. 3.11A; 3.12A-B); large symmetrical body scales (Fig. 3.11A); and smaller, slightly asymmetric spined posterior scales (Figs. 3.11D,F). All scales are extremely thick, curved to conform to the surface of the cell, have a prominent proximal rim and a secondary layer with papillae and large pits or pores. The papillae are actually situated on a smoother secondary surface (Figs. 3.12C-D). Tiny base plate pores can often be discerned with SEM on the undersurface of scales as well as within the large pits (Fig. 3.11B). Apical scales are subcircular and asymmetric due to the left distal border being raised and projected forward (Figs. 3.11A; 3.12A-B). The projected borders of the apical ring of scales forms a crown around the flagellar opening (Asmund and Kristiansen, 1986). The shield, including the forward projecting lobe, is covered with closely spaced papillae; from 8 to 12 large pits are scattered over the shield (Figs. 3.12A-C).

The surfaces of body scales are usually covered with regularly spaced, rather large papillae (Figs. 3.11A-B,E). From comparison of SEM and TEM images it appears that the papillae are hollow. Areas or patches of the surface of the scale are often lacking papillae exposing the secondary layer of silica (Figs. 3.11C; 3.12B-D). However, the templates of where the papillae should have formed, or were and fell off, can be observed at higher magnification (Figs. 3.11B; 3.12D). Thus, the use of such a characteristic in distinguishing between the two taxa is questionable. In a few populations the body scales on a given cell had two rather well defined rows of from two to six large pits. Each row on a given scale usually has a different number of pits (Figs. 3.11A,E; 3.12A). However, on other cells scales were found with few or no pits (Figs. 3.11C; 3.12C-E).

The posterior scales become progressively smal-

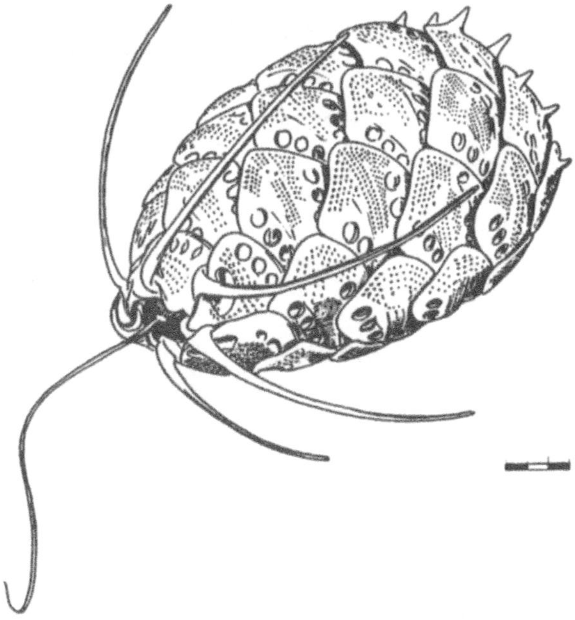

Fig. 3.10. Three-dimensional reconstruction of a whole cell of *Mallomonas lychenensis*. Scales are precisely arranged on the cell such that their longitudinal axes are perpendicular to the longitudinal axis of the cell.

ler in size; the extreme of which are about one-fourth the size of the body scales (Figs. 3.11D,F). Usually, one (Figs. 3.11A,F; 3.12A) and sometimes two or more (Fig. 3.11D) papillae situated on the right distal border become elongated into stout spines. On some cells body scales were found with spines (Fig. 3.11E). One population of cells was found where posterior scales lacked spines (Fig. 3.12E).

Cells of all populations were believed to have bristles (Figs. 3.12A,E), although in EM preparations they were not always found. Bristles are smooth, strongly curved and become tapered to a point. In live preparations the bristles often conform to the curvature of the cell making them difficult to detect (Fig. 3.12A); however, if they become orientated outwards they curve distinctly away from the cell (Fig. 3.12E).

According to Harris and Bradley (1960) and Asmund and Kristiansen (1986) *M. lychenensis* and *M. allorgei* can be distinguished using the following characteristics:

1) The surface papillae of scales of *M. allorgei* are

A
B
C
4.72KX 18.6KV 5.UM 122085
U OF CT
D
8.92KX 18.7KV 5.UM 122085
U OF CT
E
F

more numerous and more regularly spaced. Areas or patches of pores may be missing on scales of *M. lychenensis*.
2) Scales of *M. lychenensis* have a greater number of pits.
3) Scales of *M. lychenensis* have two rows of pits that usually extend more than half of the length of the scale and converge at both ends like a pair of lips (Harris and Bradley, 1960). Scales of *M. allorgei* have pits aligned in the shape of a V extending over less than half the length of the scale. Scales of *M. allorgei* also have a short row of pits positioned distally to the V structure. The pits of scales of *M. allorgei* are more irregularly arranged.
4) Posterior scales of *M. allorgei* have fewer and smaller spines.
5) Apical scales of *M. allorgei* lack the forward projecting bulge.
6) Cells of *M. allorgei* apparently lack bristles.

The number and arrangement of pits on body scales varied greatly between populations and even on a given cell. The most common arrangement was of pits aligned more or less in two rows. However, trying to decide if the rows extended less than half the length of the scale, or more, and whether they remained divergent (V-shaped) or began to converge (lip-like) was often difficult because a mixture of conditions were often found. Cells with body scales that had pits forming a V structure as described for *M. allorgei* also had features characteristic of *M. lychenensis*, such as long spined papillae, apical scales with projecting lobes and bristles. Based on the latter characteristics the cells would be more correctly identified as *M. lychenensis*.

Scales with few or no pits were often found; these scales do not fit the description for either taxon. However, such scales were often associated with other features characteristic of *M. lychenensis*. Body scales lacking pits have been previously illustrated for *M. lychenensis* (e.g. Fig. 20A of Asmund and Kristiansen, 1986). Scales with areas or patches of missing papillae were also common. However, the templates for missing papillae were often seen at higher magnification, suggesting that perhaps the papillae were present and subsequently lost. Thus, such a feature should not be used as a valid taxonomic characteristic.

All populations observed in this study were believed to have bristles, a characteristic of *M. lychenensis*. Cells with small spines or lacking spines altogether were associated with apical scales that had bristles and forward projecting lobes. In summary, the characteristics currently utilized to separate *M. lychenensis* from *M. allorgei* need to be reconsidered and a combination may be in order. In my opinion identifications based on isolated scales should be considered as tentative at best.

Sectio Punctiferae Asmund & Kristiansen 1986

Series Punctiferae Momeu & Péterfi 1979

Mallomonas punctifera Korshikov

Synonyms: *Mallomonas elegans* var. *pulchella* Kisselev
Mallomonas reginae Teiling
Mallomonas pulchella (Kisselev) Cronberg & Kristiansen

Cells are ellipsoidal to ovoid, small, covered with bristles, and divided into three distinct regions, each composed of a unique type of scale (Figs. 3.13; 3.14A-B). The anterior end consists of a single ring of asymmetric, triangular shaped scales orien-

←

Fig. 3.11. Mallomonas lychenensis. Fig. A. Whole cell depicting the arrangement of scales in spiral rows, the pattern of overlap and orientation of the scales, and the distribution of rear scales with spines. Note the spacing of the large pits and the patches devoid of papillae on individual scales. Scale bar is 5μm. Fig. B. High magnification of the cell in Fig. A. Note the templates for papillae in the patches devoid of papillae and the pores within several of the large pits. Scale bar is 1μm. Fig. C. Whole cell. Note the density and distribution of papillae, the enlargement of papillae into small spines and the presence of large pits on the shield. Scale bar is 5μm. Fig. D. Close-up of the posterior region of the cell pictured in Fig. C. Note the smaller rear scales, the pointed nature of the papillae and the distribution of short spines. Scale bar is 5μm. Fig. E. Whole cell. Note the positioning of the many large pits into two rows; the row along the side of the scale facing the posterior of the cell (right side) has more pits. Sale bar is 5μm. Fig. 6. Close-up of the posterior region of a cell. Note the alignment of scales and the relatively stout spines. Scale bar is 2μm.

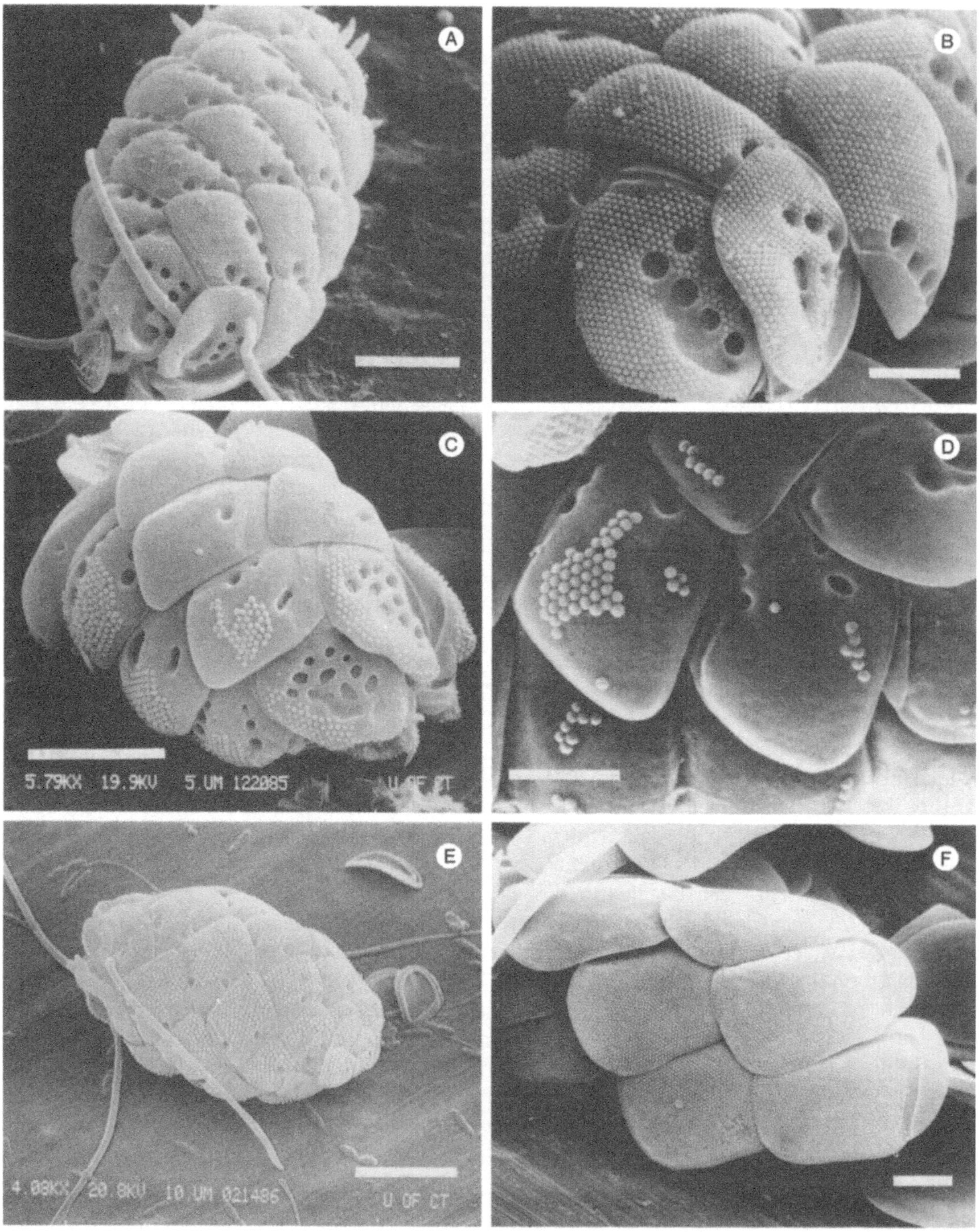

Figs. 3.12 A-E. Mallomonas lychenensis. Fig. A. View of a whole cell from the anterior end. Note the apical scales and bristles. Scale bar is 5μm. Fig. B. High magnification of the anterior end of a cell. Note the raised and forward projecting left distal border of the apical scale. Scale bar is 2μm. Fig. C. View of a cell from the anterior end. Note the lack of papillae on regions of body scales, and the large number of pits and distal bulge on the apical scales. Scale bar is 5μm. Fig. D. High magnification of scales lacking a large number of papillae; the templates for missing papillae can be seen. Scale bar is 2μm. Fig. E. Whole cell with features of both *M. lychenensis* and *M. allorgei* (see text). Scale bar is 5μm. Fig. F. *Mallomonas* cf. *bronchartiana.* Group of body scales each with a wide proximal rim and covered with closely spaced papillae. Note the V-shaped area along the rim that is devoid of papillae. Scale bar is 2μm.

tated with their longitudinal axes parallel to that of the cell; the anterior ring of scales forms a crown (Asmund and Kristiansen, 1986) or collar (Takahashi, 1978) from which the flagellum emerges (Fig. 3.13). Each apical scale is drawn out into a forward projecting spine (Figs. 3.14A,D; 3.15A). The posterior rim always extends along the left convex side of the scale and spine (Figs. 3.15A-B) and usually, but not always, along the right concave side of the scale and spine (Fig. 3.15B). A submarginal rib is present on the right hand (concave *in sensu* of Asmund and Kristiansen, 1986) side of the scale and often extends forward to the spine (Figs. 3.15A-B); the rib may be straight (Fig. 3.15B) or wavy (Fig. 3.15A). An average of seven apical scales per cell was observed. Bristles, about half as long as those associated with body scales, emerge from the right hand side of each apical scale just below the distal spine (Figs. 3.14B; 3.15B). The presence of apical bristles is in contrast to observations made by Takahashi (1978).

Each apical scale is positioned in the collar or crown such that its left hand margin is overlapped by the scale to its left and it, in turn, overlaps the scale to its right (Figs. 3.14B,D; 3.15A). In addition, the posterior margin of each apical scale is sandwiched between two body scales positioned in the ring behind the apical most ring (Figs. 3.14D; 3.15B).

The middle section of the cell is composed of body scales aligned in spiral rows with their longitudinal axes perpendicular to the axis of the cell (Figs. 3.13; 3.14A,B,D). Body scales are roughly quadrate with a rounded posterior end and a somewhat squared distal end (Figs. 3.15A-B). Each body scale has a posterior rim encircling half of the scale, two roughly parallel submarginal ribs and a small dome. The U-shaped opening from which the bristle emerges is often recurved and always positioned facing the right (see the undersurface of body scales in Figs. 3.14B,D; 3.15A) such that the bristle emerges at a right angle to the longitudinal axis of the scale and parallel to the axis of the cell. A row of teeth lines the distal end of the dome above the opening from which the bristle emerges.

Body scales lying in adjacent rows are often staggered such that the domes of scales in one row are aligned close to the middle of scales in the rows on each side (Fig. 3.14B). Such an arrangement of body scales allows for a greater freedom of rotation of the bristles. Approximately the proximal third of each body scale is overlapped by the scale positioned behind it in the same spiral row and by the scales in the spiral row above it (Figs. 3.14B,D). Only the dome and distal area of the shield of each body scale are exposed on an intact cell (Figs. 3.14B,D).

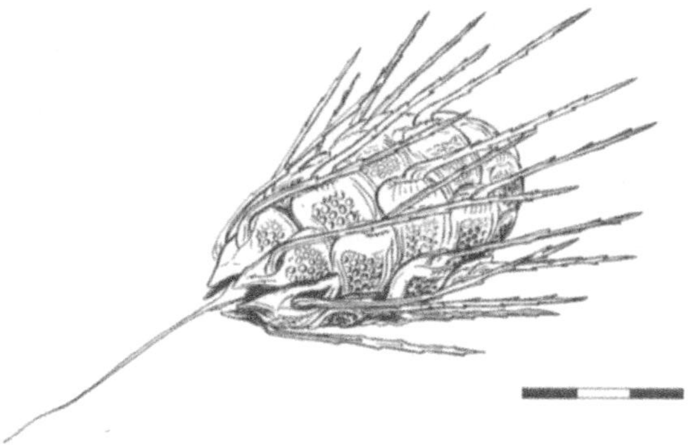

Fig. 3.13. Three-dimensional reconstruction of a whole cell of *Mallomonas punctifera* depicting the distribution of scales and bristles. Scales are positioned in spiral rows such that their longitudinal axes are perpendicular to the longitudinal axis of the cell. Note the arrangement of apical scales and serrated bristles.

Domed body scales integrade into smaller, domeless posterior or rear scales, each of which has a row of protuburences along the distal margin (Figs. 3.14C,E,F). The rear scales are primarily orientated with their longitudinal axes perpendicular to that of the cell, however, become more spirally arranged in order to cover the conical proximal end (Fig. 3.14C).

All scales have a perforated base plate with fewer pores along the perimeter and a secondary reticulum of large pores between the two submarginal ribs. The secondary layer may be less developed on apical scales (Fig. 3.15B) and is often lacking on the smaller rear scales (Figs. 3.14C,E,F).

Bristles are long, straight, thick and composed of three longitudinally aligned and serrated ridges; the ridges are spaced 120° apart (Figs. 3.14A,C; 3.15B). Bristles associated with apical scales are shorter in length (Fig. 3.13).

The details of the morphology of scales are easily observed with either TEM or SEM. The 'ear-like' extensions, seen with TEM, on each side of the dome (Asmund and Kristiansen, 1986) are observed with SEM to be the result of the continuation

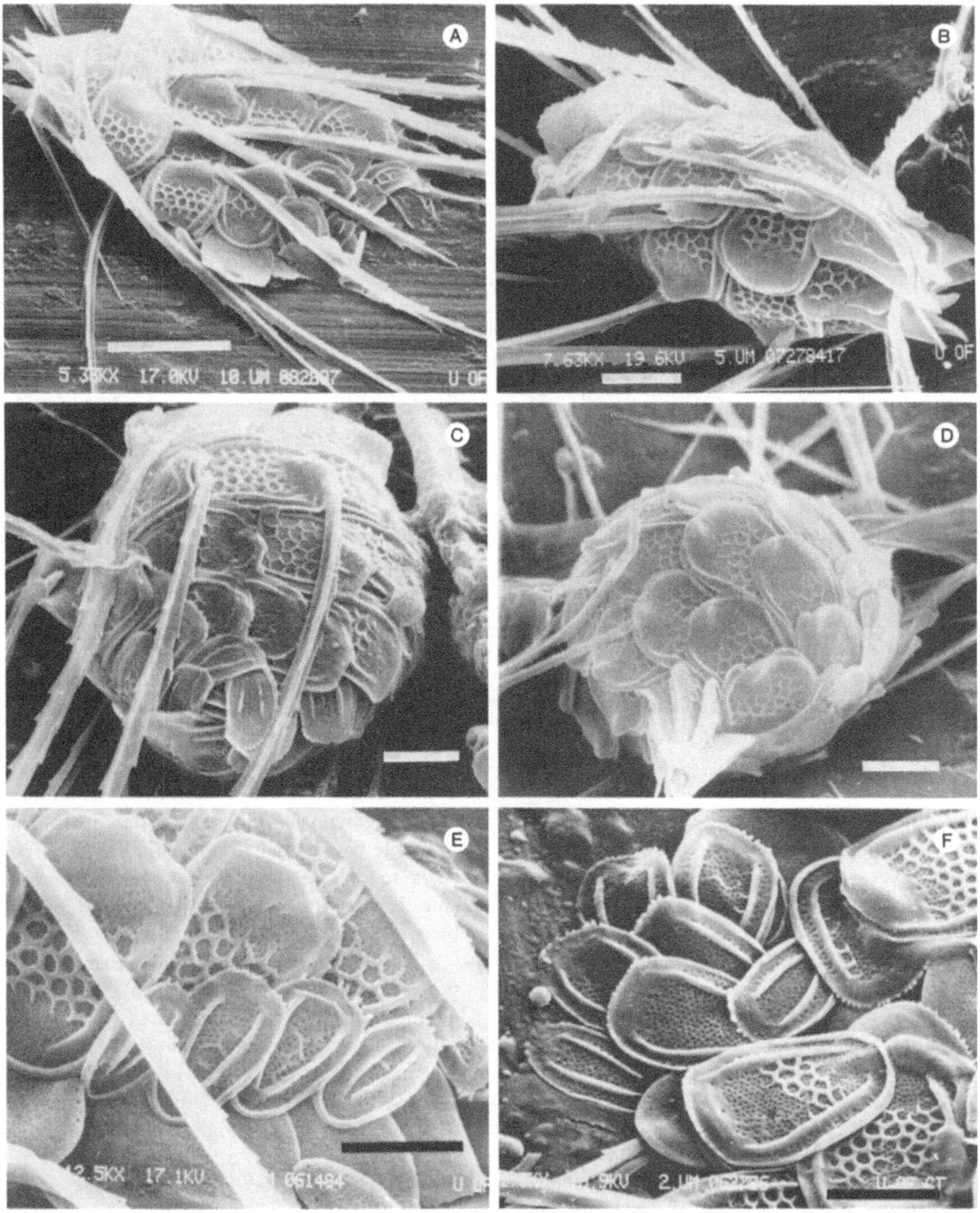

Fig. 3.14. Mallomonas punctifera. Figs. A-B. Whole cells depicting the arrangement of scales and bristles and the orientation of the apical scales around the flagellar opening (Fig. B). Scale bars are 5μm and 2μm, respectively. Fig. C. Posterior view of a whole cell. Note the spiral arrangement of the small rear scales. Scale bar is 2μm. Fig. D. View of a whole cell from the anterior end. Note the arrangement of the apical most ring of scales. Scale bar is 2μm. Figs. D-F. Posterior regions of cells with small, domeless rear scales. Some of the smaller scales lack secondary reticulation on the shield. Scale bars are 2μm.

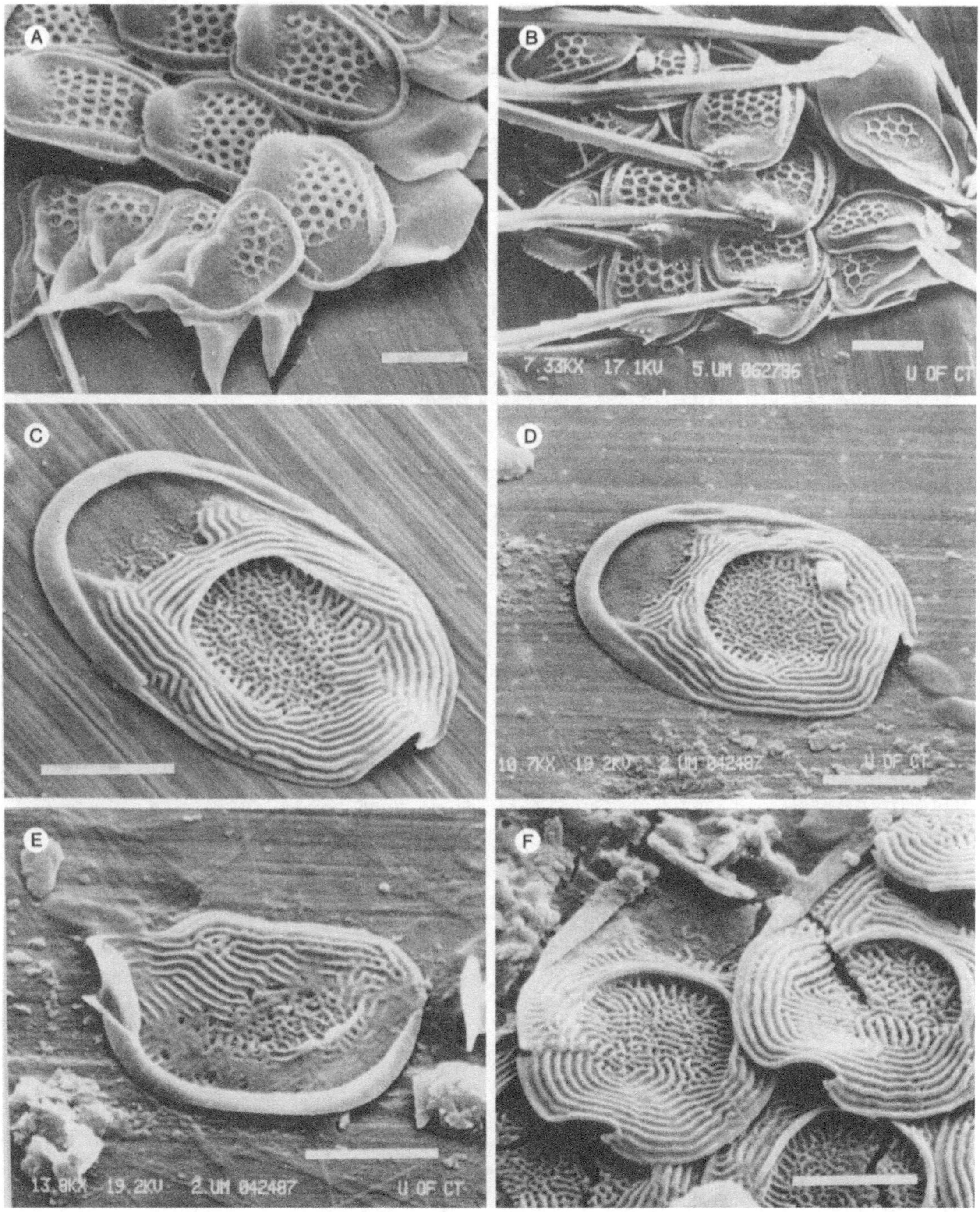

Figs. 3.15 A-B. Mallomonas punctifera. Fig. A. Apical and domed body scales. Note the forward projecting spine on each apical scale. Scale bar is 2μm. Fig. B. Anterior half of a cell. Note the attachment of the bristles to the scales, the row of teeth lining the distal portion of the dome and the serrated ridges of the bristles. Scale bar is 2μm. Figs. C-F. *Mallomonas* sp. 1. Figs. C-D. Body scales depicting the fingerprint-like pattern of ribs. Scale bars are 2μm. Fig. E. Asymmetrical apical scale. Scale bar is 2μm. Fig. F. Group of body scales. Note the opening for the bristle. Scale bar is 2μm.

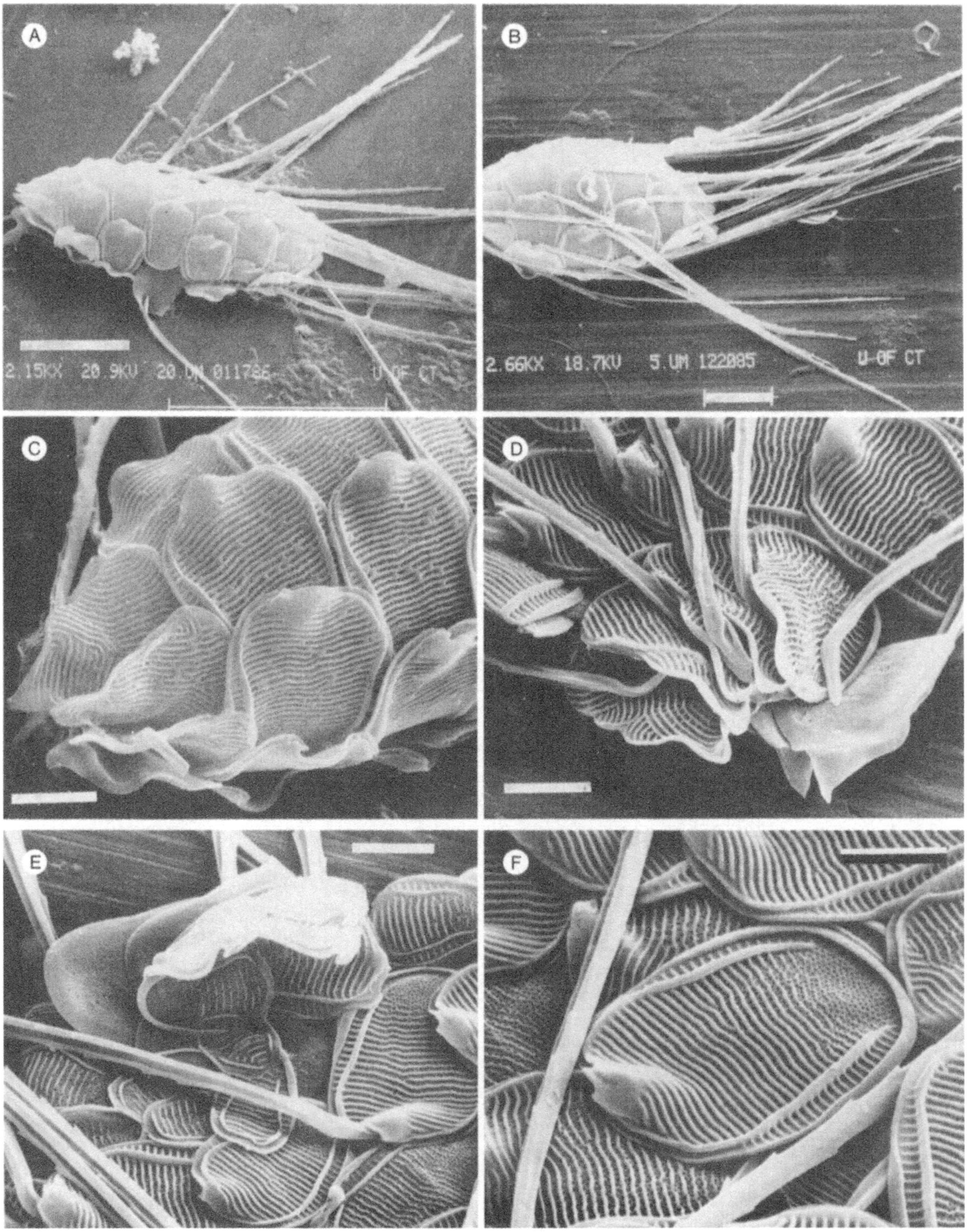

Fig. 3.16. Mallomonas transsylvanica. Figs. A-B. Whole cells depicting the arrangement of scales and bristles. Note the long, thick body bristles, each with three serrated ridges. Scale bars are 10μ and 5μm, respectively. Figs. C-D. The apical ends of two cells showing the asymmetric nature of the apical scales, the transition between apical and body scales (Fig. C) and the attachment of bristles (Fig. D). Note the three serrated ridges on each bristle. Scale bars are 2μm. Fig. E. Domed body scales and smaller posterior domeless scales. Scale bar is 2μm. Fig. F. Body scale showing the dome, submarginal rib, the distribution of transverse ribs on the shield and flanges, and the lack of ribs on the proximal end. Scale bar is 2μm.

of the submarginal ribs around the dome. The presence of the third ridge on the bristles is difficult to detect (e.g. Takahashi, 1978) especially with TEM, because of the 120° spacing. The alignment and positioning of the scales and bristles on the cell surface can only be discerned with SEM.

Mallomonas transsylvanica Péterfi & Momeu

Synonyms: *Mallomonas reginae* var. *glabra* Bourrelly
Mallomonas glabra (Bourrelly) Asmund

Most morphological characteristics of cells, scales and bristles of *Mallomonas transsylvanica* are identical to those of *Mallomonas punctifera* (Figs. 3.16A-B). The alignment and arrangement of apical, body and posterior domeless scales are essentially as described above for *M. punctifera.*

The primary difference between the two species is the morphology of the secondary layer of the scales. Rather than a reticulum of large pores, scales of *M. transsylvanica* are ornamented with closely spaced transverse ribs (Figs. 3.16B-F). The transverse ribs, sometimes wavy and/or branched (Figs. 3.16E-F), extend from the distal end of the scale to the posterior region of the shield where they are lacking or weakly developed (Fig. 3.16F). As noted by Asmund and Kristiansen (1986), the ribs are interrupted by the dome (Fig. 3.16F). Transverse ribs, which are usually aligned with those on the shield, also extend across each flange.

The pattern of ribs is much more irregular and thumbprint like on apical scales (Fig. 3.16C-D). Each apical scale has a slightly raised lateral protrusion along the right concave side from which the bristle emerges (Figs. 3.16C-D). The submarginal ribs on posterior domeless scales often bend and connect along the distal end, forming a single U-shaped rib (Fig. 3.16E). The transverse ribs of the smaller, domeless scales are less developed or lacking (Fig. 3.16E). The teeth lining the dome and the distal margin of rear scales of *M. punctifera* are less obvious on *M. transsylvanica.*

Several populations of cells were observed with body bristles almost twice the length of the cells (Fig. 3.16B). In all populations the apical most bristles were much shorter than the body bristles (Fig. 3.16B). All bristles consisted of three ridges of serrations as described for *M. punctifera*; this is in contrast with observations made by Asmund and Kristiansen (1986).

Mallomonas sp. 1

Several collections had numerous isolated scales from an unknown taxon (Figs. 3.15C-F) that could not be described because whole cells were not observed. Nonetheless, because the scales were quite unique, a short description is provided.

Two types of scales, body and apical scales, were observed. Body scales are oval-elliptical and possess a perforated base plate, a proximal rim encircling about one-half of the scale, a small slightly raised dome and an elaborate series of ribs (Figs. 3.15C-D). The distal three-quarters of the scale is covered with a secondary covering of ribs. The proximal region is unornamented and would presumably be much more electron transparent than the remainder of the scale. A raised ridge transverses the shield just below the middle region and continues along both margins to the anterior flanges. The ridge represents a structure equivalent to a submarginal rib. A series of parallel and more or less concentric ribs covers the raised ridge, anterior flanges and dome and circumscribes a sunken central region composed of a highly reticulated meshwork of ribs (Figs. 3.15 C,D,F). The overall pattern formed by both sets of ribs resembles a fingerprint. The dome is indistinct with a small opening from which the bristle emerges (Fig. 3.15F).

Apical scales are more circular in outline, but asymmetrical due to a forward projecting triangular shaped apex (Fig. 3.15E). The left margin of the apex consists of a winged rib that becomes fused with the proximal rim. The raised transverse ridge is reduced or absent and the distinction between parallel and reticulated ribs is not as obvious as on the body scales.

Although bristles were not observed attached to scales, all samples containing *M.* sp. 1 had bristles with three serrated ridges, similar in structure to those of *M. punctifera* and *M. transsylvanica*. *M.* sp. 1 also resembles these latter two taxa in having similar shaped apical scales, body scales with similar shaped domes and bristle openings, and in lacking secondary markings in the posterior region. Body scales of *M.* sp. 1 also bear a slight resemblance to those of *M. peronoides* in having a raised thickened transverse area separating less electron dense regions. However, scales of *M. peronoides* lack a dome and have quite different ornamentation. Further work will be needed to be able to fully describe *M.* sp. 1 and determine its affinity to other species of *Mallomonas.*

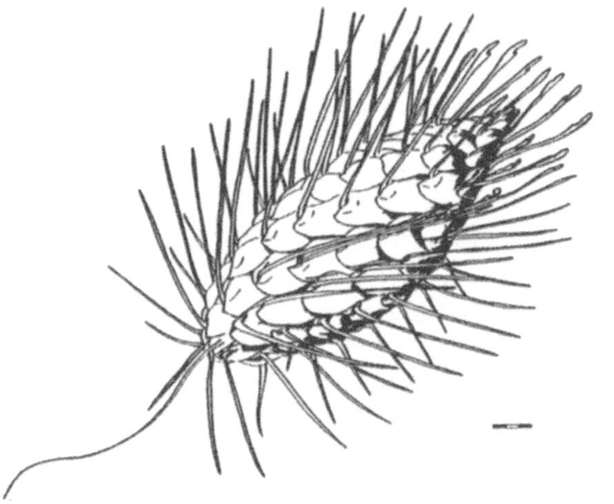

Fig. 3.17. Three-dimensional reconstruction of a whole cell of *Mallomonas hamata* depicting the distribution of scales and bristles. Scales lack domes and V-ribs and are positioned approximately 45° to the longitudinal axis of the cell. Note the distribution of the longer needle-like and the shorter hooked bristles.

Sectio Heterospinae Momeu & Péterfi 1979

Series Hamatae Asmund & Kristiansen 1986

Mallomonas hamata Asmund

Cells are oblong-ovate or ellipsoidal and are covered with two distinct types of bristles (Figs. 3.17; 3.18A,B,D).

Body and posterior (rear) scales are present, however, specialized apical scales are lacking. Body scales are ovate, with a narrow proximal rim that lines about one-half of the perimeter of the cell, and a slightly raised oval shaped dome (Fig. 3.18E). The dome is smooth and positioned to the right side of the longitudinal axis of the scale, resulting in an asymmetric morphology (Figs. 3.3A; 3.18E-F). The pores of the base plate are evenly spaced, lie in longitudinal or diagonal rows (Fig. 3.18E), and are often larger in diameter along the margins and in the posterior of the shield. The distal end of each scale usually has a short row of minute teeth (Fig. 3.18E). Additional siliceous structures are lacking. Rear scales are smaller and rounder in outline, and usually lack a clearly defined dome (Fig. 3.18C).

Using TEM, Asmund and Kristiansen (1986) made reference to a rib along the posterior border of the dome. Under SEM it is clear that such a rib is not present on the top surface of the scale and is most likely equivalent to the posterior rim of the concave opening of the dome when viewed from underneath (arrow, Fig. 3.18F).

Bristles are of two types (Figs. 3.18A,D). Long, needle-like bristles usually cover all but the posterior end of the cell and are associated only with body scales. Smaller, hooked shaped bristles are restricted to the posterior of the cell, but, can be associated with either body scales or the smaller rear scales (Figs. 3.18A,D). The needle bristles have a smooth shaft that is roughly the same diameter throughout (Figs. 3.18A-B), a foot that is positioned at a 20° to 90° angle to the shaft (Fig. 3.18A), and a blunt distal tip lined with teeth (Fig. 3.3B). The smaller bristles, known as hooked bristles, have a 'C' shaped or cleft like distal portion connected to a thin shaft (Figs. 3.3A; 3.18F). A ring of teeth lines the lower edge of the cleft and

→

Fig. 3.18. Mallomonas hamata. Fig. A. Whole cell depicting the arrangement and orientation of scales and the distribution of both long needle-like bristles and short hooked bristles. Scale bar is 5μm. Fig. B. Remains of the siliceous covering of one cell. Note the length and nature of the needle-like bristles and the shallow cavity of the dome. Scale bar is 10μm. Fig. C. Close-up of the posterior end of a cell. The rear most scales are smaller, more rounded and arranged with their longitudinal axes more or less parallel with the longitudinal axis of the cell. Scale bar is 2μm. Fig. D. Partial remains of one cell showing the distribution of the hooked scales along the posterior end of the cell. Scale bar is 5μm. Fig. E. Body scales depicting the proximal rim, evenly spaced base plate pores and the slightly raised dome. Note the small row of teeth along the distal margin (arrow). Scale bar is 2μm. Fig. F. Body scales and hooked bristles. Note the 'C' shaped distal portion and thin shaft of each bristle. The arrow depicts the posterior rim of the dome viewed from the undersurface of the scale; this structure appears as a rib in TEM micrographs. Scale bar is 2μm.

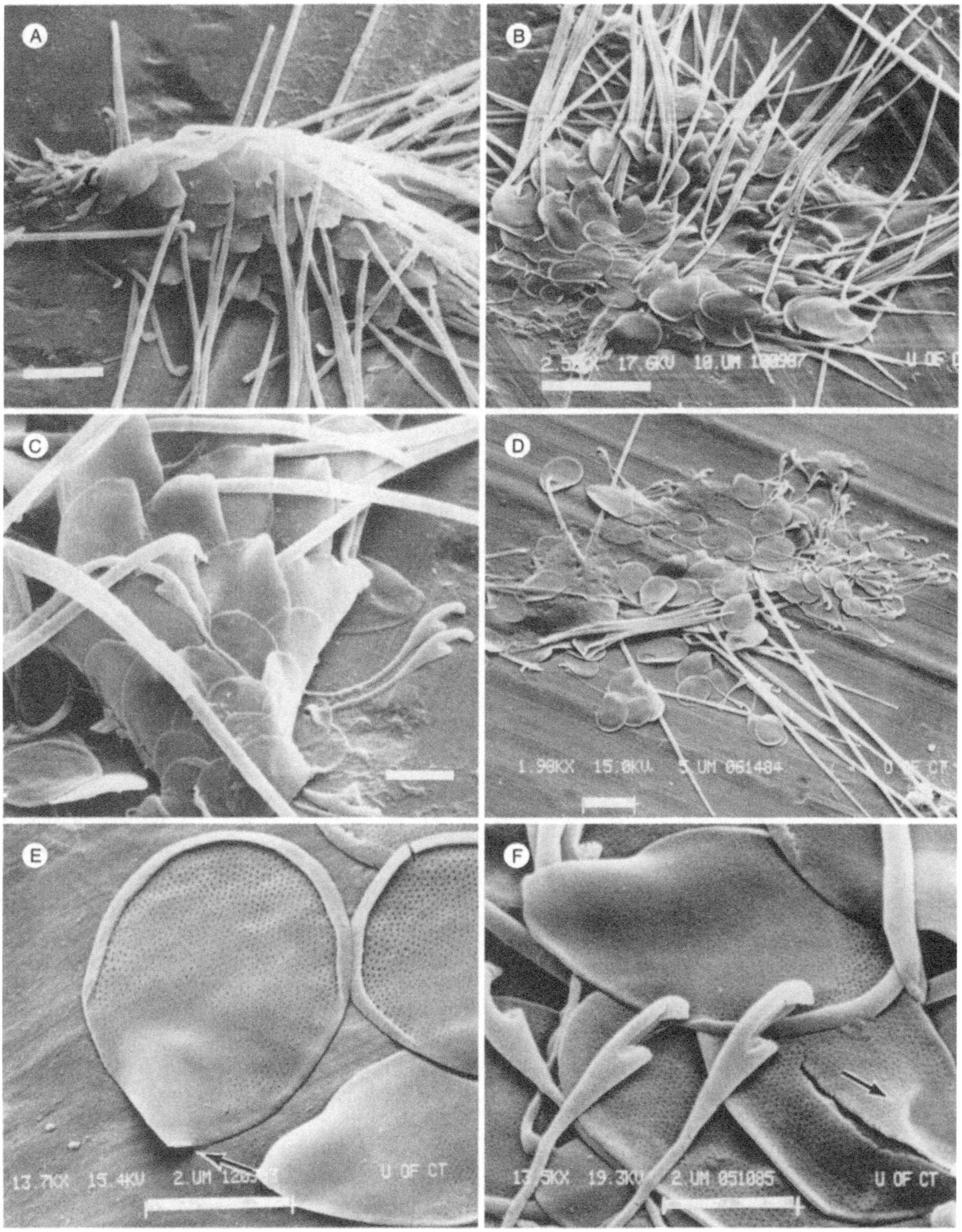
A
B
2.5 KX 17.6KV 10.UM 1 987
U OF C
C
D
1.90KX 15.0KV 5 UM 061484
U OF CT
E
13.7KX 15.4KV 2.UM 120
U OF CT
F
13.5KX 19.3KV 2.UM 051085
U OF CT

the upper portion is marked with ribs (Figs. 3.3A; 3.18F).

Normally, the number of needle bristles on a given cell is greater than the number of hooked bristles; over 80 needle bristles have been observed on a given cell. However, in several populations cells with hooked bristles distributed over one-third to one-half of their posterior were observed; some cells had over 65 hooked bristles (Fig. 3.18D).

Body scales are spirally arranged on the cell surface with their longitudinal axes positioned at approximately a 45° angle to the longitudinal axis of the cell (Fig. 3.18A). Such an arrangement of the scales results in the bristle openings of the domes being positioned such that all face the posterior of the cell; this, in turn, allows the bristles to become parallel with the cell when the organism is swimming. The extreme posterior body scales are often not positioned as above, but, with their longitudinal axes being parallel with that of the cell (Fig. 3.18C).

All features of the siliceous armour can be determined with either SEM or TEM. Thus, identification with either instrument is easily accomplished. By observing cell shape, the distribution of the two clearly distinct types of bristles and the characteristically shaped body scales, identification is also possible with LM. Identifications cannot be based solely on the presence of the hooked bristles since other species in the Sectio Heterospinae Momeu and Péterfi have similar bristles. Identification based on body scale architecture is possible.

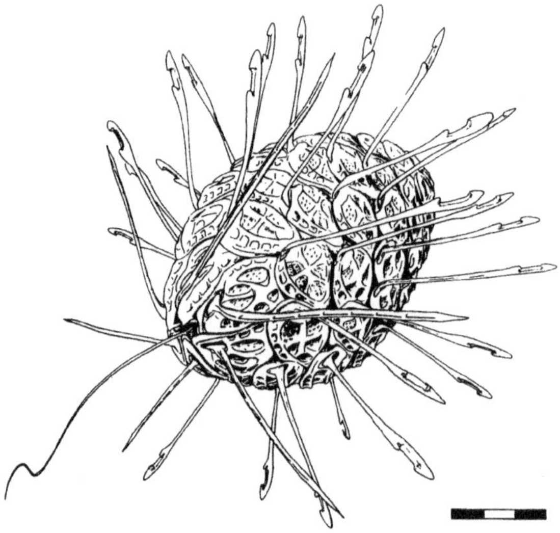

Fig. 3.19. Three-dimensional reconstruction of a whole cell of *Mallomonas canina* depicting the distribution of scales and bristles. Scales are positioned in spiral rows with their longitudinal axes at approximately a 45° angle to the longitudinal axis of the cell.

Mallomonas hamata could be confused with *Mallomonas americana* Dürrschmidt which is identical in all respects to *M. hamata* except for the presence of a posterior submarginal rib with struts on the body scales. To date, *M. americana* has been observed only from several localities in South America (Dürrschmidt, 1983b) and described with TEM. Thus, it is unclear if this taxon could be distinguished from *M. hamata* using SEM or LM. If the submarginal rib is a surface structure, as is true of other species in the Sectio Heterospinae, their separation using SEM would be possible. It is unclear if the submarginal rib will be discernible with LM.

Series Heterospinae Asmund & Kristiansen 1986

Mallomonas canina Kristiansen

Cells are small, spherical to slightly ovoid and are covered with short bristles of one or two types (Figs. 3.19; 3.20A,C).

A few smaller, domeless and less ornamented posterior scales are present (Fig. 3.20B), however, the majority of scales are larger, obovate and domed (Figs. 3.20B,D). All scales have an upturned proximal rim encircling one-half to two-thirds of the scale, a thickened distal margin, a base plate perforated with small pores often aligned in rows, and a continuous submarginal rib (Fig. 3.20E). The V-rib is broadly U-shaped with equally thick forward projecting arms that fuse on the distal end of the dome forming the continuous submarginal rib (Figs. 3.20A,E). The distal half of the submarginal rib is marked with small rounded papillae (Fig. 3.20E); by observing thinly silicified scales it appears that the distal half of the submarginal rib is constructed by the fusion of individual papillae.

The surface of the scale circumscribed by the submarginal rib is ornamented with a series of prominent papillae and a reticulum of ribs (Fig. 3.20). Four primary ribs radiate at 90° angles from a central point (Fig. 3.20E). One primary rib,

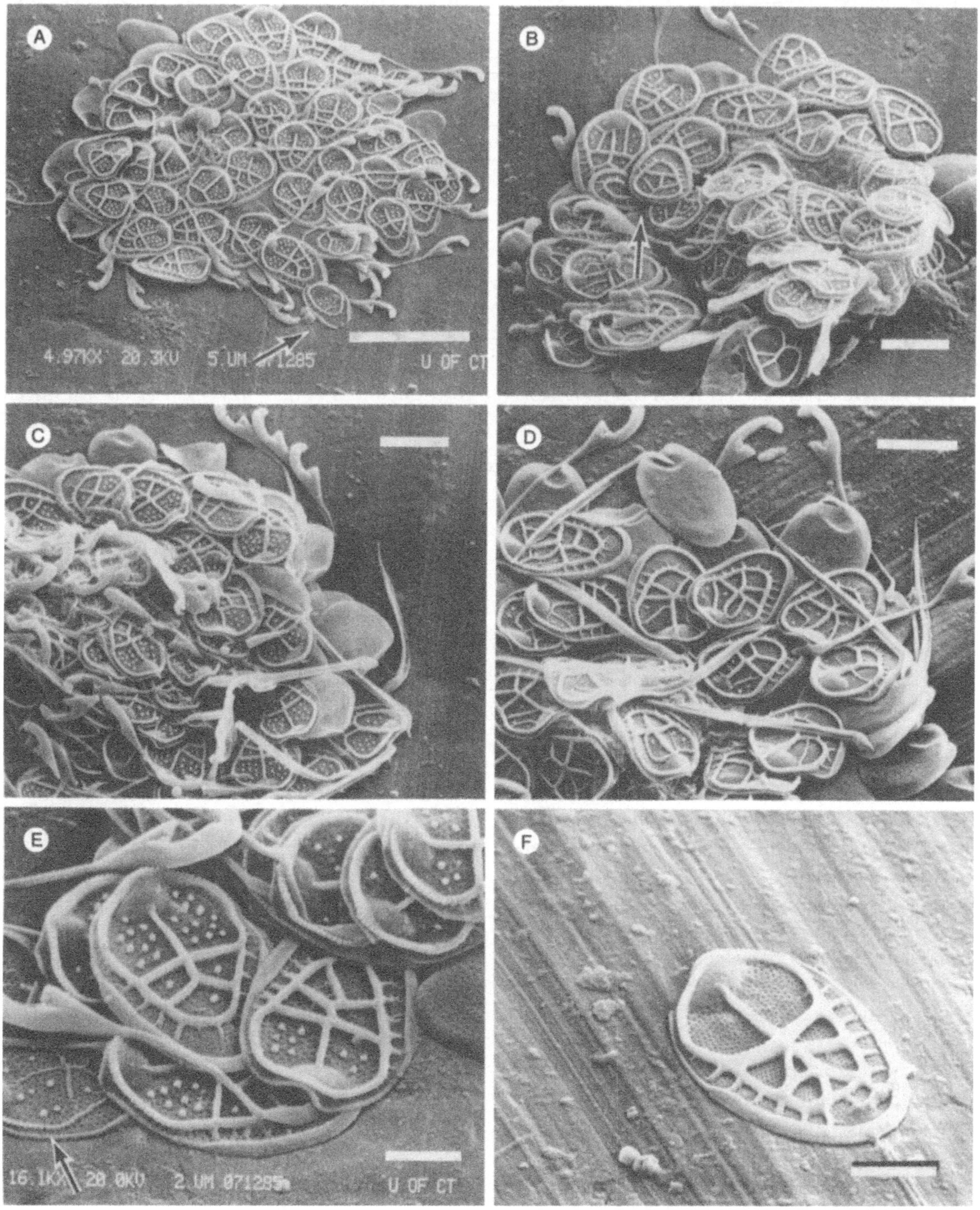

*Figs 3.20A-E. **Mallomonas canina.*** Figs. A-B. Siliceous remains of whole cells depicting body scales and hooked bristles. Note the smaller, domeless posterior scales (arrow). Scale bars are 5μm and 2μm, respectively. Figs. C-D. Groups of body scales with greater (Fig. C) or fewer (Fig. D) shield papillae. Note both needle and hooked bristles. Scale bars are 2μm. Fig. E. Close-up of body scales denoting the small dome, the continuation of the submarginal rib over the dome and the shallow bristle opening. Note the apparent fusion of papillae (arrow) to form the distal half of the submarginal rib. Scale bar is 1μm. Fig. F. ***Mallomonas hindonii.*** Body scale. Note the lack of surface papillae and the overall close resemblance to body scales of *M. canina*. Scale bar is 1μm.

referred to as the longitudinal apical rib, runs forward, parallel with the longitudinal axis of the scale, and terminates at the base of the dome. Two primary ribs radiate transversely at 180° across the scale and fuse with the submarginal rib: both of these ribs form the transverse rib. The fourth primary rib, directed towards the posterior of the scale, is almost always forked or branched at least once; each branch is fused with the submarginal rib (Fig. 3.20E). Any of the four primary ribs may have additional branches (Fig. 3.20D) or be lacking (Fig. 3.20A). From one (Fig. 3.20D) to 35 (Fig. 3.20A) papillae, mostly distributed on the anterior half of the scale, have been observed. A series of short ribs radiate at 90° angles from the posterior half of the submarginal rib (V-rib portion) onto the flange (Fig. 3.20D). All ribs and papillae are usually situated in areas between the rows of base plate pores.

Domes are small, smooth and without a distinct posterior border (Figs. 3.20C,E). The inverted U-shaped cavity from which the bristle emerges is shallow and slightly offset to the right of center (when viewed with the distal end of the scale on top).

Needle and/or hooked bristles of approximately equal length may be found (Figs. 3.20C-D). Needle bristles are stout and with a longitudinal ridge running the length of the shaft. The hooked bristles are as described for *Mallomonas hamata* with a swollen 'C' shaped distal end attached to a thin and delicate shaft (Fig. 3.20). In this study cells with only hooked bristles or both hooked and needle bristles were observed. Because of the large difference in thickness between the thin shaft and the expanded distal hook, the latter are often observed on living cells as forming a ring around the cell proper.

Body scales are spirally arranged on the cell surface with their longitudinal axes positioned at about a 45° to 60° angle with the longitudinal axis of the cell (Fig. 3.19); the more spherical the cell becomes the less obvious is the arrangement of scales.

Mallomonas heterospina Lund

Cells are small, spherical, obovoid or ovoid and are usually covered with short bristles of two types (Fig. 3.21A).

Scales are oval to slightly obovate, with an upturned posterior rim encircling one-half to two-thirds of the scale, a thickened distal margin, a relatively small dome, a base plate perforated with evenly spaced pores and a dense reticulum of ribs on the shield (Figs. 3.21C-F). A continuous submarginal rib and transverse rib are present (Fig. 3.21F), however, are not always clearly defined due to the high degree of ribbing on the shield (Figs. 3.21C-D). As observed on scales of *Mallomonas canina*, the distal portion of the continuous submarginal rib, as well as the transverse rib, are often marked with small teeth (Figs. 3.21C,E,F). The distal border of the dome above where the bristle emerges is also marked with teeth (Fig. 3.21E). The area of the shield enclosed by the continuous submarginal rib is covered with a reticulum of ribs of various thicknesses (Fig. 3.21). In the proximal region thick ribs often form large meshes or pores which, in turn, are subdivided by less prominent ribs (Figs. 3.21E-F). A distinct longitudinal apical rib is not present. Short ribs extend at 90° angles onto the flange along the proximal two-thirds of the continuous submarginal rib (Figs. 3.21C-D); the latter may be branched (Fig. 3.21F).

Domes are small, account for 10.4% of the total surface area of the scale and, except for the continuation of the submarginal rib, are usually smooth (Figs. 3.21B-D). Domes with additional ribs have been observed (Figs. 3.21E-F; Wee, 1982; Cronberg, 1980; Asmund and Kristiansen, 1986). The inverted U-shaped cavity from which the

→

Fig. 3.21. Mallomonas heterospina. Fig. A. Siliceous remains of a whole cell with both hooked and needle bristles. Note the pattern of overlap of the body scales. Scale bar is 5μm. Fig. B. Close-up of body scales in Fig. A. Note the continuation of the serrated submarginal rib over the dome. Scale bar is 2μm. Fig. C. Domed scales and a hooked bristle. Note the reticulation of ribs and the small, unornamented dome. Scale bar is 1μm. Fig. D. High magnification of a body scale in Fig. C. Note the arrangement of base plate pores. Scale bar is 1μm. Fig. E. Body scale with a series of thick ribs which in turn are subdivided by less prominent ribs. Note the serration (arrow) lining the distal perimeter of the dome. Scale bar is 1μm. Fig. F. Body scales and bristles. Note the ribs on the domes. Scale bar is 2μm.

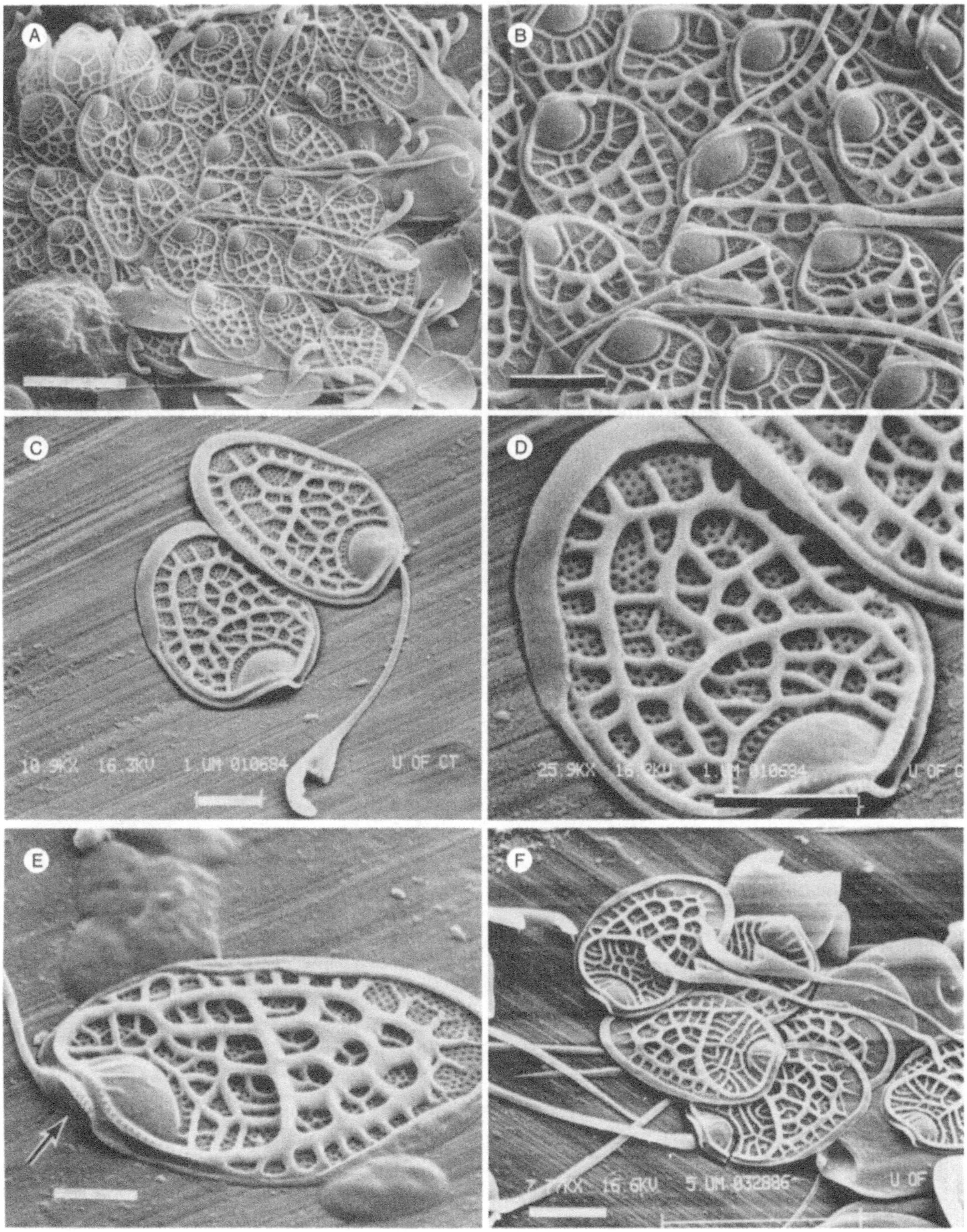
A
B
C
D
E
F
10.9KX 16.3KV 1.UM 010684 U OF CT

bristle emerges is distinct (Figs. 3.21E-F) and positioned to the right of center (when viewed with the distal end of the scale on top) (Fig. 3.21C). A few small, domeless scales may be observed.

Cells have both needle and hooked bristles (Fig. 3.21A). Hooked bristles are short, have a thin and delicate shaft and are sculptured similar to those described for *Mallomonas hamata*. Cells of *M. heterospina* are most often reported to have needle and hooked bristles more or less restricted to the anterior and posterior ends of the cell, respectively (Asmund, 1956; Takahashi, 1978; Asmund and Kristiansen, 1986). Cells with only needle bristles, but not cells with only hooked bristles, have also been observed (Asmund and Kristiansen, 1986). In this study only cells with both bristle types were observed. Although a few populations had cells where the hooked bristles were predominantly arranged on the posterior end of the cell, most populations had cells where both bristle types were scattered over the cell. As discussed for *Mallomonas canina* the expanded distal tips of the hooked bristles are often seen in live preparations as a ring around the cell proper.

Body scales are arranged in spiral rows on the surface of the cell with their longitudinal axes aligned at a 45° to 60° angle with the longitudinal axis of the cell. Each scale is overlapped by the scale positioned below it in the same spiral row and by scales positioned in the spiral row above it (Figs. 3.21A-B).

Mallomonas hindonii Nicholls

Although whole cells were not observed, they are reported to be small, similar in size to those of *M. canina*, ellipsoidal and covered with bristles (Nicholls, 1982). Except for specialized apical scales, all scales are similar in structure.

Body scales are elongate-obovate and have a perforated base plate with evenly spaced pores, a small dome, a narrow proximal rim encircling about one-half to two-thirds of the scale and a continuous submarginal rib (Fig. 3.20F). The V-rib is broadly rounded with arms that are continuous with the anterior submarginal ribs, which in turn fuse on the dome forming a single continuous rib. The surface of the scale circumscribed by the submarginal rib is ornamented with a group of five to seven thick ribs that usually appear to originate from a central point. Two of these ribs radiate outwards at about a 140° to 180° angle across the shield and fuse with the submarginal rib; these two ribs form the transverse rib that serves to divide the area circumscribed by the submarginal rib in half (Fig. 3.20F). A third rib, the longitudinal apical rib, extends forward parallel with the longitudinal axis of the scale and terminates at the base of the dome; this rib serves to divide the anterior half of the scale into two relatively large unornamented areas. The remaining two to four ribs radiate towards the posterior of the scale and fuse with the submarginal rib. These latter ribs may be branched. A series of short, sometimes forked, struts radiate from the outer margin of the submarginal rib onto the posterior flanges. Besides the ribs, additional ornamentation (e.g. papillae) are lacking. The ridge of the distal portion of the submarginal and transverse ribs are papillate as described for *M. canina*. The dome is small, smooth and symmetrically positioned.

The apical most scales are more circular in outline, but with a very broad and highly asymmetric dome. The left side of the dome is drawn out into a pointed apex and the longitudinal apical rib crosses the dome and fuses with the submarginal rib.

Bristles are short, 6.5 to 8.5 μm in length (Nichols, 1982), needle-like, ribbed and have a single subapical tooth.

Mallomonas multiunca Asmund

Cells are small, spherical to slightly ovoid, and covered with short hooked bristles.

The majority of scales are ovoid with an upturned posterior rim encircling one-half to two-thirds of the scale, a base plate perforated with relatively large evenly spaced pores, a broad dome, and a series of large meshes on the shield formed by a reticulum of thick ribs (Figs. 3.22E-F). The continuous submarginal rib, less conspicuous than in most species within the series, crosses the shield proximal to the dome; as such, *Mallomonas multiunca* is the only species in the series where the submarginal rib does not extend over the dome. Two (sometimes three) prominent, large anterior meshes or 'eyes' are formed by the fusion of the longitudinal apical rib with the distal end of the submarginal rib (Figs. 3.22E-F). All other meshes

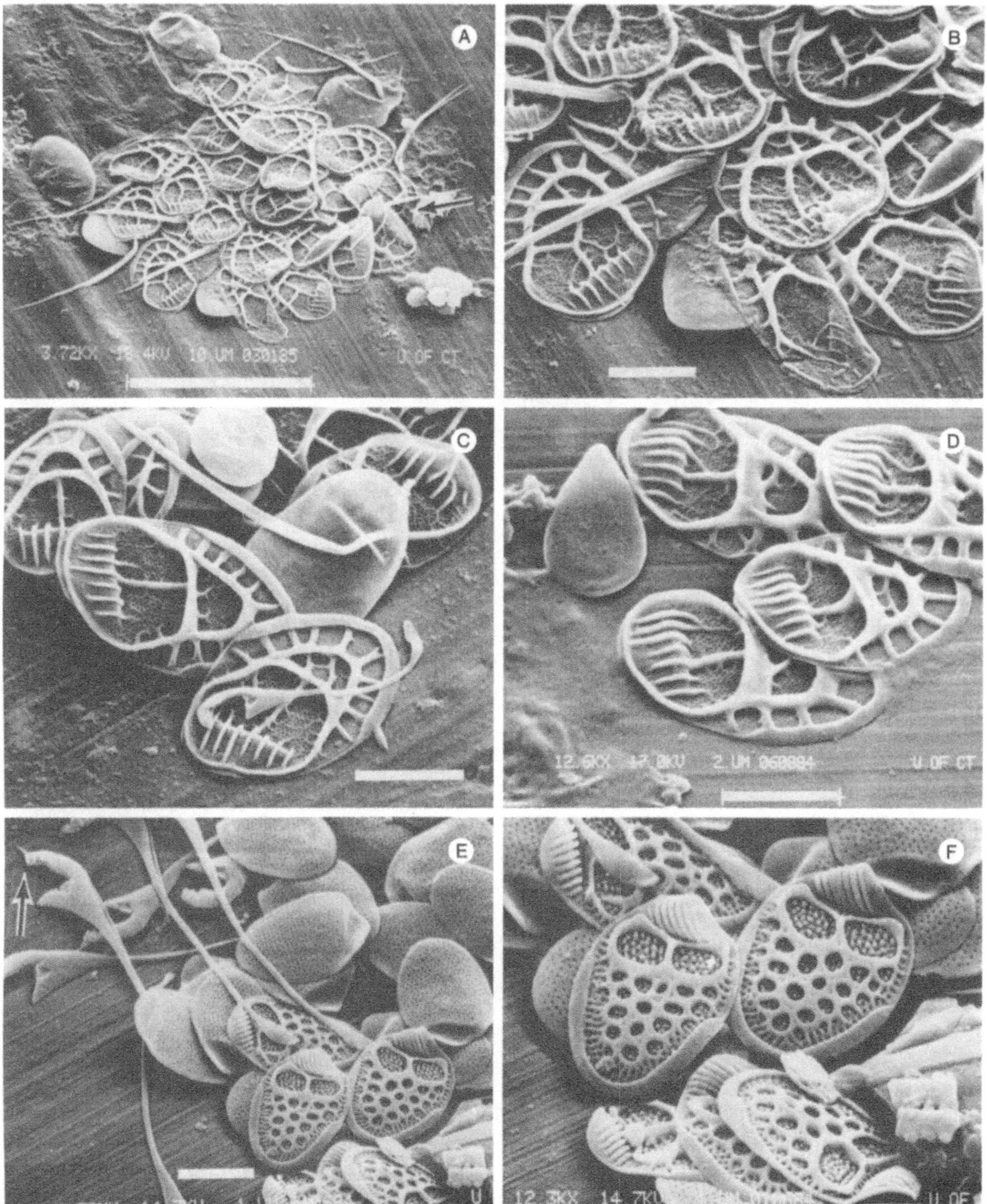

Fig. 3.22A-D. Mallomonas pugio. Fig. A. Siliceous remains of a whole cell denoting body scales, apical scales (arrow) and needle bristles. Scale bar is 10μm. Fig. B. Close-up of body scales from Fig. A. Scale bar is 2μm. Figs. C-D. Body scales depicting the distribution of ribs on the shield, flange and dome. The hooked bristle in Fig. C belongs to *M. hamata*. Scale bars are 2 μm. Figs. E-F. *Mallomonas multiunca*. Fig. E. Body scales and bristles. Note the distribution of base plate pores, the large eye-like meshes, the smaller meshes, the papillae originating from the base plate and the sculptured dome. Also, note the broad hook and extended distal tip (arrow) of the bristles. Scale bar is 2μm. Fig. 6. Close-up of the body scales in Fig. E. Scale bar is 1μm.

are significantly smaller in diameter. The base plate within each mesh is furnished with papillae; smaller ribs are not found (Figs. 3.22E-F). A series of short ribs radiates from the submarginal rim towards the base of the dome; however, ribs radiating from the posterior part of the submarginal rib onto the flange are either lacking or rare and short. A few domeless scales are often found on each cell; these may be larger or smaller than the domed scales. The smaller domeless scales are restricted to the posterior end of the cell.

Domes are relatively small and broad, cover approximately 10 to 12% of the surface area of the scale, and are marked with a series of oblique parallel ribs (Figs. 3.22E-F). The inverted U-shaped cavity from which the bristle emerges is shallow, broad and positioned to the right of center.

The hooked bristles differ from those of other species within the Sectio Heterospinae in having a broader hook and an extended distal tip (Fig. 3.22E).

Scales are arranged and positioned on the cell in a fashion similar to other taxa within the series.

Mallomonas pugio Bradley

Cells are small, nearly spherical and are covered with short needle bristles approximately half the length of the cell. Body scales are oval with an upturned posterior rim encircling half of the perimeter of the scale, an evenly spaced perforated base plate, well developed flange ribs and a continuous submarginal rib (Figs. 3.22C-D). The transverse and longitudinal apical ribs (see the description for *M. canina*) are the most prominent ribs found within the area circumscribed by the continuous submarginal rib (Fig. 3.22D); it is common to find two longitudinal ribs or a single one that is branched on a given scale (Fig. 3.22B). Additional smaller struts of lesser stature radiate at right angles from the larger ribs (Fig. 3.22D). The area within the continuous submarginal rib is unevenly divided by the transverse rib such that the distal portion is much larger (Fig. 3.22D).

Domes are large, broad with a straight posterior edge and marked with about 5 to 8 evenly spaced parallel ribs (Figs. 3.22C-D). The longitudinal apical rib(s) of the shield are continuous with ribs on the dome. Domes are often asymmetrically positioned (e.g. see Fig. 3.22A) and lack a clearly defined inverted U-shaped structure from which the bristle emerges.

Specialized apical scales are asymmetric with a large dome that is drawn out or extended on its left side into a point (arrow, Fig. 3.22A). The longitudinal apical rib extends across the dome dividing it in half (Fig. 3.22A); several smaller dome ribs may or may not be present.

Cells with only needle bristles were found. Bristles have a single subapical tooth and a ridge running the length of the shaft (Figs. 3.22A-C).

Body scales are spirally arranged on the cell surface with their longitudinal axes positioned at about a 45° to 60° angle with the longitudinal axis of the cell.

Similarities and Differences Between Species Within the Series Heterospinae

There are currently six species within the series Heterospinae; *Mallomonas canina, M. heterospina, M. harrisiae, M. pugio, M. hindonii* and *M. multiunca.* Distinguishing between the species, especially if whole cells are available, is not difficult and can be done with either SEM or TEM. However, some confusion could be encountered when identifying taxa within this Series if a decision must be based on isolated body scales. Because most of these species live in well defined habitats and, as such, represent valuable indicator organisms (see below) it is critical that proper identifications are made. As a result, a general overview of the characteristics common to the Series as a whole and a discussion of the more difficult comparisons is presented.

Cells of all taxa are small, often between 10 μm and 20 μm in length (Table 4.5), ovoid to spherical, and covered with short bristles that are approximately half the length of the cell. The majority of scales on a given cell are similar and oval to obovate in shape; very few specialized scales are produced. Two features common to all species are the presence of the continuous submarginal rib and the transverse rib that divides the area circumscribed by the continuous submarginal rib. Species differ in the degree and type of secondary ornamentation on the shield and in the prominence of the transverse rib.

The transverse rib is most obvious on scales of

Mallomonas pugio (Fig. 3.22), *M. canina* (Fig. 3.20) and *M. hindonii* (Fig. 3.20), less obvious on scales of *M. heterospina* (Fig. 3.21) and *M. harrisiae*, and the least apparent on scales of *M. multiunca* (Fig. 3.22). The area of the scale distal to the transverse rib on scales of *M. canina, M. hindonii, M. pugio* and *M. multiunca* is divided by an additional rib originating midway along the transverse rib and running parallel to the axis of the scale; this rib is referred to herein as the longitudinal apical rib. The longitudinal apical rib is usually not apparent on scales of *M. heterospina* and *M. harrisiae* where the distal portion of the shield is covered with a rather dense reticulum of ribs (Fig. 3.21). However, the longitudinal apical rib is apparent on the specialized apical scales of *M. harrisiae*.

Scales of *M. multiunca* and *M. pugio* have broad domes with a series of parallel ribs, while those of *M. canina* and *M. hindonii* are small, smooth and usually lack ribs. The longitudinal apical rib of *M. multiunca* fuses with the submarginal rib just below the proximal border of the dome forming two distinctive pores or 'eyes' (Fig. 3.22). In *M. pugio*, the longitudinal apical rib may branch and is always continuous onto the dome (Fig. 3.22). The longitudinal marginal rib of scales of *M. canina* and *M. hindonii* terminates at the proximal border of the dome and is rarely branched.

The most difficult distinctions to make within the Series Heterospinae are between *Mallomonas canina* and *M. hindonii, M. pugio* and *M. harrisiae* and *M. harrisiae* and *M. heterospina.*

Mallomonas canina vs. *Mallomonas hindonii*

Isolated body scales of *M. canina* and *M. hindonii* can be very difficult to separate. Scales from both taxa have transverse and longitudinal apical ribs, similar domes and weakly formed flange ribs. In addition, the ribs within the continuous submarginal rib often appear to originate from a central point.

The most obvious difference in scale ornamentation is the presence of papillae on scales of *M. canina*. However, some cells of *M. canina* have scales with only a few papillae (e.g. one) making a proper identification often difficult (e.g. Fig. 3.20D). Thus, when possible, additional features should also be used. *M. hindonii* has asymmetrical apical scales with pointed distal ends, and only needle bristles with a single subapical tooth. *M. canina* lacks specialized apical scales and most cells have hooked bristles. Lastly, on scales of *M. hindonii* the transverse rib is often slightly V-shaped (Fig. 3.20).

Mallomonas pugio vs. *Mallomonas harrisiae*

Both of these taxa have specialized apical scales that are similarly shaped and ornamented, needle bristles and a series of parallel ribs on the dome. Body scales can be distinguished in that those of *M. pugio* have a more prominent transverse rib, a distinct longitudinal apical rib, a broader dome and significantly less ribbing on the shield.

Mallomonas harrisiae vs. *Mallomonas heterospina*

Isolated body scales from these two species could also be difficult to separate. Scales from both taxa are similarly shaped and have a high degree of shield reticulation, especially on the distal end. The best character to identify isolated body scales is the presence of parallel dome ribs on *M. harrisiae*. However, although scales of *M. heterospina* have been reported to lack dome ribs (Takahashi, 1978; Asmund and Kristiansen, 1986), they have been observed (e.g. Fig. 3.21F; plate XVIII of Wee, 1982; Fig. 28E-F of Asmund and Kristiansen, 1986).

If whole cells are present other distinguishing features are the specialized apical scales of *M. harrisiae* and the presence of hooked as well as needle bristles on *M. heterospina*.

Sectio Akrokomae Asmund and Kristiansen 1986

Mallomonas akrokomos

Cells of *Mallomonas akrokomos* are narrow, spindle-shaped with a rounded anterior end, a tapered posterior tail that may bend slightly, and an apical tuft of bristles (Figs. 3.23; 3.24A).

Scales are of three types: apical, domed, bristle bearing scales; domeless body scales; and elongated posterior (caudal) scales. The base plate of all scales has a proximal patch of pores and fewer, irregularly spaced pores of lesser diameter on the distal end. On body and caudal scales the posterior patch of pores is more obvious when viewed from the

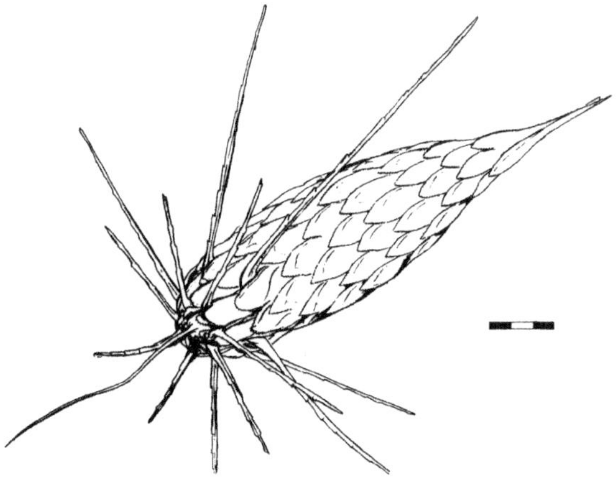

Fig. 3.23. Three-dimensional reconstruction of a whole cell of *Mallomonas akrokomos* depicting the distribution of scales and bristles. Scales are positioned parallel to the longitudinal axis of the cell and bristles are restricted to the anterior end of the cell. Note the spindle shape of the cell and the elongated caudal scales.

undersurface because the diameters of the pores are larger than they are on the top surface of the scale (compare body scales on Fig. 3.24C and caudal scales on Fig. 3.24D). This difference in pore size between the two surfaces of a scale is easily seen with SEM, but not with TEM. The area of the base plate with the patch of pores is elevated slightly above the plane of the scale (Fig. 3.24F); this feature can only be detected when the undersurface of a scale is viewed. On apical scales the diameters of the pores are equal on both surfaces (Fig. 3.24E).

All scales have a posterior rim and a row of teeth along the distal most margin (Fig. 3.24). Apical and body scales have a submarginal row of regularly spaced pores that lines the length of the proximal rim (Figs. 3.24C,E,F); this row of pores can be seen using either TEM or SEM, although with SEM it may be difficult to view from the top surface if the pores lie under the rim.

Apical scales are oblong with a U-shaped submarginal rib; the submarginal rib and the posterior rim both extend further along the concave edge of the dome resulting in an asymmetric shape (Fig. 3.24E). Body and posterior scales are symmetric in outline and lack both a dome and submarginal rib. Body scales are rhombic with a pointed distal end. Body scales grade into the narrower and longer spatula shaped posterior scales (Figs. 3.24A,D).

Scales are arranged in spiral rows with their longitudinal axes parallel to that of the cell and their distal ends facing the anterior of the cell. The proximal end of a scale, including the dense patch of pores, is overlapped by the distal end of the scale(s) located in the spiral row below it and by the scale positioned behind it within the same spiral row (Fig. 3.24A). Thus, although the pattern of overlap of scales within a given spiral row is identical to all other species of *Mallomonas*, adjacent spiral rows are overlapped in an opposite manner (see Chapter 2). Perhaps the organic substance which binds the scales together is exuded out through the pores in the patch and submarginal row. The overlapping pattern of the scales is clearly observed with SEM; however, using TEM the pattern could appear opposite if the scales are viewed upside down.

Bristles are delicate, needle-like with a 'conspicuous slit' (Asmund and Kristiansen, 1986) that originates at the base and usually extends one-third to one-half the length of the bristle (Fig. 3.24A). There are two distinct types of bristles. One is shorter, attached to the apical most ring of scales, with two rows of serrations; the other is longer, attached to the second ring of apical scales, with a single row of teeth. Typically, the longer uniserrate bristles are as long as the cell (Figs. 3.23; 3.24A).

Sectio Striatae Asmund and Kristiansen 1986

Series Striatae Momeu & Péterfi 1979

Mallomonas asmundiae (Wujek & van der Veer) Nicholls

Synonym: *M. cratis* var. *asmundiae* Wujek & van der Veer

Cells are ovoid to spherical (Asmund and Kristiansen, 1986) and completely covered with slightly curved bristles (Figs. 3.25; 3.26A). Except for the rare presence of a small domeless posterior scale (Fig. 3.26E), all scales are domed and similar in size. Scales are spirally arranged with their longitudinal axes positioned at a 70° to 90° angle to the longitudinal axis of the cell (Figs. 3.25; 3.26A) and are overlapped as described for *M. striata*.

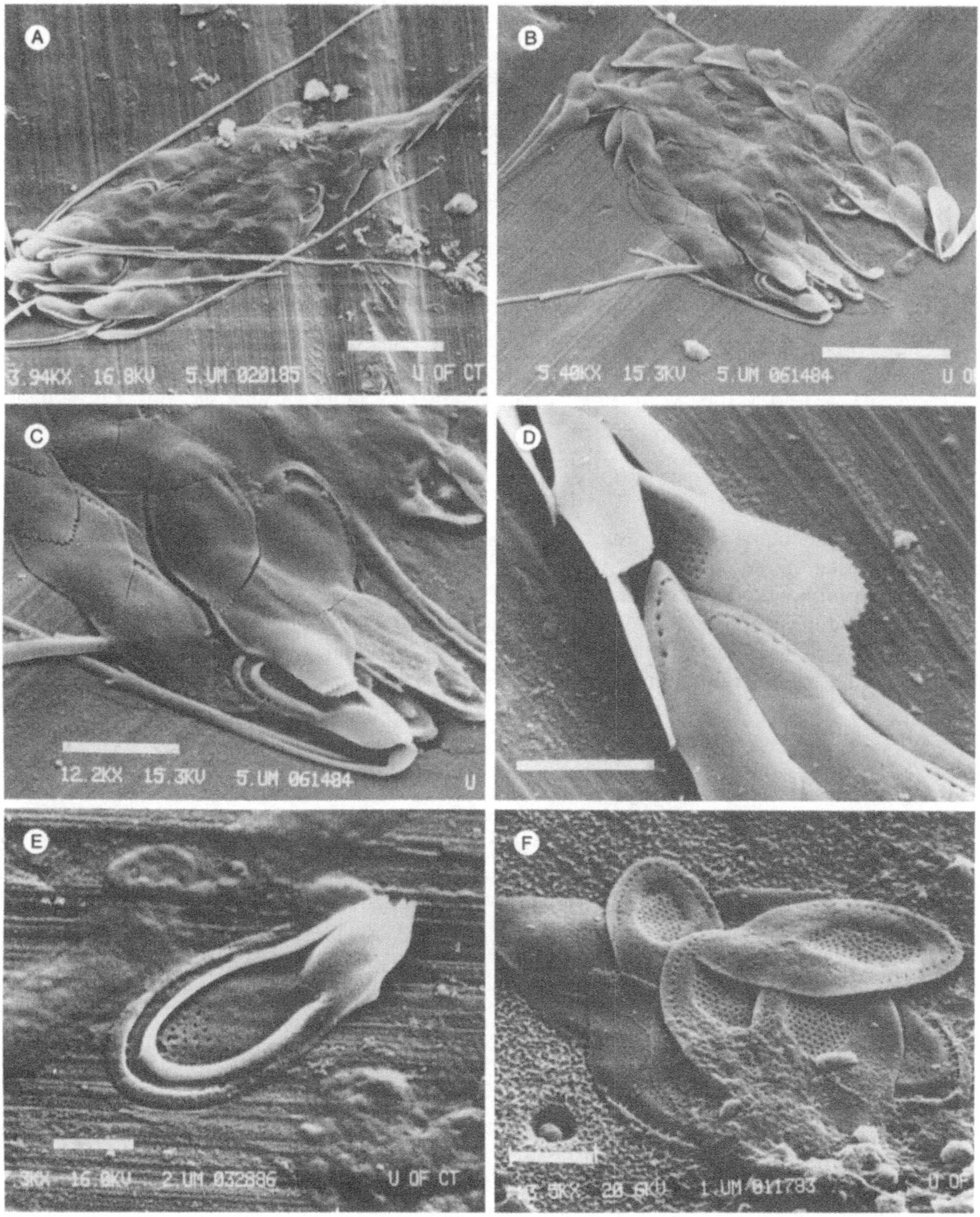

Fig. 3.24. Mallomonas akrokomos. Figs. A-B. Whole cells depicting the spindle-like shape, tapered posterior tail, apical tuft of bristles and the overlapping of scales. Note the shorter bristles with two rows of teeth and the longer ones with a single row of teeth. Scale bars are 5μm. Fig. C. Close-up of anterior end of the cell in Fig. B. Note the serrated distal margins of the scales. Scale bar is 2μm. Fig. D. Close-up of the posterior end of a cell illustrating the narrower and more elongated scales. Scale bar is 2μm. Fig. E. Isolated bristle-bearing apical scale. Note the asymmetry of the scale and the presence of both a dome and a submarginal rib. Scale bar is 1μm. Fig. F. Body scales. The proximal patch of base plate pores and the uniseriate row of pores lining the posterior flange are clearly observed, especially when viewed from the undersurface of the scale.

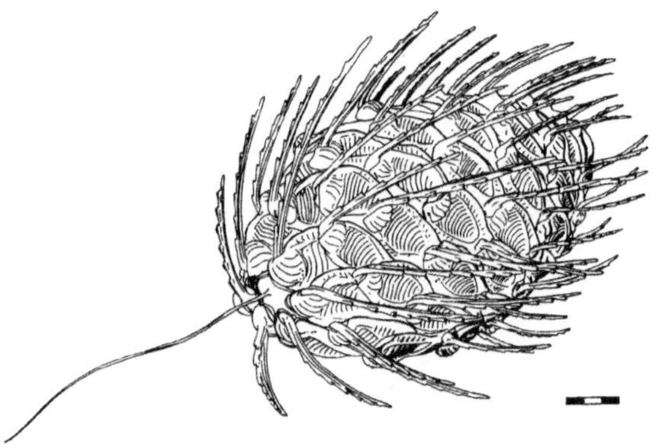

Fig. 3.25. Three-dimensional reconstruction of a whole cell of *Mallomonas asmundiae* depicting the distribution of scales and bristles. Scales are positioned in spiral rows with their longitudinal axes at approximately a 45° to 60° angle to the longitudinal axis of the cell. The serrated bristles cover the cell.

Scales are elongate-obovoid with strong lateral incurvings, a posterior rim encircling one-half of the perimeter, a pronounced and strongly hooded V-rib and a large dome (Figs. 3.26C-D). The distal ends of the arms of the V-rib curve and become continuous with the anterior submarginal ribs (Table 3.2), the latter of which may be slightly winged and terminate along the sides of the dome (Figs. 3.26C-E). The dome, shield and posterior flange are ornamented with neatly arranged, equally spaced, sets of parallel ribs (Figs. 3.26C-E). The shield, which extends along the sides of the dome (Asmund and Kristiansen, 1986), is transversed with 9 to 18 ribs (Figs. 3.26C-D); the anterior most ribs may be continuous with those of the dome (Fig. 3.26C). About 30 to 35 ribs radiate at right angles from the V-rib onto the posterior flange (Fig. 3.26D), but, do not extend under the posterior rim. The anterior flanges are very narrow and lack ornamentation (Figs. 3.26C-D). Scales with twice the compliment of ribs, as described by Nicholls (1987a), were not observed.

Domes are marked with 6 to 11 parallel ribs which in turn are arranged parallel (Fig. 3.26D) or at an oblique angle (Fig. 3.26C) with the longitudinal axis of the scale. Short perpendicular struts often connect neighbouring ribs (Fig. 3.26B). As described for *M. cratis*, the bristles emerge from under the dome through a strongly U-shaped opening (Figs. 3.26C,E) positioned slightly to the right of center. Nicholls (1987a) observed on TEM micrographs that along the posterior margin of the dome the curved portion of the ribs could be mistaken as a series of struts radiating onto the shield; such scales could be mistaken as being from *M. corcontica*. This is not a problem with SEM.

Bristles are strongly ribbed, curved and serrated along the convex rib. Each point along the serration consists of 2 or 3 teeth (Figs. 3.26B,E,F), the distal most cluster being close to the terminal end of the bristle.

Mallomonas cratis Harris & Bradley

Cells are ovoid to cylindrical, often two times (Wee, 1982) or more as long as wide and completely covered with curved bristles (Fig. 3.27A). All scales possess a dome and are spirally arranged with their longitudinal axes orientated at approximately a 70° angle to the longitudinal axis of the cell (Fig. 3.27A). On the longer more cylindrical cells, scales are positioned closer to a 90° angle, especially in the posterior region. The overlap of scales is identical to that described for *M. striata* such that only the dome, anterior flanges and a portion of the shield are exposed on intact cells (Fig. 3.27A).

Scales are narrow ovoid (Takahashi, 1978) with slight lateral incurvings, and have a posterior rim encircling one-half of the perimeter, and a strongly hooded, roundly-angled V-rib with straight arms that extend almost to the margin of the scale. Scales usually lack well developed anterior submarginal ribs and the margin of the hood in the angle of the V-rib is serrated (Table 3.2; Figs. 3.27B-C). The shield, flanges and dome are all marked with ribs (Table 3.2). There is a progression from relatively straight, to slightly U-shaped, to strongly U-shaped ribs with a shift from the posterior of the shield, to the anterior of the shield and to the dome, respectively (Fig. 3.27C). The shield consists of about 15 to 18 equally spaced, parallel or concentric ribs; the ribs on the anterior part of the shield are continuous with those of the anterior flanges (Figs. 3.27C-D). The ridges of ribs in the posterior region of the shield are often lined with tiny serrations (Fig. 3.27C). The posterior flange has 16 to 23 short ribs that are spaced wider apart than those of the shield.

The dome is small, covering 12% of the surface area of the scale, and marked with strongly U-shaped ribs (Figs. 3.27C-D). The ribs extending

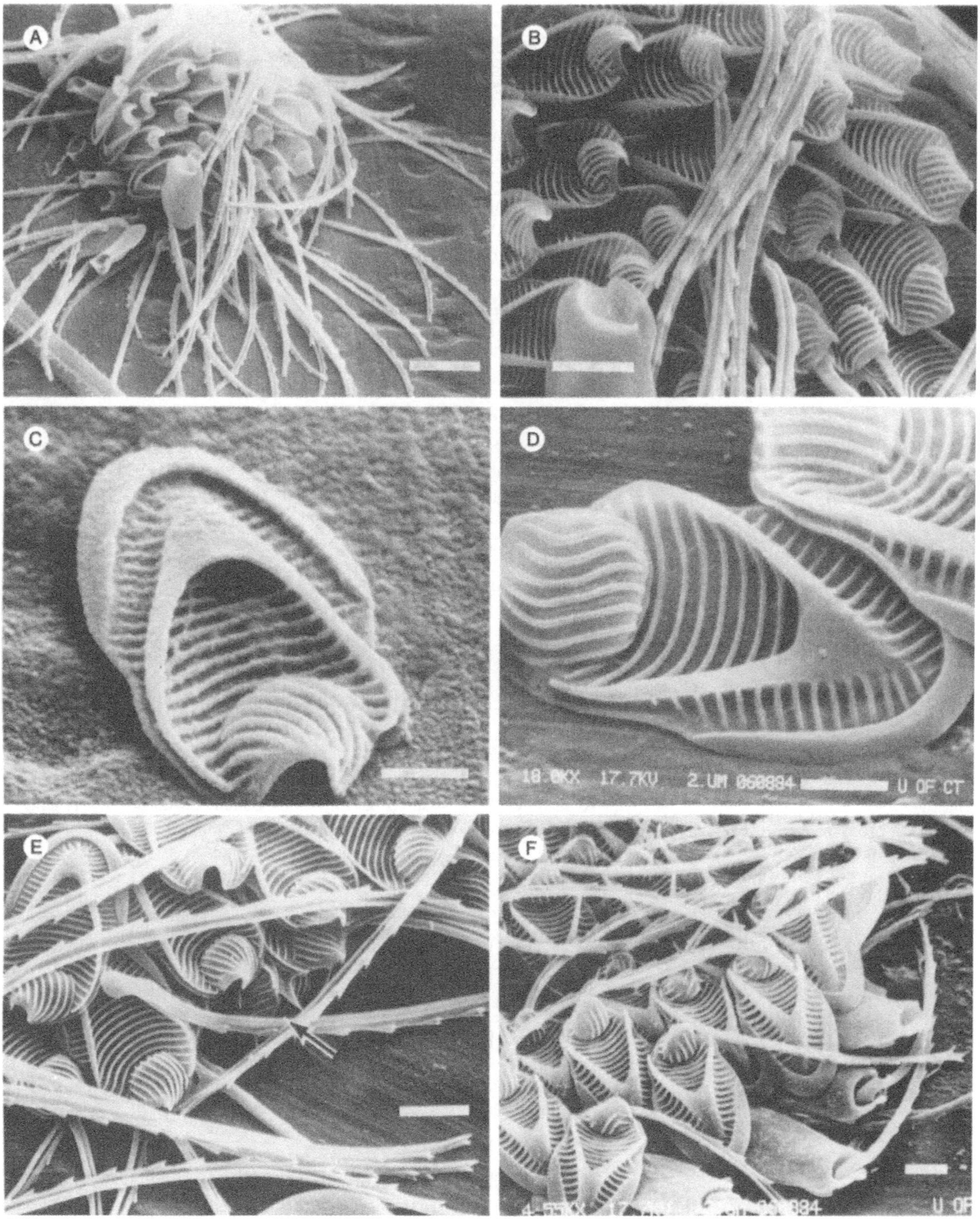

Fig. 3.26. Mallomonas asmundiae. Fig. A. Whole cell showing arrangement of scales and bristles. Scale bar is 5 μm. Fig. B. Close-up of Fig. A. Note the pattern of scale overlap, the short perpendicular struts connecting adjacent ribs on some domes and the serrations of the bristles. Scale bar is 2 μm. Figs. C-D. Isolated body scales depicting surface ornamentation. Note the arrangement of ribs, the lateral incurvings, the slightly winged submarginal ribs, the bristle opening and the large domes. Scale bars are 1 μm. Fig. E. Close-up of scales and bristles. Note the small domeless posterior scale (arrow) and the serrations of the bristles. Scale bar is 2 μm. Fig. F. Body scales and bristles. Note the attachment of the bristle to the undersurface of the dome. Scale bar is 2 μm.

Table 3.2. Summary of the major diagnostic features observable with SEM used in this study to separate closely related taxa of the Series Striatae and Corconticae. This table represents an updated and modified version of one published by Asmund and Kristiansen (Table 2, p. 61, 1986). Key: ASR = anterior submarginal rib.

Taxon	Lateral incurvings	Anterior submarginal ribs	V-rib/Anterior submarginal rib juncture	Ribs on anterior flange	Ribs on posterior flange	Ribs on shield
Striata v. striata	+	+, rather long	Arms of V-rib extend to scale margin, separate from ASR.	3–6, equally spaced, parallel ribs at 60° angle with ASR.	Usually 7–12 widely spaced ribs that do not reach posterior border.	7–12 equally spaced, parallel ribs.
v. serrata	+	+, rather long	As above.	As above.	As above.	As above.
Cratis	slight or absent	–	Sides of V-rib straight, extend to scale margin.	3–7, equally spaced, parallel ribs, continuous with ribs on shield.	Usually 16–23 ribs spaced further apart than those of shield.	10–18 equally spaced, parallel ribs. Anterior ones continuous with flange ribs.
Asmundiae	+, pronounced	+	Arms of V-rib curve and become continuous with ASR.	Usually lacking.	Usually 30–35 equally spaced, ribs that do not reach posterior border.*	9–18 equally spaced, parallel ribs. Anterior most rib restricted to extensions of shield along sides of dome.
Corcontica	+, pronounced	–	Arms of V-rib straight, extend to scale margin.	Series of ribs arranged roughly parallel to scale margin.	Usually 20–25 ribs, somewhat irregularly arranged.	9–12 equally spaced, parallel ribs. Few cross-connecting struts present.
Sphagniphila	+, slight	+	Arms of V-rib straight, extend to scale margin.	Lacking.	28–40 ribs, extend to proximal border.	8–16 slightly curved, parallel ribs.
Pseudocratis	–	+, may be winged	Arms of V-rib curve and become continuous with ASR.	Lacking.	Around 16, many incompletely developed.	10–15 equally spaced, parallel ribs.
Flora	–	+	Arms of V-rib curve and become continuous with ASR. ASR's long, curve around dome.	8–10 short, parallel ribs at 60° angle with ASR.	16–20, equally spaced.	9–13 equally spaced, parallel ribs.

* Nicholl's (1987) reported scales with twice the normal allotment of ribs.

Table 3.2. continued

Markings on dome	Ribs radiating from posterior of dome onto shield	Presence of bristles with lance shaped tips	Detail of bristle teeth	Other
Ribs parallel, U-shaped or obliquely arranged. Cover dome.	–	–	Single subapical tooth, lack serrations.	
As above.	–	–	Unilaterally serrated, pointed teeth.	
Strongly U-shaped ribs often connected with smaller perpendicular ribs.	–	–	Unilaterally serrated, teeth pointed or in groups of twos or threes.	Edge of hood at the angle of V-rib with teeth. Progression of ribs from straight to slightly U-shaped to strongly U-shaped as move further form posterior of shield to anterior of dome.
6–11 equally spaced parallel ribs orientated with or at a slight angle to the longitudinal axis of scale. Short cross-connecting struts may be present.	–	+, sometimes	Unilaterally serrated with teeth in groups of 2 or 3.	Shield extends along both sides of the dome. On some scales dome ribs continuous with those of shield extensions.
Series of irregularly arranged ribs, some continuous with ribs on anterior flange.	+	+	Unilaterally serrated, teeth pointed; sometimes bifurcate.	Ribs of dome, anterior portion of the shield and posterior flange irregularly arranged compared with other species.
2–5 straight or bent ribs.	–	+	Unilaterally serrated, pointed teeth.	
4–6 parallel ribs orientated at slightly oblique angle with longitudinal axis of scale.	–	–	Unilaterally serrated, teeth usually pointed.	
Strongly curved, almost U-shaped.	–	–	Unilaterally serrated, teeth usually pointed.	Edge of hood at the angle of V-rib with teeth. Flower-like structure in posterior region of shield.

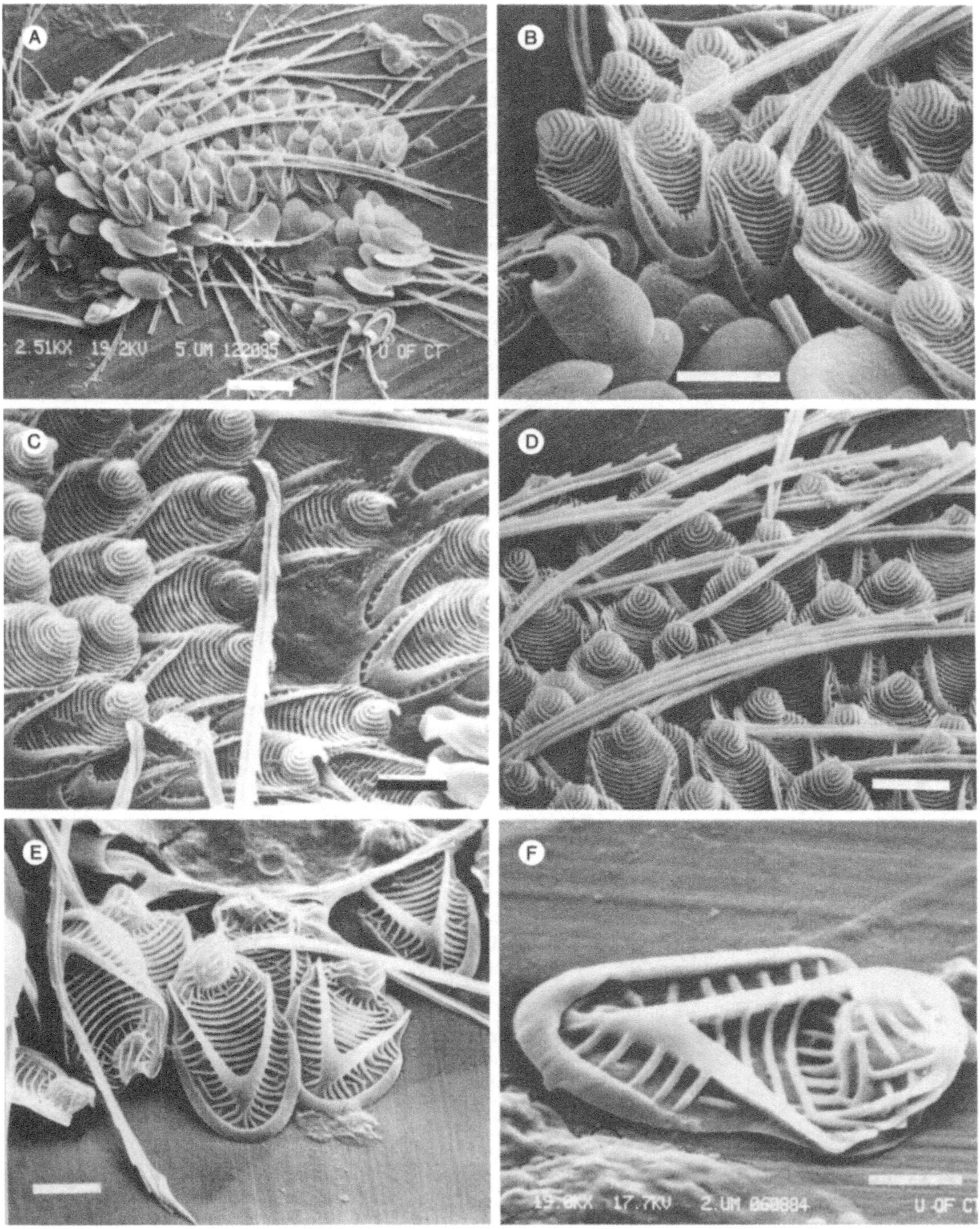
A
B
C
D
E
F
2.51KX 19.2KV 5.UM 122085
U OF CT
19.0KX 17.7KV 2.UM 060884
U OF CT

along the posterior border of the dome are spaced closely to those on the shield, often giving the appearance in TEM micrographs that a thick rib encircles the dome (e.g. Fig. 34C of Asmund and Kristiansen, 1986). Domes of apical scales are large and often marked with a series of parallel, instead of U-shaped, ribs orientated obliquely across the dome and inter-connected with smaller struts (Fig. 3.27B) (Asmund and Kristiansen, 1986).

Bristles are strongly ribbed, curved, and unilaterally serrated along the convex rib (Figs. 3.27A,D). The teeth of the serration are most often in groups of 2 or 3, with the distal most group being close to the terminal end of the bristle, yielding a somewhat blunt appearance (Fig. 3.27A). Bristles emerge from under the dome via a strongly U-shaped opening positioned to the right of center (Fig. 3.27C).

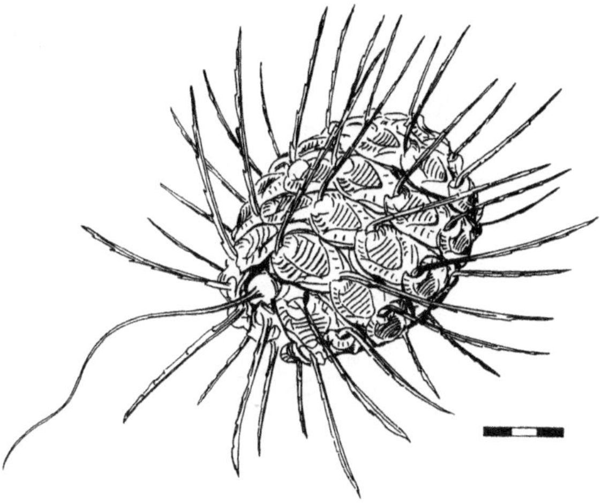

Fig. 3.28. Three-dimensional reconstruction of a whole cell of *Mallomonas striata* depicting the distribution of scales and bristles. Scales are positioned in spiral rows with their longitudinal axes at approximately a 45° to 60° angle to the longitudinal axis of the cell.

Mallomonas striata Asmund

Cells are small, ovoid and completely covered with short, slightly curved bristles (Figs. 3.28; 3.29A). All scales possess a dome and are spirally arranged with their longitudinal axes orientated at a 45° to 60° angle with the longitudinal axis of the cell (Figs. 3.28; 3.29A). The right posterior flange of each scale is overlapped by the scale positioned behind it in the same spiral row. The left posterior flange is overlapped by the scale(s) positioned in the spiral row above it. On an intact cell only the dome, anterior flanges and the anterior part of the shield of each scale are exposed (Figs. 3.28; 3.29A-B).

Scales are ovate with lateral incurvings, have a narrow posterior rim encircling one-half of the scale, a small dome and an acutely angled V-rib with arms that extend to the perimeter of the scale (Figs. 3.29C,E,F). The anterior submarginal ribs form acute angles with the arms of the V-rib; the former are not continuous with the latter. The anterior submarginal ribs are often continuous with ribs on the dome. The anterior flanges, shield, posterior flange and usually the dome are all marked with ribs. The anterior flange is wing-like and transversed with 3 to 6 parallel struts, each originating from the anterior submarginal rib at a 60° angle (Figs. 3.29C,E). The shield is marked with a series of 7 to 12 parallel, equally spaced, transverse ribs (Figs. 3.29C,E,F). About 7 to 12 struts radiate from the V-rib at right angles onto the posterior flange; these ribs are wider spaced than those on the anterior flange and shield and do not extend under the posterior rim. Most specimens had U-shaped, parallel or irregularly arranged ribs on the domes; domes with a reticulated pattern (Asmund, 1959; Kristiansen, 1975) or smooth in nature (Wee, 1982) have also been reported. Scales with aberrant or deviating rib patterns have been observed (Wujek and Hamilton, 1973; Asmund and Kristiansen, 1986). The group

←

Fig. 3.27A-D. Mallomonas cratis. Fig. A. Whole cell showing the arrangement of scales and bristles. Scale bar is 5 μm. Figs. B-D. Close-ups depicting the details of scales and bristles and the arrangement of the scales on the cell. Note the pattern of ribs on the scales, the apparent lack of anterior submarginal ribs on most scales, and the teeth along the margin of the hood at the base of the V-rib. Fig. B. represents a close-up of the anterior end of the cell in Fig. A. Note the larger domes with more or less parallel and inter-connected ribs. The strongly U-shaped opening from which the bristle emerges can be seen on several scales in Fig. C. Scale bars for Figs. B-D are 2 μm. Figs. E-F. *Mallomonas corcontica.* Fig. E. Body scales and bristles. Note the lateral incurvings, the large V-rib, the lack of anterior submarginal ribs and the irregularly arrayed ribs on the dome and anterior part of the shield. A serrated bristle with a lance-shaped distal end is also shown. Scale bar is 2 μm. Fig. F. A smaller apical scale. Scale bar is 1 μm.

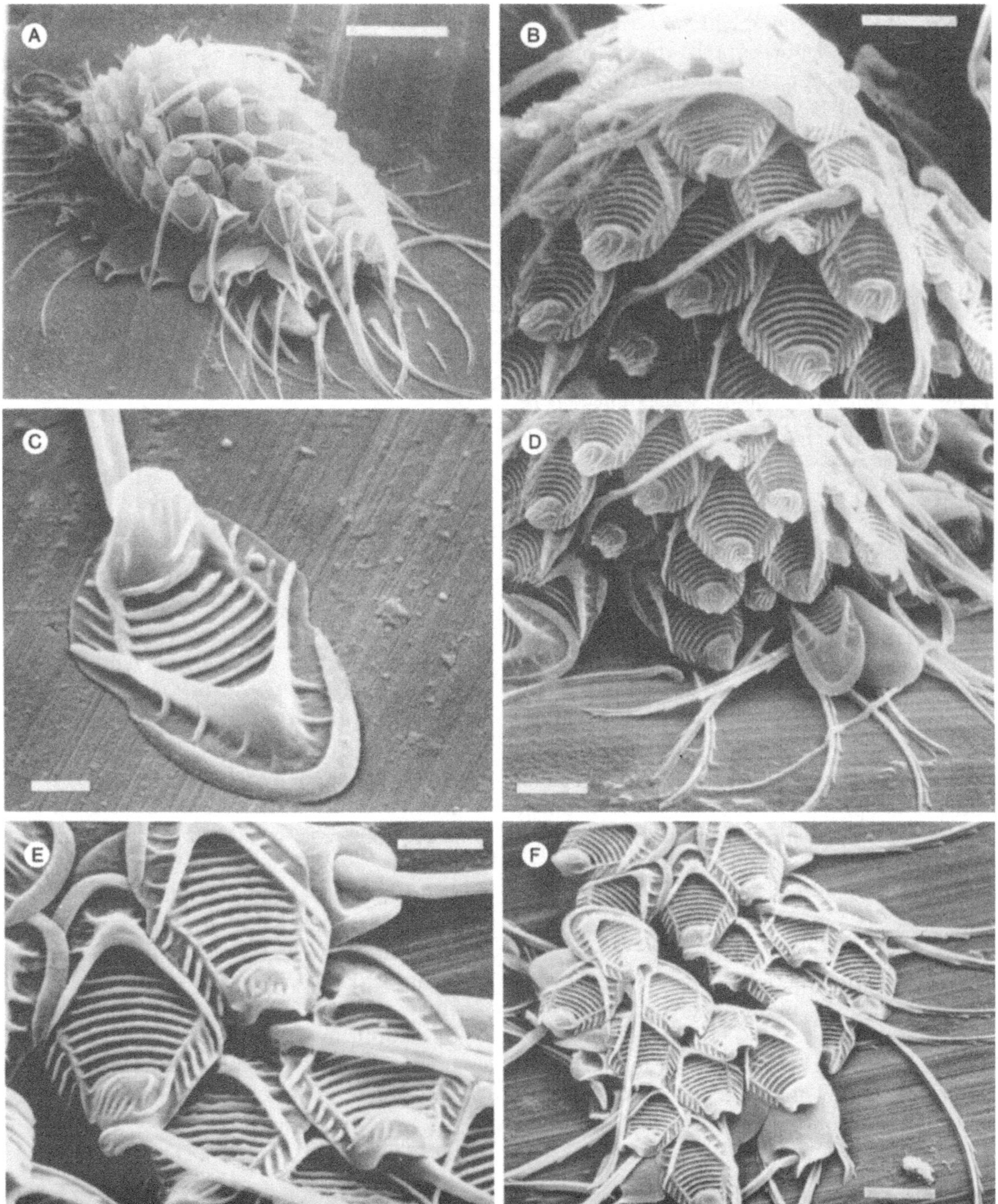

Fig. 3.29. Mallomonas striata var. *serrata.* Fig. A. Whole cell covered with short, slightly curved bristles. Note the orientation and pattern of overlap of the scales. Scale bar is 5 μm. Fig. B. Close-up of an intact cell illustrating the pattern of overlap of the scales. Scale bar is 2 μm. Fig. C. Body scale. Note the lateral incurvings, expanded anterior flanges, the prominent V-rib with arms extending to the perimeter of the scale and the ornamentation of the ribs on the shield and flanges. Scale bar is 1 μm. Fig. D. Scales and bristles. Bristles are curved, ribbed and unilaterally serrated. Note the U-shaped ribs on the domes. Scale bar is 2 μm. Fig. E. Close-up of scales and bristles. Note the emergence of the bristles and the somewhat parallel ribs on the dome. Scale bar is 1 μm. Fig. F. Scales and bristles. Note the position of the proximal end of the bristle under the dome. Scale bar is 2 μm.

of pores located on the posterior part of the shield are most often observed with SEM only from the under side of the scale.

The anterior most ring of scales usually consists of 5 scales each slightly smaller in size and with a larger dome. This ring of scales surrounds the flagellar pore and each scale is orientated with its longitudinal axis roughly parallel with that of the cell axis.

Two varieties of *M. striata* have been described, differing primarily in the structure of the bristles. Bristles of var. *striata* (Asmund, 1959) are small, curved and with a single subapical tooth on the convex edge. Those of var. *serrata* (Harris and Bradley, 1960) have a unilateral serration of small teeth along the convex edge and are strongly ribbed (Figs. 3.29A,D,F). In both varieties bristles emerge from a U-shaped opening of the dome positioned slightly to the right of center (Fig. 3.29F).

In this study var. *serrata* was more common than var. *striata*. Although cells of var. *serrata* were originally described as being slightly smaller (Harris and Bradley, 1960) than those of var. *striata*, the ranges for both varieties overlapped considerably in this study.

Sectio Leboimianae Asmund & Kristiansen 1986

Mallomonas cristata Dürrschmidt

Cells are ovoid to ellipsoidal with bristles radiating from each end (Fig. 3.30A). A few bristles may also originate from the middle section of the cell. The body scales are arranged in spiral rows and aligned with their longitudinal axes at a 60° to 90° angle to the longitudinal axis of the cell. Each scale is overlapped by the scale positioned behind it in the same spiral and by the scales situated above it (Fig. 3.30A).

Scales are either domed or domeless. Domed scales are primarily restricted to the ends of the cell, but a few may also be found in the middle portion (Fig. 3.30A). Body scales are oval to obovate, symmetrical and possess a perforated base plate, proximal rim, a widely hooded and angled V-rib with straight arms that extend to the margin of the scale, and lateral incurvings (Figs. 3.30B-C). Anterior submarginal ribs are lacking or at best poorly developed. The anterior flanges are broad and unmarked except for a few small papillae (Fig. 3.30C). Most scales appear to have a thin perforated secondary layer covering the shield. The posterior flange is ornamented with pores. Domes are large and prominent and usually marked with a few small papillae (Fig. 3.30C). A single short rib may also be found on top of the dome and the distal margin is slightly protruded forward. The apical and the rear most scales are asymmetrical and slightly more circular in outline than body scales (Dürrschmidt, 1981). Rear scales may possess a short subapical spine due to the extension of one of the arms of the V-rib (Asmund and Kristiansen, 1986).

Bristles range in size from 4 μm to 12 μm, are bent and have a shaft that consists of three equally spaced ridges. One of the ridges is either serrated (Fig. 3.30C) or smooth with a single subapical tooth (Fig. 3.30B). Cells up to 33 μm in length, considerably larger than previously reported in the literature (Asmund and Kristiansen, 1986), were observed.

Mallomonas paludosa Fott

Cells are elongate or ellipsoidal and covered with bristles (Fig. 3.30D). Scales are arranged in spiral rows with their longitudinal axes perpendicular to the longitudinal axis of the cell (Fig. 3.30D). Each scale is overlapped by the scale positioned behind it in the same spiral row and by scales in the spiral row above it.

Scales vary little in their morphological appearance along the cell; there is little distinction between apical, body and posterior scales (Fig. 3.30D). Scales are rather large, slightly bowed in order to conform to the surface of the cell, and consist of a perforated base plate, a proximal rim, a large dome and a prominent V-rib (Figs. 3.30E-F). The arms of the V-rib curve and become continuous with the anterior submarginal ribs, which in turn terminate along the lateral sides of the dome (Figs. 3.30E-F). The anterior flanges are thin and unornamented. The height of the V-rib structure may reach 0.5 μm above the base plate and the anterior submarginal ribs often curve outward and appear winged (Fig. 3.30F). The submarginal rib structure is often asymmetric with the left side being more strongly curved.

The dome, shield and posterior flange are all

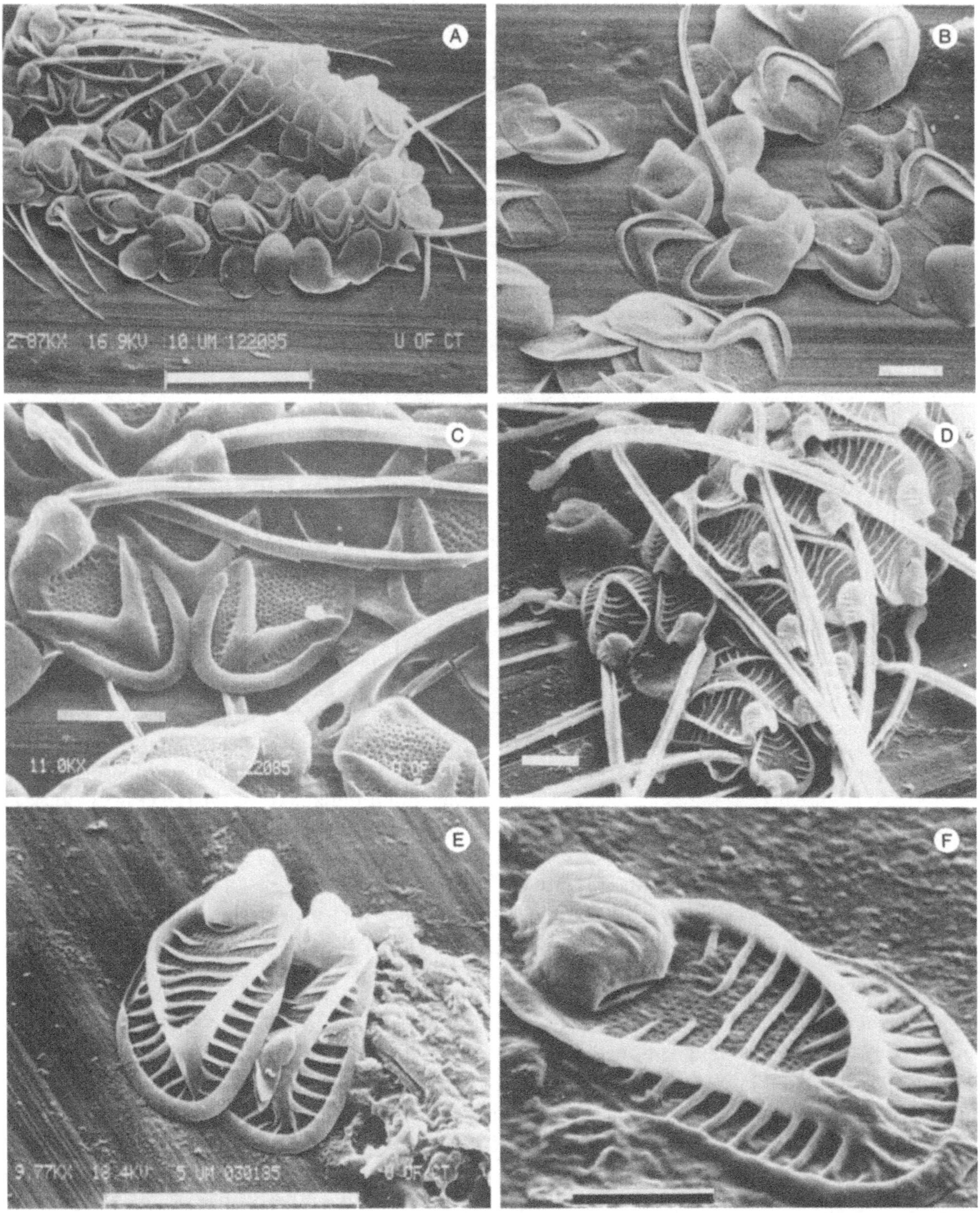

Fig. 3.30A-C. Mallomonas cristata. Fig. A. Whole cell showing the arrangement of domed and domeless scales, and the distribution of bristles primarily from each end of the cell. Scale bar is 10 μm. Fig. B. Body scales. Note the widely hooded and broad V-rib, the lateral incurvings and the lack of anterior submarginal ribs. Scale bar is 2 μm. Fig. C. Body scales. Note the structure of the base plate and dome, and the serrated bristles. Scale bar is 2 μm. Figs. D-F. *Mallomonas paludosa.* Fig. D. Posterior portion of an intact cell. Note the arrangement of scales and the three rows of serrations on each bristle. Scale bar is 2 μm. Figs. E-F. Body scales depicting the distribution of shield, flange and dome ribs, the elliptical hole along the posterior border of the dome and the continuation of the arms of the V-rib with the anterior submarginal ribs. Scale bars are 5 μm and 2 μm, respectively.

ornamented with sets of parallel ribs (Figs. 3.30D-F). The ribs that transverse the shield are rather widely spaced and aligned with those on the posterior flange (Fig. 3.30F). The ribs of both the shield and flange have a greater height at the points where they adjoin the submarginal rib structure (Fig. 3.30E). The distal most ribs on the shield are often incomplete or interrupted (Figs. 3.30E-F).

The domes are large and have a prominent elliptical shaped hole along the proximal border (Figs. 3.30E-F). The surface of the dome is usually ornamented with ribs that are arranged in a parallel manner and often restricted to one side. The border surrounding the inverted U-shaped opening from which the bristle emerges is strongly protruded (Figs. 3.30D-E).

The shaft of each bristle is thick and consists of three equally spaced longitudinal ridges (Fig. 3.30D). Each ridge is serrated with rather small teeth. Apical bristles are shorter in length than body bristles.

Sectio Mallomonas

Series Alpinae Asmund and Kristiansen 1986

Mallomonas alpina Pascher & Ruttner emend. Asmund & Kristiansen

Synonyms: *Mallomonas tonsurata var. alpina* (Pascher & Ruttner) Krieger
Mallomonas monograptus Harris and Bradley

Cells are broadly to elongate ellipsoidal and have bristles of two lengths that are primarily restricted to the anterior half of the cell (Figs. 3.31A-B).

Four types of scales are present: apical domed, domed body, domeless body and rear scales with small spines. All scales consist of a base plate perforated with large, evenly spaced pores, a narrow proximal rim encircling one-half of the scale and a V-rib. On most scales the arms of the V-rib extend forward and are continuous with the anterior submarginal ribs. As observed with *Mallomonas tonsurata*, the arms of the V-rib on some apical domed scales extend to the posterior of the dome; these scales lack anterior submarginal ribs. The ridges of the submarginal ribs are usually curved outwards (Fig. 3.31C). A small patch of closely spaced pores with smaller diameters is often present at the base of the V-rib. Isolated scales are virtually identical to those of *Mallomonas areolata* (see below), but differ from those of *Mallomonas tonsurata* in lacking a secondary reticulum over the shield and flange. Domes are larger on the apical most scales, and usually smooth or sculptured with longitudinal or oblique ribs (Fig. 3.31A).

All bristles are unilaterally serrated with sharp (pointed) teeth and lack a long tapered distal tip. The apical most bristles, associated with the apical most ring of scales, are shorter in length and stout. Body bristles are slender and considerably longer in length.

The overlapping and positioning of scales on the cell are similar to that described for *Mallomonas tonsurata*. Comparisons with other closely related taxa are summarized below.

Mallomonas alpina, originally described as a variety of *Mallomonas tonsurata* (Pascher, 1913), was considered a juvenile form of *Mallomonas monograptus* by Harris and Bradley (1960). The TEM micrographs of related scales of *M. monograptus* published by Harris and Bradley (1960) are identical with those of *M. alpina*; however, mature cells of the former were illustrated as being covered with bristles. The species *Mallomonas alpina*, originally described by Pascher and Ruttner, was emended to include *Mallomonas monograptus* by Asmund and Kristiansen (1986).

Mallomonas areolata Nygaard

Cells are most often ellipsoidal or elongate-ellipsoidal with bristles distributed over the anterior half of the cell. As such, they can easily be confused with other closely related taxa, especially *Mallomonas alpina* (see below).

The types of scales are as described for *M. alpina* and their arrangement on the cell similar to that described for *Mallomonas tonsurata*. All scales possess a base plate with evenly spaced rather large pores, a narrow proximal rim, a V-rib, and lack a secondary layer. Body scales tend to be larger than those of *M. alpina*, broad relative to their length, rhomboidal and with obtusely-angled V-ribs (Asmund, 1959; Nicholls, 1982; Asmund and

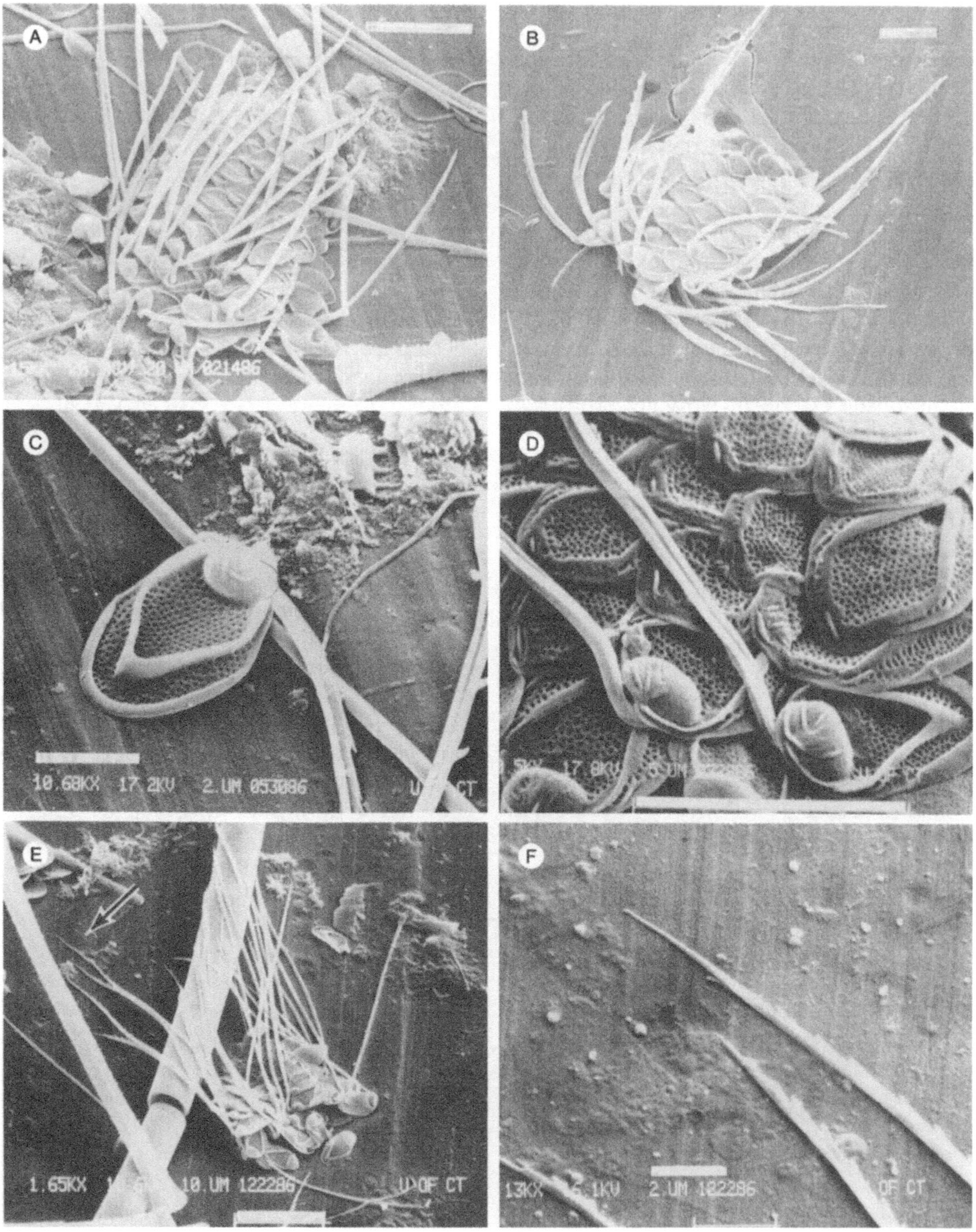
A
B
C
10.68KX 17.2KV 2.UM 053086
D
E
1.65KX
10.UM 122286
F
2.UM 122286
OF CT

Kristiansen, 1986). The rhomboidal shape results in a smaller anterior submarginal rib on the right hand side. However, other domed body scales and domeless body scales are narrower, more symmetrical, and with an acutely angled V-rib, similar to scales of *M. alpina* (Asmund and Kristiansen, 1986). The apical most domed scales are more strongly asymmetrical than are the domed body scales. On many scales a patch or cluster of tiny pores is often observed at the base of the V-rib (Nicholls, 1982); however, this feature was not always present and is also common on body scales of other species (e.g. *M. alpina* and *M. elongata*).

Bristles are unilaterally serrated and of two sizes. The shorter ones restricted to the anterior part of the cell, are thicker with the distal most tooth of the serration being close to the bristle tip. The longer bristles are serrated only along the distal portion and possess a long tapered tip extending beyond the distal most tooth (Figs. 3.31E-F).

Cells of *M. areolata* were rare and distinguished from other species primarily on the basis of bristle morphology.

Mallomonas elongata Reverdin

Cells are large with a mean size of 40.2 μm · 15 μm and elongate to ellipsoid in shape (Figs. 3.32; 3.33A). The anterior one-half to two-thirds of the cell possesses bristles of varying lengths; the posterior end is rounded and devoid of bristles.

Three types of scale are present; domed scales (Figs. 3.33B; 3.34A), symmetrical domeless scales (Fig. 3.33D), and asymmetrical domeless rear scales with short spines (Fig. 3.34D). All scales are large relative to most other species (Table 4.2). Each scale consists of a perforated base plate, a narrow posterior rim that encircles about one-half of the cell and a prominent V-rib that is raised high above the base plate and curved over the

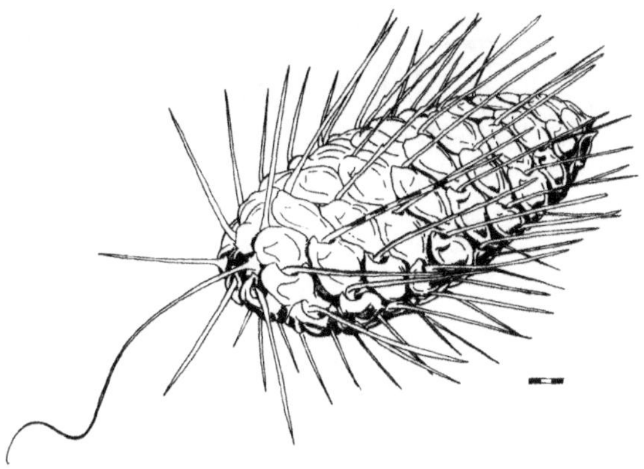

Fig. 3.32. Three-dimensional reconstruction of a whole cell of *Mallomonas elongata* depicting the distribution of scales and bristles. Scales are positioned in spiral rows with their longitudinal axes at approximately a 45° to 60° angle to the longitudinal axis of the cell. The serrated bristles are distributed over the anterior half of the cell.

shield, forming a hood (Figs. 3.33D-E; 3.34A-D). The base of the V-rib is acutely or more often obtusely angled and the arms extend forward, curve and become continuous with the anterior submarginal ribs (Figs. 3.33D; 3.34A,B,D). The ridges of the V-rib are curved inward towards the shield; however, the ridges of the anterior submarginal ribs are curved outward (Figs. 3.34A,B,D), yielding an image that is usually identifiable with LM. On some scales the ridges of the V-rib and submarginal ribs are lined with minute teeth (Fig. 3.34D).

Body scales are broadly rhomboidal (Figs. 3.33D; 3.34A). All scales lack a secondary layer on their shield and flanges. The small base plate pores, often arranged in transverse rows on the shield, are unevenly distributed and clearly less dense on the distal portion of the shield (Figs. 3.34A,B,D); most often pores are missing from the distal most rows. The uneven distribution of pores is a unique characteristic that can be used to

←

Fig. 3.31A-C. Mallomonas alpina. Fig. A. A whole cell showing the distribution and pattern of overlap of domed and domeless scales and bristles. Scale bar is 10 μm. Fig. B. Intact portion of the anterior half of a cell. Note the apical tuft of short bristles and the longer body bristles. Both types of bristles are serrated. Scale bar is 5 μm. Fig. C. Domed scale. Note the relatively large base plate pores and the slightly winged anterior submarginal ribs. Scale bar is 2 μm. Fig. D. *Mallomonas cyathellata* var. *cyathellata*. A group of domed and domeless scales. Note the irregularly shaped pores on the shield, the continuation of the left anterior submarginal rib around the dome and the ribs on the dome and distal portion of the domeless scales. Scale bar is 5 μm. Figs. E-F. *Mallomonas areolata*. A portion of a cell denoting the morphology of scales and bristles. Note the long tapered tip (arrow) extending beyond the distal most tooth on the body bristles. Scale bar is 10 μm. Fig. F. Close-up of the distal tips of bristles. Scale bar is 2 μm.

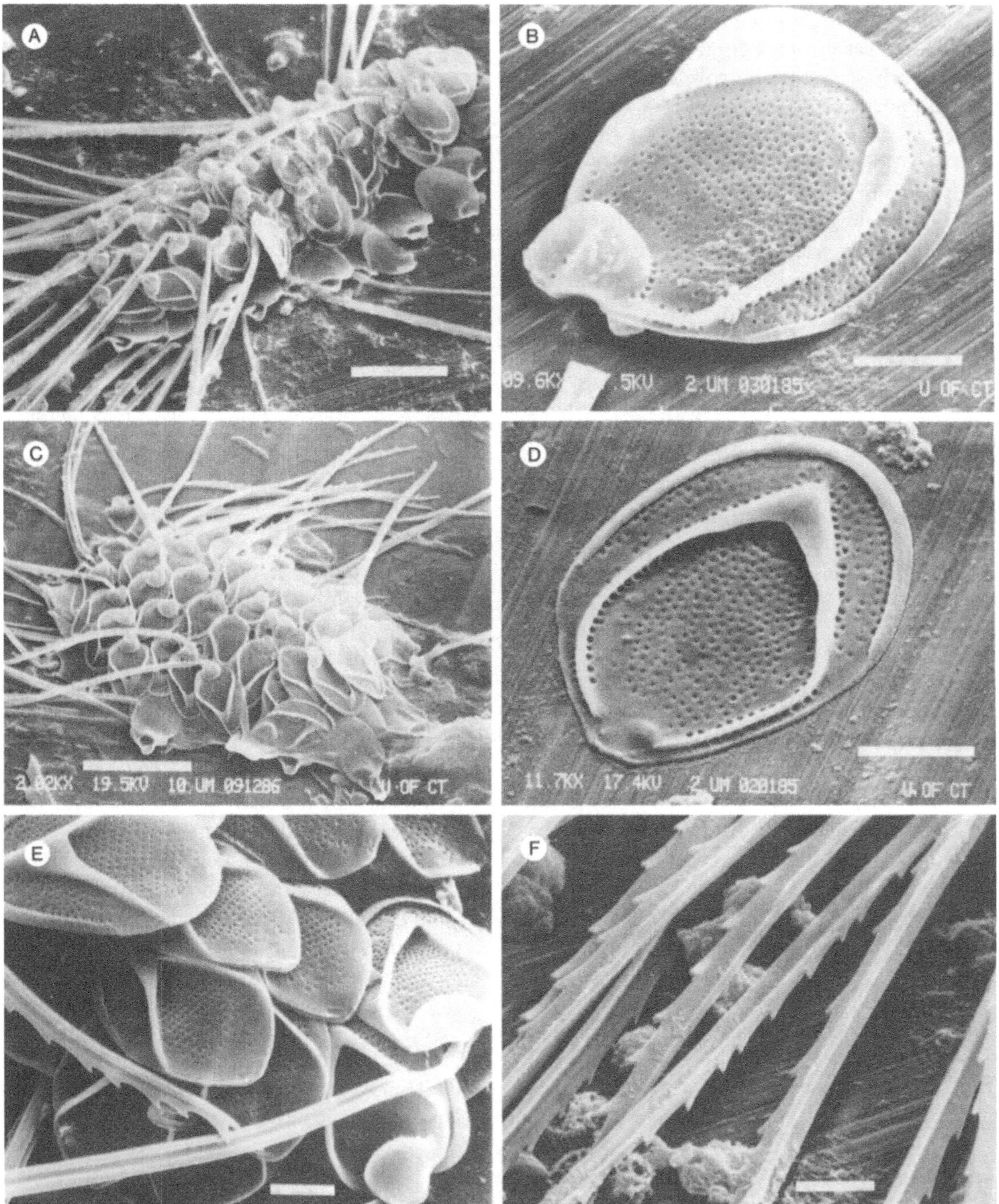

Fig. 3.33. Mallomonas elongata. Fig. A. Whole cell denoting the elongated shape and the arrangement and positions of domed scales, domeless scales and bristles. Scale bar is 10 μm. Fig. B. Domed body scale. Note the distribution of base plate pores, the dome and the slightly winged anterior submarginal ribs. Scale bar is 2 μm. Fig. C. Group of body scales, still somewhat aligned in spiral rows, and their associated bristles. Scale bar is 10 μm. Fig. 4. Domeless body scale. Note the continuation of the arms of the V-rib with the anterior submarginal ribs. Scale bar is 2 μm. Fig. 5. Body scales with fewer base plate pores in the distal region. Scale bar is 2 μm. Fig. 6. The strongly ribbed shafts of bristles. Each shaft is serrated along one ridge and each tooth of the serration is, in turn, toothed. Scale bar is 2 μm.

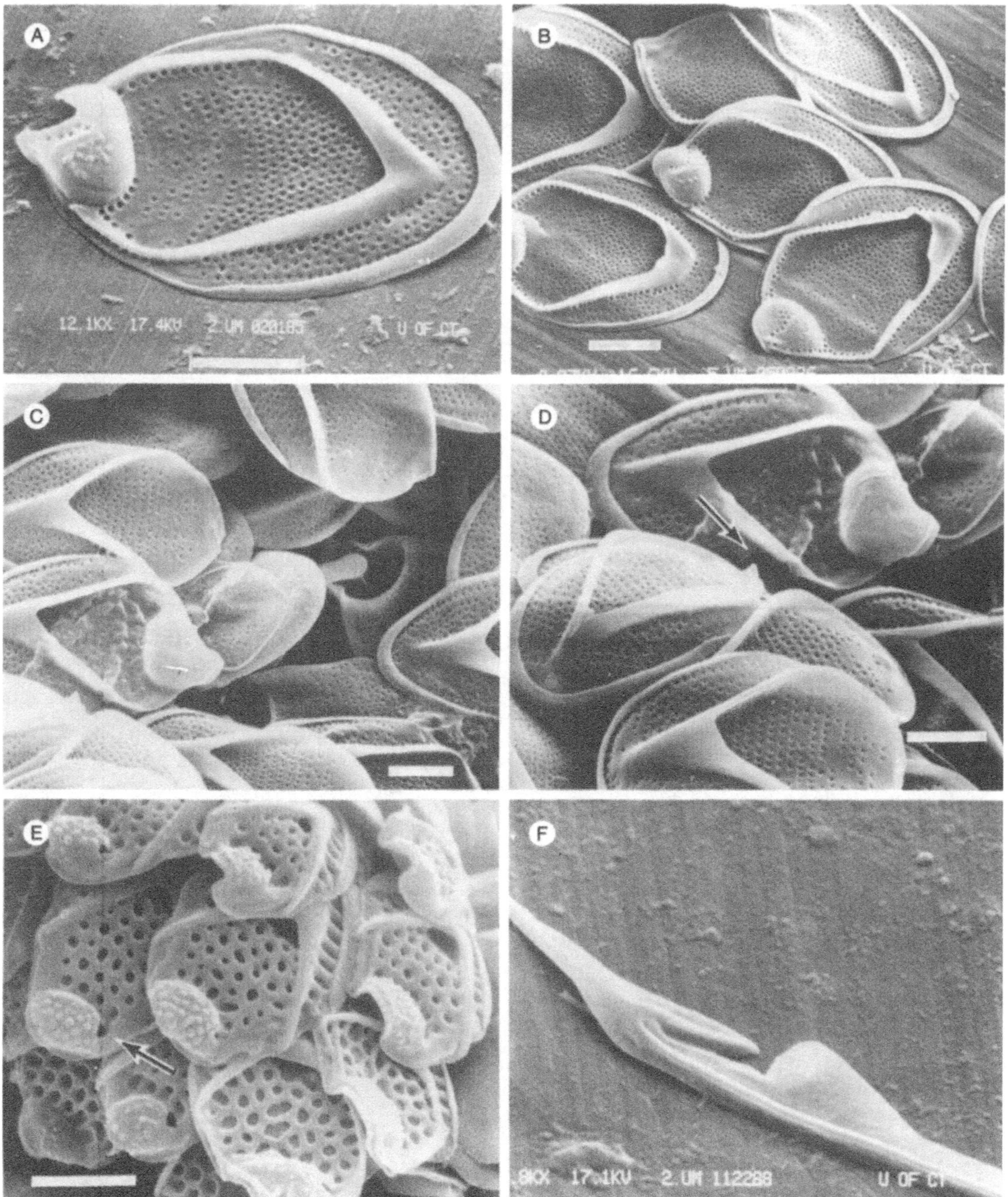

Fig. 3.34. Mallomonas elongata. Figs. A-B. Body scales with small base plate pores and larger pores aligned in chains along the bases of the V-rib, anterior submarginal ribs, base of the dome and the medial axis of the dome. Note the distribution of smaller pores into transverse rows on the distal portion of the shield; some pores within each row are missing. Scale bars are 2 μm. Figs. C-D. Body scales. Note the small spines on a scale in Fig. D (arrow). Scale bars are 2 μm. Figs. E-F. *Mallomonas crassisquama.* Fig. E. Close-up of an intact cell denoting the arrangement of scales. Note the papillated domes and the hole located on left side of the dome (arrow); the latter hole was found on each domed scale. Scale bar is 2 μm. Fig. F. Distal end of a helmet bristle. Scale bar is 2 μm.

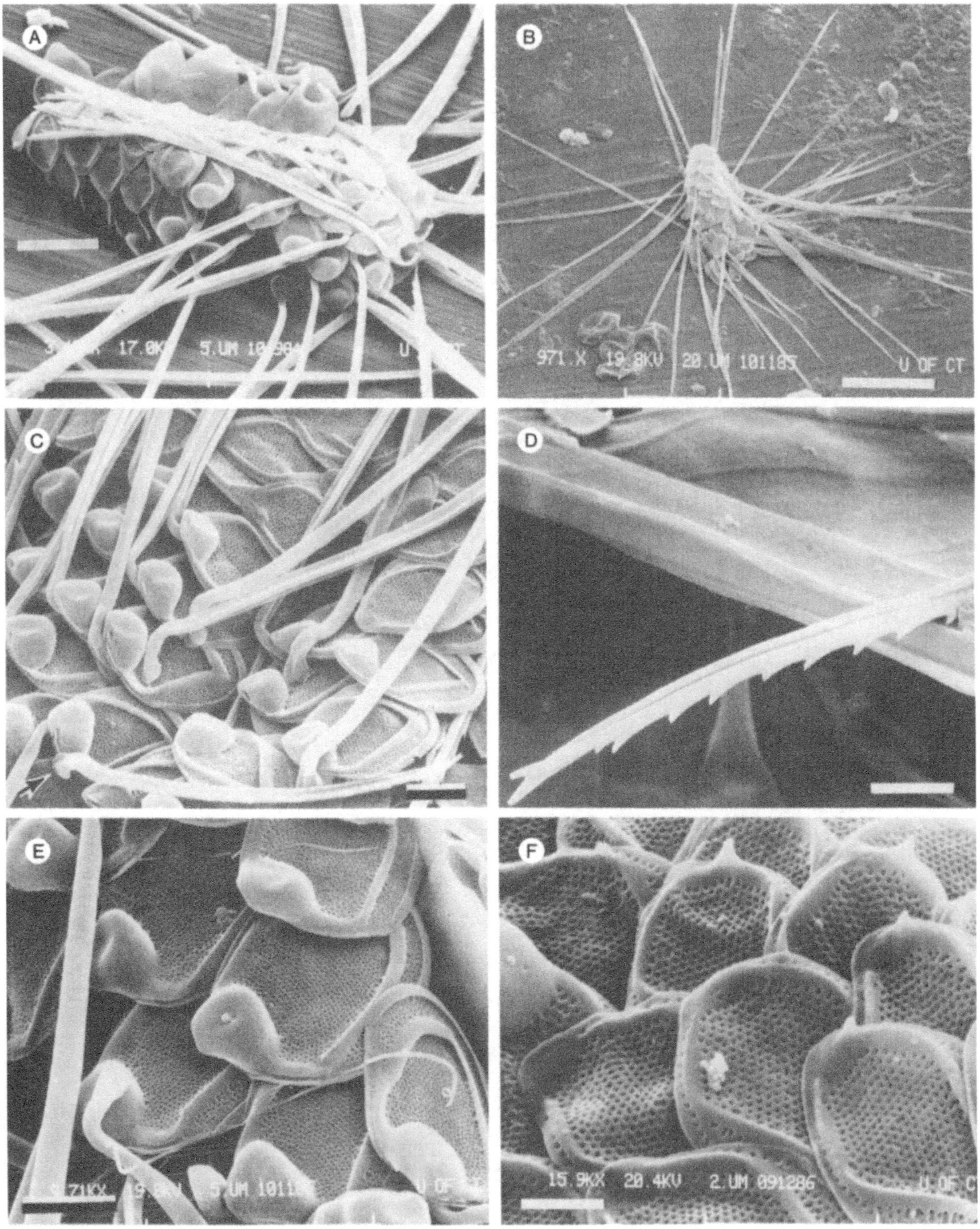
A
17.0KV
5.UM
B
971.X
19.8KV
20.UM
101185
U OF CT
C
D
E
9.71KX
5.UM
F
15.9KX
20.4KV
2.UM
091286

identify isolated scales with EM. A patch of closely spaced and smaller pores positioned at the base of the V-rib, similar to ones described for *M. areolata* (Asmund and Kristiansen, 1986; Nicholls, 1982) has been observed on a few scales (Fig. 3.34A). If the patch lies under the hood of the V-rib it would be missed by SEM. Such a patch has not been previously reported for *M. elongata* (Asmund, 1959; Nicholls, 1982; Asmund and Kristiansen, 1986), however, a similar patch can be seen under the hood of the scale shown in Fig. 43a of Asmund and Kristiansen (1986).

Scales from cells in many populations have a second type of pore, much larger, and aligned in distinct chains along the base of the V-rib and anterior submarginal ribs, the proximal base of the dome, the anterior margin of the dome opposite of where the bristle emerges and along the medial axis of the dome (Figs. 3.33D; 3.34A-B).

Domes are prominent with a sharply concave cavity, but are also small, covering only 11.4% of the surface area of the scale (Figs. 3.34A-B). The concave opening of the dome, from which the bristle emerges, is always positioned on the right hand side of the dome and is often angular and projected forward causing the scale to be asymmetric. Rear scales are also often asymmetric due to a small spine positioned on the right hand submarginal rib (Fig. 3.34D).

Bristles are slightly tapered, unilaterally serrated and strongly ribbed. Each bristle consists of three equally spaced ribs that run the length of the shaft, but, only one is provided with serrations (Fig. 3.33F). Each tooth of the serration is, in turn, serrated. The distal tip of the bristle is blunt, lacks a distinct subapical tooth and often ends in a small bifurcation. Anterior bristles are shorter than body bristles; the latter often as long as the cell.

Scales are arranged in longitudinally aligned spiral rows (Figs. 3.32; 3.33A,C). The longitudinal axis of each scale is positioned at about a 60° to 90° angle to the longitudinal axis of the cell. Each body scale is overlapped by the scale positioned behind it in the same spiral row and by the scales in the spiral row above it (Figs. 3.32; 3.33C). As a result, only part of the shield and dome (if present) of each scale is exposed on intact cells.

Series Tonsuratae Asmund and Kristiansen 1986

Mallomonas corymbosa Asmund & Hilliard

Cells are large, elongate to ellipsoidal and possess two types of bristles that originate from the anterior one-third (Fig. 3.35A) to three-forths (Fig. 3.35B) of the cell.

A progression of four types of scales can be found from the anterior to the posterior of the cell: domed apical; domed body; domeless body; and domeless rear scales with spines (Fig. 3.35A). All scales are relatively large, have a base plate with evenly spaced small pores, a posterior rim encircling one-half of the scale, an acutely angled V-rib and a thin continuous secondary layer (Figs. 3.35C,E,F). The V-rib is sometimes more broadly angled. The pores of the secondary layer are evenly spaced and larger in diameter than those of the base plate. The secondary layer is often lacking at the base of the V-rib forming a 'window' (Asmund and Hilliard, 1961).

Domes are prominent with a sharply concave cavity, smooth, and cover approximately 21% of the surface area of the scale (Figs. 3.35C,E). As is true with other species within the Sectio, the bristles emerge from a concave opening on the right side of the dome. On the apical most scales the V-rib extends to the posterior of the dome (Fig. 3.35C), whereas on body scales the arms of the V-rib bend 40° to 70° to form anterior submarginal ribs. The left handed submarginal rib often extends about one-half way around the lower edge of the dome forming a rib (Fig. 3.35C). The posterior

←

Fig. 3.35. Mallomonas corymbosa. Fig. A. Whole cell depicting the elongate-ellipsoidal shape and the distribution of domed scales, domeless scales and bristles. Note that about half of the length of the cell is covered with domed scales. Scale bar is 5 μm. Fig. B. Whole cell with long bristles. Scale bar is 20 μm. Fig. C. Anterior end of a cell showing the overlapping of the scales and emergence of bristles. The arrow points to the foot of an apical bristle. Most other bristles represent the longer body bristles. Note the smooth unornamented domes and the spatula-shaped proximal foot of the bristle; the latter forms approximately a 90° angle with the shaft. Scale bar is 2 μm. Fig. D. Distal end of a long serrated bristle. Note that the distal most tooth points in the opposite direction from the others. Scale bar is 2 μm. Fig. E. Close-up of domed body scales. Scale bar is 2 μm. Fig. F. Posterior scales with short spines. Scale bar is 1 μm.

most scales become progressively smaller, are asymmetric, and possess a small spine.

Scales are spirally arranged with their longitudinal axes aligned at a 45° to 60° angle to the longitudinal axis of the cell (Fig. 3.35A). The position and overlap of scales on the cell are as described for *Mallomonas tonsurata*. On some individuals, it is common for the smaller asymmetric scales to be positioned more or less at right angles to the longitudinal axis of the cell.

Two types of bristles are found on a given cell (Asmund and Hilliard, 1961). Both types of bristles usually consist of three equally spaced ridges that run the length of the shaft (Figs. 3.35A,C,D). The apical most scales possess short bristles with a single row of serrations along the length of one ridge of the shaft (Fig. 3.35C). Some of the serrations may be slightly recurved. Bristles originating from domed body scales are stout and much longer in length. The distal one-half to two-thirds of one of the ridges is serrated (Figs. 3.35A,D); the proximal end of the shaft lacks serrations (Fig. 3.35C). The distal most tooth is distinctive in that it points in the opposite direction from the others (Fig. 3.35D). The apices of the teeth and the tips of the bristles are often bifurcated (Fig. 3.35D); both pointed and bifurcated teeth can be found on the same bristle.

Mallomonas cyathellata var. *cyathellata* Wujek & Asmund

Specimens of *M. cyathellata* var. *cyathellata* were observed only once. The sizes and morphologies of both cells and scales, the arrangement of scales and bristles on the cell, the presence of two sizes of bristles and the thick secondary layer on all scales are similar to that of *M. tonsurata* (Wujek & Asmund, 1979).

Scales of *M. cyathellata* var. *cyathellata* differ from other taxa in the series in having a secondary layer with irregularly shaped pores (Fig. 3.31D). Domed scales have a larger, prominent dome that is usually marked with longitudinally or obliquely positioned ribs (Asmund and Kristiansen, 1986). The left anterior submarginal rib continues around the distal margin of the dome. Domeless scales have one or more irregularly positioned ribs on their distal ends (Fig. 3.31D).

According to the original description (Wujek and Asmund, 1979), the rearmost scales possess goblet-like protuburences; cells observed in this study lacked such appendages, but one specimen had several rear scales with small spines. In this sense, the specimens were similar to var. *kenyana* Wujek and Asmund, however, they have been identified herein as var. *cyathellata* because of the presence of the distinctive ribs on the domed and domeless scales.

Bristles are unilaterally serrated and of two types. The anterior most ones are shorter, thicker and serrated along the entire length of the shaft. The body bristles are longer, may lack serrations along the proximal end and have a furcated tip; the furcated tip is sometimes the result of a subapical tooth positioned on the side of the shaft opposite of the serration, similar to bristles of *Mallomonas corymbosa*.

Mallomonas galeiformis Nicholls

Cells are relatively small (Table 4.5), obovate in shape, possess two different types of bristles that radiate from the anterior half of the cell and small posterior spines (Figs. 3.36A-C). Scales are arranged in spiral rows and aligned with their longitudinal axes forming a 60° to 90° angle with the longitudinal axis of the cell (Fig. 3.36C). Each scale is overlapped by the scale positioned behind it within the same spiral row and by the scale(s) in the spiral row above it (Fig. 3.36C).

The scale coat consists of four types of scales; anterior domed, domed body, domeless body and posterior spined scales. All scales have a perforated base plate, a wide proximal rim, a broad U-shaped V-rib and, most often, transverse shield ribs (Figs. 3.36D-F; 3.37A-B). Domed scales are circular to slightly elliptical in outline, have a large pronounced dome and a V-rib that often forms an extensive canopy over the shield (Figs. 3.36D-E). The arms of the V-rib extend to the base of the dome and become fused with relatively short anterior submarginal ribs (Figs. 3.36D-E; 3.37A). The anterior submarginal ribs continue around the distal most portion of the dome forming a ridge (Figs. 3.36C-D; 3.37A). Often, additional (inverted) U-shaped ridges lie above the one formed by the submarginal rib (Figs. 3.36D; 3.37A). Additional ribs, usually orientated parallel with the longitudinal axis of the cell, are also present on the dome. Each scale

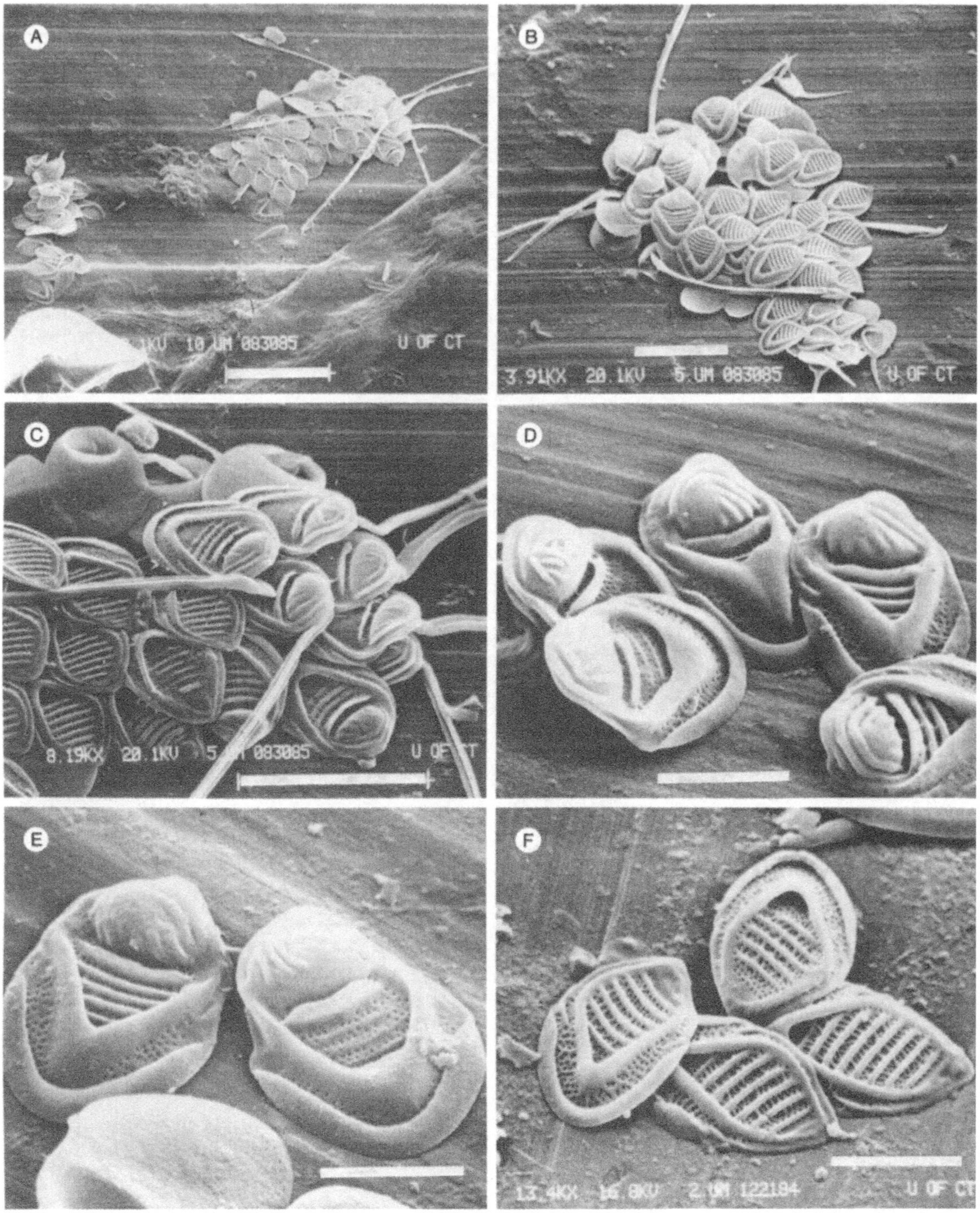

Fig. 3.36. *Mallomonas galeiformis.* Figs. A-B. Whole cells with domed scales, domeless scales, rear scales with spines and two types of bristles. Scale bars are 10 μm and 5 μm, respectively. Fig. C. Anterior end of the cell in Fig. A. Note the orientation and pattern of overlap of the scales and the apical most domed scales, each with a single prominent transverse rib. Scale bar is 5 μm. Fig. D. Scales with large, prominent and highly ribbed domes. The transverse shield rib located directly behind the dome is the most pronounced. Scale bar is 2 μm. Fig. E. Domed scales with multiple (left) and a single (right) transverse shield rib. Scale bar is 2 μm. Fig. F. Domeless body scales. Scale bar is 2 μm.

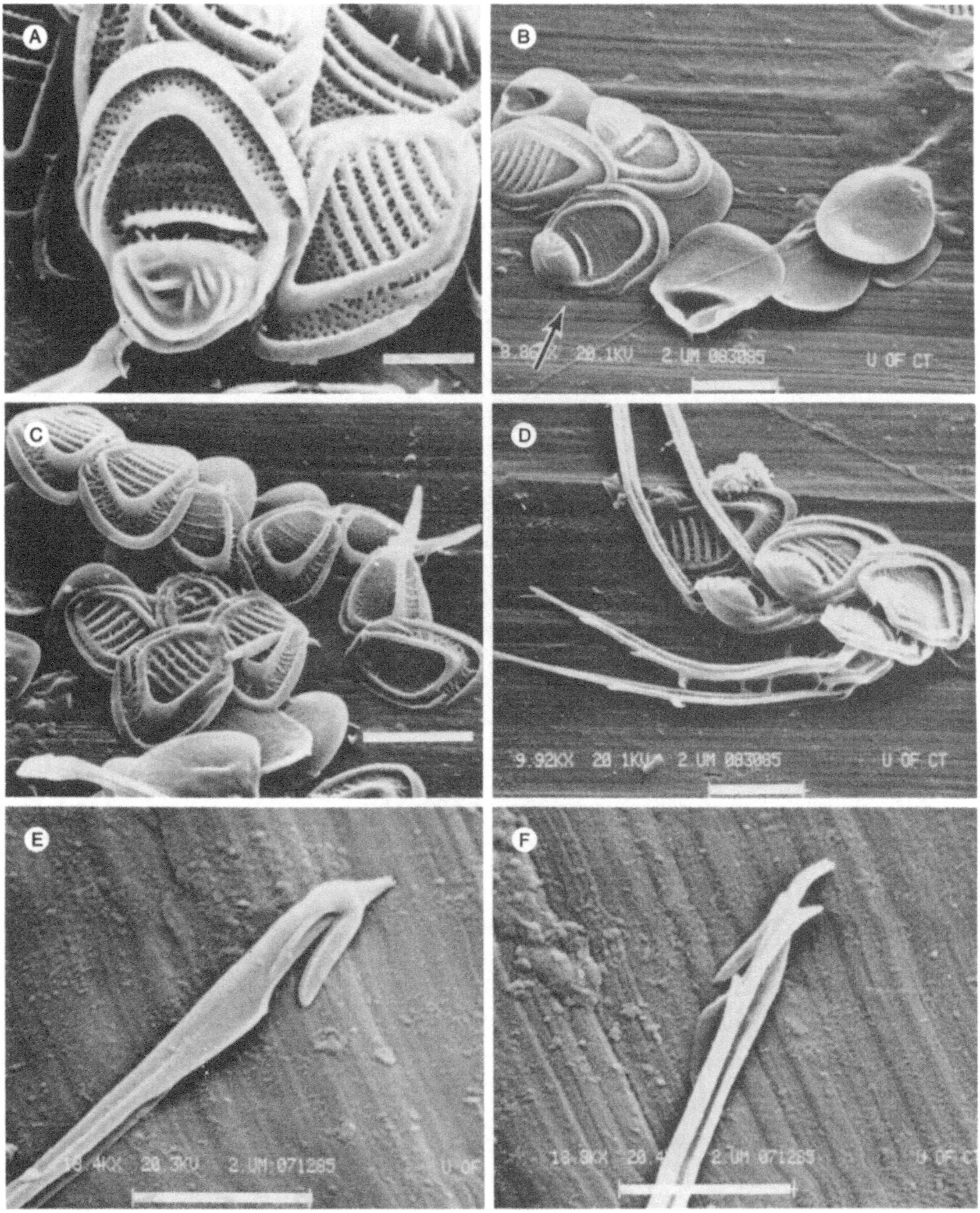

Fig. 3.37. Mallomonas galeiformis. Fig. A. Domed scale with a single transverse shield rib and a large highly ribbed dome, and a domeless scale. Note the continuation of the short anterior submarginal ribs around the distal margin of the dome. Scale bar is 2 μm. Fig. B. Domed and domeless scales. Note the larger domed scale with a single transverse shield rib (arrow). Scale bar is 2 μm. Fig. C. Posterior rear scales with short spines. Note the secondary ribbing on the posterior flange and the lack of transverse shield ribs. Scale bar is 2 μm. Fig. D. Short, thick, strongly ribbed, serrated bristles each with a bifurcate distal end and a proximal recurved hook. Scale bar is 2 μm. Figs. E-F. Distal ends of helmet bristles without (Fig. E) and with (Fig. F) a single subapical tooth on the side opposite of the cleft. Scale bars are 2 μm.

of the anterior most ring generally has a single pronounced transverse shield rib located directly behind the dome (Figs. 3.36C,E; 3.37A-B). Additional shield ribs, if present, are much less pronounced. The domed body scales positioned behind the apical most ring, have from one to five transverse shield ribs (Figs. 3.36C-E).

Domeless body scales, which comprise the majority of the cell armour, are slightly more elliptical than the domed scales and have up to eight transverse ribs (Figs. 3.36C,F). The anterior submarginal ribs are continuous with the arms of the V-rib and become fused in the distal end. The rearmost scales are small, asymmetrical in shape and design, usually lack transverse shield ribs, and possess a single spine ranging in length from 0.2 μm to 3.5 μm (Figs. 3.36B; 3.37C). Posterior flanges are either smooth and lacking of secondary ornamentation (e.g. Fig. 3.36E), or more often with a reticulation of ribs (Figs. 3.36F; 3.37C).

Cells have two types of bristles. Anterior domed scales possess short (mean length of 7.4 μm), thick, strongly ribbed, serrated bristles that terminate with a bifurcate tip (Fig. 3.37D). The proximal-most tooth is recurved. Domed body scales possess longer (mean length of 12.1 μm), ribbed, helmet bristles that may or may not be serrated (Figs. 3.37E-F). The cleft of the helmet is unusual in that it is narrow and runs parallel to the shaft of the bristle (Fig. 3.37E). Each helmet bristle has a pointed distal tip and a single subapical tooth on the side opposite of the cleft (Fig. 3.37F); the subapical tooth may be lacking (Fig. 3.37E).

Several characteristics of *M. galeiformis* are similar to those of *M. costata* Dürrschmidt, *M. intermedia* Kisselev em. Péterfi and Momeu, *M. trummensis* Cronberg and *M. portae-ferreae* Péterfi and Asmund (Siver, 1988b). Siver (1988b) and Nicholls (1988a) discuss in detail the similarities and differences between *M. galeiformis* and each of these taxa. A brief summary is given here. Isolated body scales and often posterior spined scales of *M. galeiformis* can be difficult to separate from those of *M. costata*. Anterior domed scales with a single transverse shield rib have not been observed for *M. costata* and thus, would serve as one basis for separating the two taxa. Scales of *M. costata* are slightly larger and commonly have better developed anterior submarginal ribs. In addition, *M. galeiformis* has helmet bristles as opposed to the lance-shaped ones of *M. costata* (Siver, 1988b).

Scales of *M. trummensis* are also quite similar to those of *M. galeiformis* (Siver, 1988b; Nicholls, 1988a). However, *M. galeiformis* differs from *M. trummensis* in dome and bristle structure, bristle distribution, the presence of posterior scales with spines and overall cell morphology (Siver, 1988b; Cronberg, 1975). The isolated domed scales with a single transverse rib of *M. galeiformis* (e.g. Fig. 3.37B) may be difficult to distinguish from those of *M. intermedia*. However, *M. intermedia* differs from *M. galeiformis* in lacking additional ribs on the shield and dome, lacking posterior spines, in the structure and position on the cell of bristles and in having larger scales.

There is also some similarity between the scales with transverse ribs of *M. galeiformis* and those of *M. portae-ferreae* var. *portae-ferreae*, however, the ribs on the latter are thinner and usually connected by cross struts. In addition, other features of the scales, bristles and cells can adequately serve to separate *M. galeiformis* from *M. portae-ferreae*. Although *M. galeiformis* can easily be separated from *M. tonsurata* with EM, whole swimming cells of the two taxa look similar when viewed with LM (Siver, 1988b).

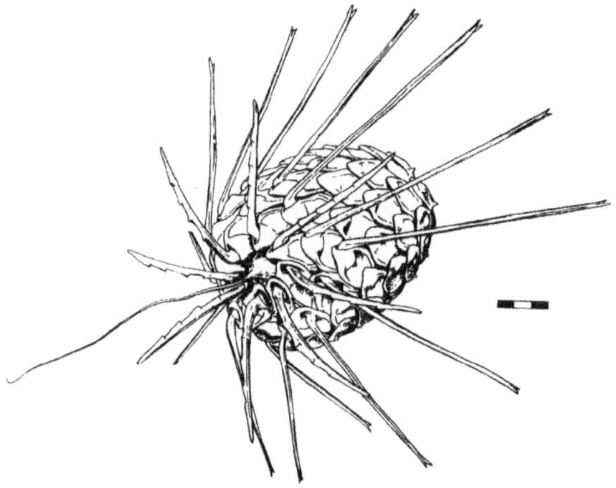

Fig. 3.38. Three-dimensional reconstruction of a whole cell of *Mallomonas tonsurata* depicting the distribution of scales and bristles. Scales are positioned in spiral rows with their longitudinal axes at approximately a 45° to 60° angle to the longitudinal axis of the cell. Note the arrangements of both the shorter serrated and longer forked bristles.

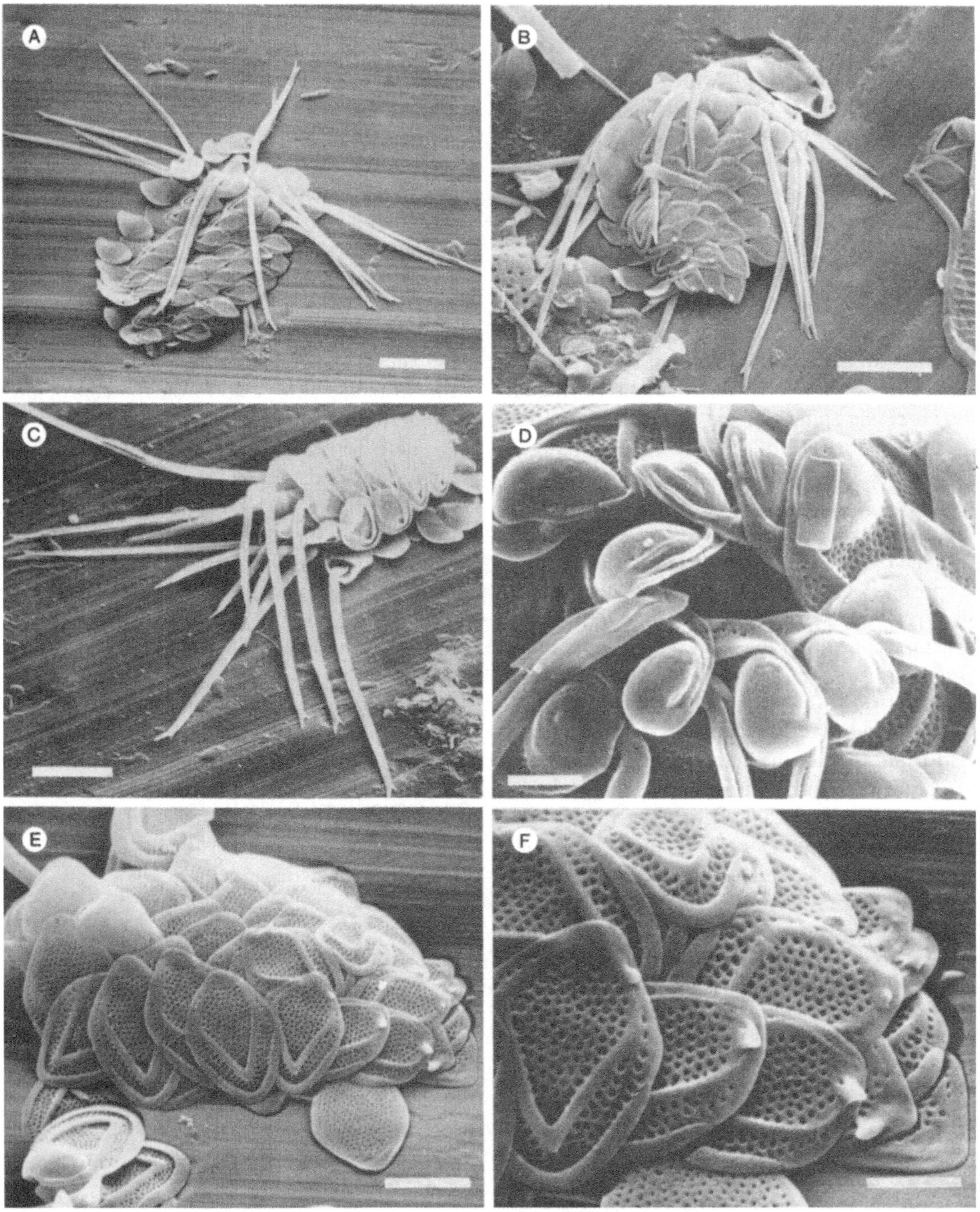
A
B
C
D
E
F

Mallomonas tonsurata Teiling emend. Krieger

Cells are small, rounded ovoid to elongate ovoid in shape, possess two types of bristles that are restricted to the anterior one-third of the cell and posterior scales with short spines (Figs. 3.38; 3.39A-C). The posterior end is often broader than the anterior.

Four types of scales can be discerned; domed apical, domed body, domeless body and domeless rear scales with spines (Figs. 3.39B-F). All scales have a base plate evenly perforated with fine pores, a posterior rim that encircles about one-half of the scale and a distinct V-rib (Figs. 3.39E-F; 3.40B-C). The sides of the V-rib are generally straight and form an acute angle (Figs. 3.39E; 3.40B-C). A reticulum of evenly spaced pores, the secondary layer (Asmund, 1959; Asmund and Kristiansen, 1986), covers the shield and flange portions of the base plate (Figs. 3.40A-C,F). The diameters of the pores of the secondary layer are larger than those of the base plate (compare the top and bottom surfaces of scales in Figs. 3.39E and 3.40B). The secondary layer is lacking at the base of the V-rib forming a 'window' (Asmund, 1959; Nicholls, 1982). As noted by Asmund (1959), a patch of small, closely spaced pores is often visible in the region of the 'window' (Fig. 3.40A) similar to what is observed in *Mallomonas areolata* and *Mallomonas alpina*. Domes are large, prominent and occupy an average of 22% of the surface area of the scale (Fig. 3.40C). On domed scales the arms of the V-rib usually extend to the base of the dome (Fig. 3.40C) whereas on domeless scales they are bent at a 30° angle and form anterior submarginal ribs (Figs. 3.40B-C). On domed scales the left arm of the V-rib continues two-thirds of the way around the lower edge of the dome producing a distinct ridge (Figs. 3.39D-E). The ridge is most prominent on the apical scale (Fig. 3.39D).

A ring of apical domed scales, usually eight in number, line the flagellar pore (Fig. 3.39D); each is overlapped 'piggy-back style' by a neighbouring apical scale and partially by a scale in the next ring (Fig. 3.39D). Body scales are arranged in spiral rows with their longitudinal axes aligned at a 45° angle to the longitudinal axis of the cell (Figs. 3.38; 3.39A-B). Each scale is overlapped by the scale positioned behind it in the same spiral row and by the scale(s) in the spiral row above it (Figs. 3.38; 3.39A). On whole cells the rim, posterior flange and V-rib of body scales are commonly covered while the shield and dome remain exposed (Figs. 3.38; 3.39B). The windows may also be covered on intact cells. Although domed body scales may be slightly asymmetric due to the shape of the dome and the uneven continuation of the V-rib, domeless body scales are symmetric in outline (Figs. 3.39E; 3.40C,F). Towards the posterior of the cell the domeless body scales become slightly smaller in size and grade into asymmetric spine bearing rear scales (Figs. 3.39B,E,F).

Two distinct types of bristles are produced. One type, associated with the apical ring of scales, is short, thick, slightly curved, and has a uniserrate row of teeth along the convex margin (Fig. 3.40C). The proximal teeth are often recurved to form hooks (Fig. 3.39B). The second type, associated with the domed body scales, is twice as long, has a non-serrated shaft and terminates in an uneven bifurcation (Fig. 3.39B). A longitudinal ridge or fold, separated from the main shaft by grooves, runs along the axis of the bristle, twists slightly and terminates in the subapical tooth (Fig. 3.39B). The longitudinal ridge is observed from only the ventral surface of the bristle; the dorsal surface is smooth (Fig. 3.39B). The main axis of the shaft extends past the longitudinal ridge and forms the distal most tip; the distal tip often has two very small teeth (Figs. 3.39B; 3.40D). On a given cell, each type of bristle is remarkedly consistent in length.

←

Fig. 3.39. Mallomonas tonsurata. Figs. A-C. Whole cells illustrating the distribution of domed scales, domeless scales and bristles, the overlapping nature of the scales and the presence of posterior scales with spines. Both the short, thick, serrated bristles and the longer, smooth forked bristles can be seen. Scale bars are 5 μm. Fig. D. Close-up of the apical most ring of scales surrounding the flagellar pore. Note the continuation of the left arm of the V-rib around the perimeter of the dome. Scale bar is 1 μm. Fig. E. Domed and domeless body scales and domeless rear scales with spines. Note the secondary layer covering the shield and flanges and the window at the base of the V-rib. The diameter of the pores in the secondary layer is larger than those of the pores in the base plate. Scale bar is 2 μm. Fig. F. Close-up of Fig. E depicting the position of the spines. Scale bar is 1 μm.

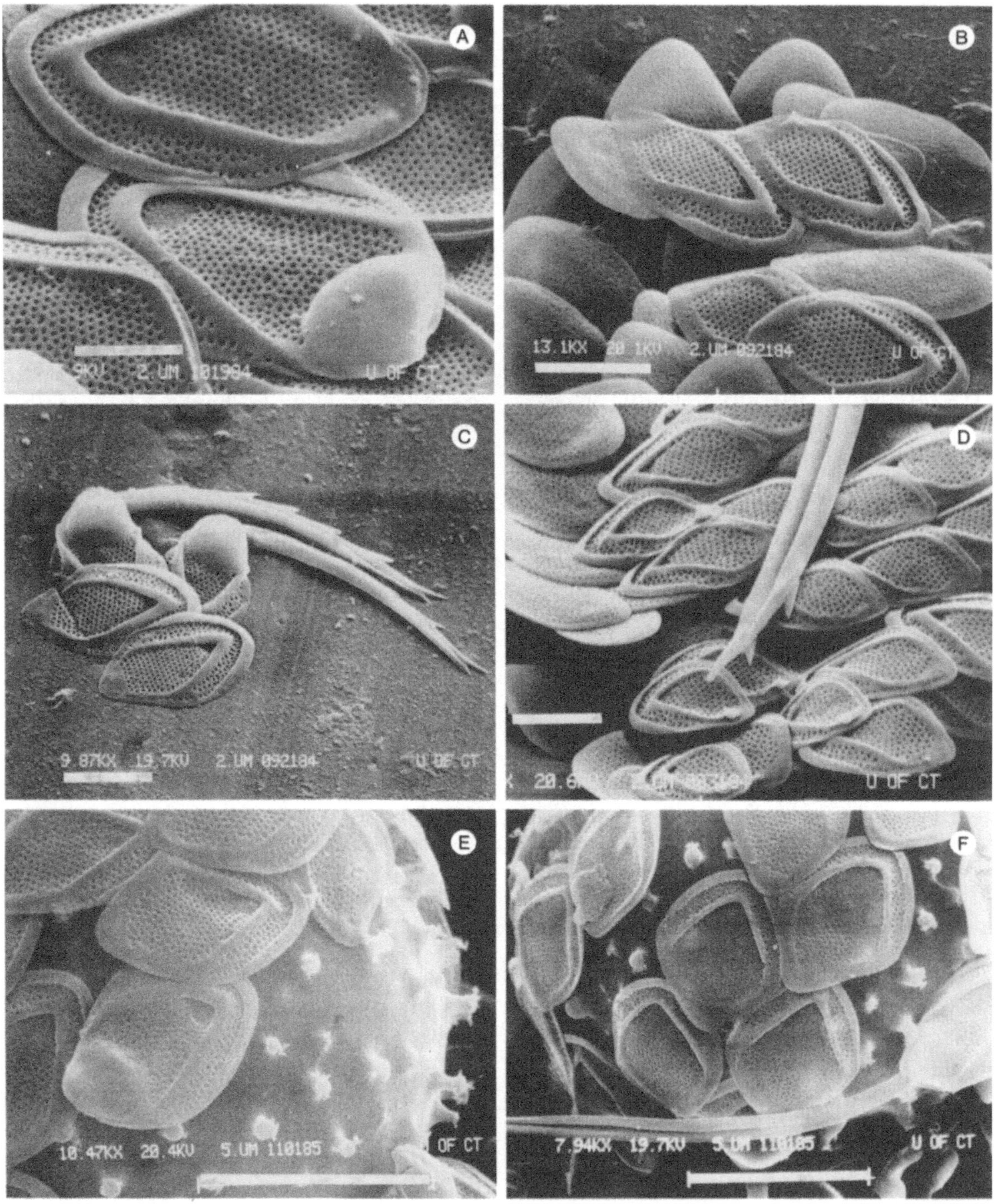

Fig. 3.40. Mallomonas tonsurata. Fig. A. Domed and domeless scales. A patch of closely spaced pores can be seen in the region of the window. Scale bar is 1 μm. Fig. B. Domeless scales. Note the smaller diameter of the base plate pores (as viewed from the undersurface of the scale) compared to the diameter of the pores in the secondary layer. Scale bar is 2 μm. Fig. C. Two apical domed scales with short, thick, serrated bristles and two domeless scales. Scale bar is 2 μm. Fig. D. Domeless scales and the distal tips of two long, bifurcated bristles. Scale bar is 2 μm. Figs. E-F. Surface detail of stomatocysts. Note the short, blunt projections with irregularly shaped arms. Scale bars are 5 μm.

Stomatocysts are spherical, 15 μm in diameter and covered with short (1μm) blunt projections (Figs. 3.40E-F). Small, irregularly shaped arms proliferate from the tops of most blunt projections.

Series Portaferreanae Asmund and Kristiansen 1986

Mallomonas portae-ferreae Péterfi & Asmund

Many characteristics of cells, scales and bristles for this species are similar to those of *Mallomonas corymbosa* making the distinction between the two virtually impossible with LM. Cells are large, cylindrical to elongate-ellipsoidal in shape and have bristles covering one-half to three-fourths of the length of the cell (Fig. 3.41A).

As described for *Mallomonas corymbosa*, a progression of four types of scales can be found along the length of the cell: domed apical; domed body; domeless body; and domeless rear scales (Figs. 3.41A-B). Like *M. corymbosa*, scales are relatively large and broad (Table 4.2), have a base plate with evenly spaced minute pores, a narrow posterior rim encircling one-half of the scale, and a well developed acutely angled V-rib (Figs. 3.41 C-E; 3.42D-F). Domes are prominent, strongly concave, usually smooth and cover 17.8% of the surface area of the scale (Figs. 3.41D-E; 3.42F). The juncture between the anterior submarginal ribs and the arms of the V-rib is curved and continuous in nature (Figs. 3.41D-E; 3.42D). The submarginal ribs are often slightly winged (Fig. 3.42F) and become greater in length with increasing distance from the anterior end of the cell.

Scales are further decorated with a secondary layer consisting of a series of parallel transverse ribs (var. *portae-ferreae* - Figs. 3.42D,F) or an irregular reticulated pattern (var. *reticulata* Gretz, Sommerfeld and Wujek - Figs. 3.41D-E); in this manner they differ from *M. corymbosa*. The transverse ribs of var. *portae-ferreae* are most often curved and usually connected by smaller perpendicular ribs (Figs. 3.42D,F). The flange of both varieties is usually marked with a secondary reticulation (Fig. 3.41E). Domed scales lacking a secondary layer on the shield and resembling scales of *M. areolata* were often found on the anterior ends of whole cells of var. *reticulata* (Figs. 3.41 B-D).

Domeless body scales are symmetrical in nature and grade into smaller, slightly asymmetrical rear scales; the latter may possess a tiny spine.

As described for other taxa of this section, scales are in spiral rows and arranged with their longitudinal axes positioned at a 45° to 60° angle to the longitudinal axis of the cell (Fig. 3.41B). The arrangement and overlap of scales is as described for *Mallomonas tonsurata*.

Both varieties have bristles of two different lengths (Péterfi and Asmund, 1972; Gretz et al., 1985); the smaller ones being associated with the apical most domed scales (Figs. 3.41A,C; 3.42C). All bristles are stout, consists of three ridges that run the length of the shaft, and are unilaterally serrated along one of the ridges (Figs. 3.41F; 3.42B). Each tooth of the serration terminates with two or three smaller teeth (Figs. 3.41C; 3.42A,D). The morphology of the distal tip of each long bristle varies. Asmund and Kristiansen (1986) described the distal portion of the longer bristles as being twisted such that the apical most tooth pointed in the opposite direction from the other teeth, and as being 'several times the length of the subapical teeth'. Bristles with distal tips as described by Asmund and Kristiansen (1986) were found in this study; however, others were not twisted and/or had distal sections equal in length to the subapical most tooth.

Taxonomic Difficulties Between Species in the Series Alpinae, Tonsuratae and Portaferreanae

The identification of species within the series Alpinae, Tonsuratae and Portaferreanae can be difficult, and often impossible, when based solely on characteristics of isolated scale(s). This is especially true if the identification is made with LM. Even with SEM or TEM micrographs of isolated scales, identifications may still be questionable. Taxonomic problems within these three series are well documented (Takahashi, 1978; Wee, 1982; Asmund and Kristiansen, 1986) and exist primarily due to the high degree of overlap in characteristics used to separate the taxa, especially those of scale size and morphology. For example, the range in size and morphology of *Mallomonas alpina* and *Mallomonas areolata* are practically identical (Asmund and Kristiansen, 1986); in addition, the range in size of cells and the types of scales produced by

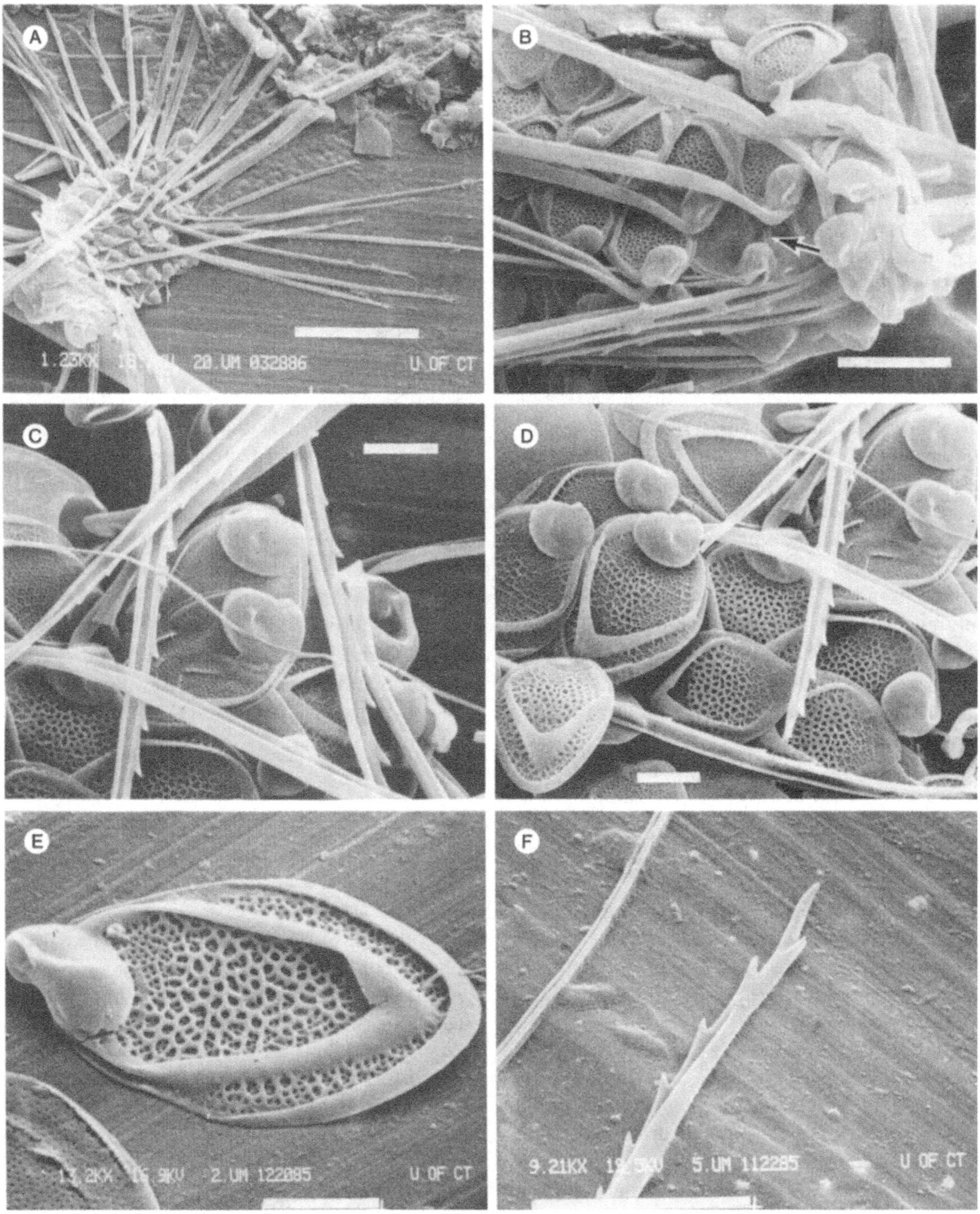

Fig. 3.41. Mallomonas portae-ferreae. Fig. A. Whole cell showing the arrangement of domed scales, domeless scales and long body bristles. Scale bar is 20 μm. Fig. B. Close-up of the anterior end of an intact cell. Note the scale (arrow) lacking a secondary reticulum. Scale bar is 5 μm. Figs. C-D. Body scales with varying degrees of secondary reticulation. Note the prominent and smooth domes and the serrations of the bristles. Scale bars are 2 μm. Fig. E. Domed body scale with a well developed dome and secondary reticulum. Note the continuation of the anterior submarginal ribs with the arms of the V-rib. Scale bar is 2 μm. Fig. F. Distal portion of a body bristle. Scale bar is 5 μm.

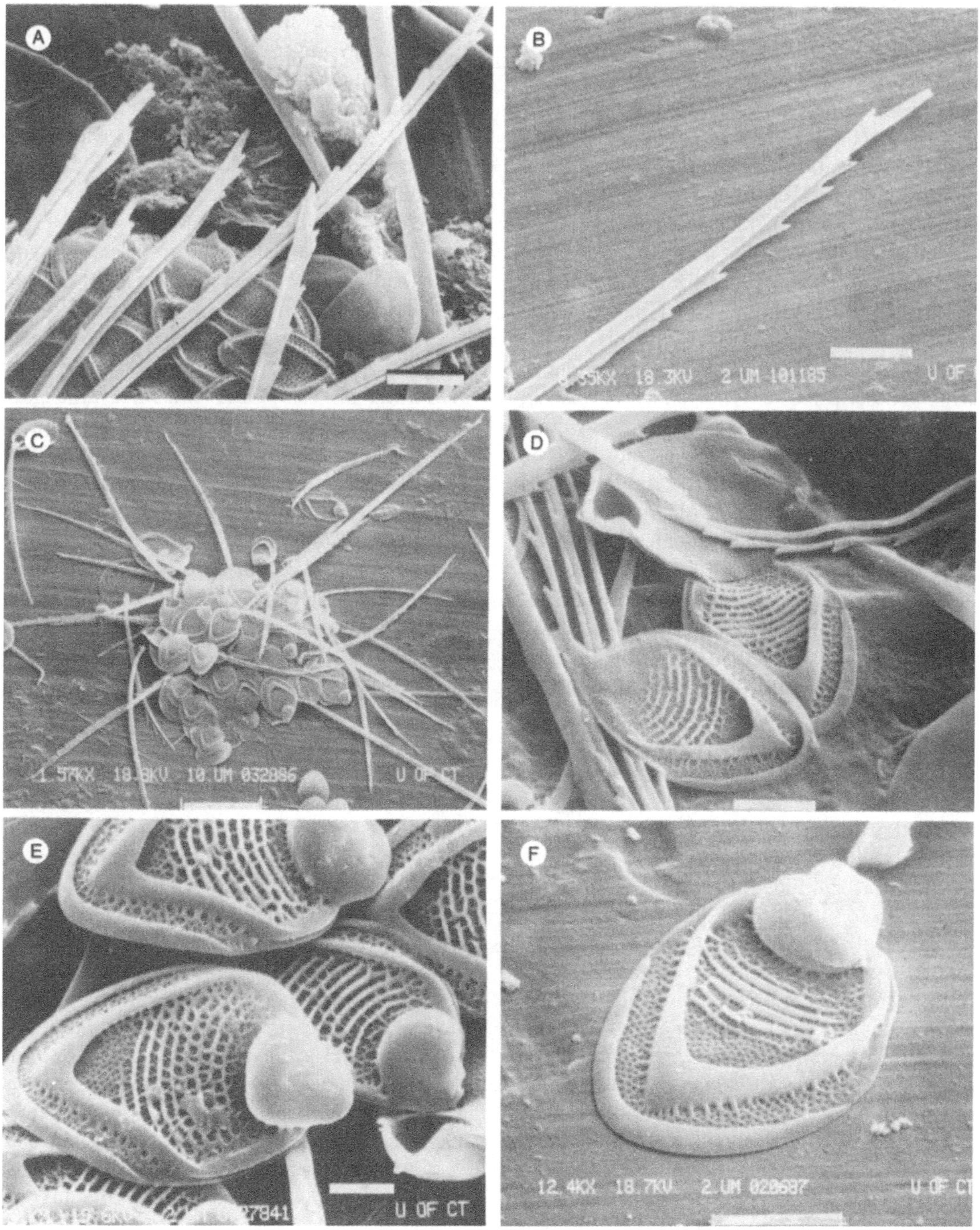

Fig. 3.42A-B. Mallomonas corymbosa. Ribbed body bristles that are serrated along the distal portion of the shaft. Note that the distal most tooth (Fig. B) is pointed in the opposite direction from the other teeth. Scale bars are 2 μm. Fig. C. *Mallomonas portae-ferreae* var. *reticulata.* Scales and bristles. Note both short and long bristles. Scale bar is 10μm. Figs. D-F. *Mallomonas portae-ferreae* var. *portae-ferreae.* Fig. D. Domeless body scales. Scale bar is 2μm. Figs. E-F. Domed body scales possessing a prominent and well defined dome and transverse shield ribs that are interconnected by smaller perpendicular struts. Scale bars are 1μm and 2μm, respectively.

both taxa are very similar (Asmund and Kristiansen, 1986). A review of the TEM micrographs of domed and domeless body scales for *Mallomonas tonsurata, M. corymbosa* and *M. cyathellata* var. *cyathellata* in the recent monograph by Asmund and Kristiansen (1986) (e.g. compare Figs. 45a and 46b of that publication) will further illustrate the taxonomic difficulties encountered where identifications are attempted from isolated scales.

Difficulties in the concept of a species in the Series Alpinae and Tonsuratae are well documented (Asmund, 1959; Asmund and Hilliard, 1961; Takahashi, 1978; Wee, 1982; Nicholls, 1982; Asmund and Kristiansen, 1986). For example in the original description of *Mallomonas corymbosa*, Asmund and Hilliard (1961) stated, '*M. corymbosa* is closely related to *M. tonsurata* and may be considered a variety of the latter.' Asmund (1959) commented that since *M. alpina* (*M. tonsurata* var. *alpina* in her publication) like scales have been observed on *M. tonsurata* cells, the 'distinction' is not always clear. Péterfi and Momeu (1976a) described a form intermediate between *M. tonsurata* and *M. alpina*. Wee (1982) described a large variation in the structure of bristles attached to '*tonsurata*-like scales' and concluded that this was a poorly understood group. Asmund and Kristiansen (1986) pointed out many similarities between *M. alpina* and *M. areolata*. In addition, many of the taxa in these Series are often found in the same habitats (Siver and Hamer, 1989; Siver, 1989a).

It is clear that valid identifications can be made by observing a combination of different characters. In particular, it is believed that the structure of the bristles can greatly aid in the identification of many of the more common species within the three Series. A comparison of the similarities and differences and the resultant problems encountered is presented below for *Mallomonas tonsurata, M. alpina, M. areolata, M. elongata, M. corymbosa, M. cyathellata* var. *cyathellata* and *M. portae-ferrae*.

The first step in the identifications of the seven species is to note their common features: a) cells lacking bristles in the posterior region; b) presence of domed and domeless scales; c) rear scales with spines (or protuberances) and; d) bristles of two size classes with the smooth ones forming an apical tuft.

Mallomonas tonsurata, M. corymbosa, M. cyathellata var. *cyathellata* and *M. portae-ferreae* differ from the other three in having scales with a secondary layer and window. In many LM preparations the window appears as a bright region and indicates the presence of a secondary layer. However, the visual absence of a window using LM does not preclude that the scale lacks a secondary layer. The secondary layer is easily detected with TEM and SEM; the diameter of the pores of the base plate (ventral surface) are smaller than those of the secondary reticulum (dorsal surface). With EM scales of *M. portae-ferreae* are separated on the basis of the characteristic shield reticulum; however, the secondary layers of *M. tonsurata, M. corymbosa* and *M. cyathellata* and *cyathellata* are similar and difficult to distinguish. The protuberances on rear scales of *M. cyathellata* var. *cyathellata* are clearly different (Wujek and Asmund, 1979) from the other three taxa.

Scales of *Mallomonas corymbosa, M. portae-ferreae* and *M. cyathellata* var. *cyathellata* are commonly larger, have smaller domes relative to the surface area of the scale and possess short submarginal ribs when compared to scales of *M. tonsurata*. However, considerable overlap exists in the dimensions and morphologies of scales for *M. tonsurata, M. corymbosa, M. portae-ferreae* and *M. cyathellata* var. *cyathellata* making taxonomic decisions based solely on these features questionable.

Cells of *Mallomonas tonsurata* tend to be smaller, more rounded, ovoid and have bristles covering a lesser area of the cell when compared to *M. corymbosa* and *M. portae-ferreae*. Based on the few available records, cells of *M. cyathellata* var. *cyathellata* are quite small but within the known range for *M. tonsurata* (see Chapter IV). However, a review of the literature shows that there is considerable overlap in the range of cell size for *M. tonsurata* and *M. corymbosa* and between *M. corymbosa* and *M. portae-ferreae*; little overlap exists however, between *M. tonsurata* and *M. portae-ferreae*. As observed with scales, the dimensions of cells cannot be used as the sole taxonomic character.

Based on the close examination of many populations of *Mallomonas tonsurata, M. corymbosa* and *M. portae-ferreae* it is concluded that the differences in the morphologies of the long bristles provide a useful means to separate the taxa. The presence of a long bristle with a smooth shaft and a single subapical tooth (forked tip) identifies the cell as *M. tonsurata*; such a bristle morphology has not been observed on cells of *M. corymbosa* or *M. portae-ferreae*. Populations of *M. tonsurata* cells with both typical types of bristles as well as a third intermediate type have been observed in this study and by other researchers (Asmund, 1959; Péterfi and Momeu, 1976a; Wee, 1982; Asmund and Kristiansen, 1986). The intermediate type of bristle is usually long with a few teeth on the shaft and a forked tip; these bristles would be virtually impossible to distinguish from those of *Mallomonas alpina* and *M. areolata*. However, the long, smooth, forked bristles typical of *M. tonsurata* have never been observed on *M. alpina, M. areolata* (see below), *M. corymbosa* or *M. portae-ferreae*.

The shafts of the long bristles of *Mallomonas corymbosa* and *M. portae-ferreae* are always uniseriate and, thus, distinguished from *M. tonsurata*. Bristles of *M. corymbosa* are separated from those of *M. portae-ferreae* by the distinct subapical tooth positioned below the distal tip on the side of the shaft opposite the serration. The distal tips of some bristles of *M. portae-ferreae* are forked (Figs. 3.41F; 3.42B), but the fork is not the result of a subapical tooth as in *M. corymbosa*; such a bristle tip can be found on cells of *M. corymbosa* and probably represents an intermediate type between the shorter serrated bristles and the longer ones with the subapical tooth. The features of the distal tips discussed above can be observed with LM, however, it must be realized that many bristles will be orientated such that the true tip features may not be seen.

Scales of *Mallomonas alpina, M. areolata* and *M. elongata* lack a secondary reticulum and window, however, as noted above, this feature is not always evident with LM. Scales and cells of *M. areolata* are generally considered to be larger than those of *M. alpina* (Asmund and Kristiansen, 1986), however, the ranges for both taxa are virtually identical. In addition, both *M. alpina* and *M. areolata* have similar scales on identical positions on the cell (see Asmund and Kristiansen, 1986 for a review). Scales of *M. areolata* are generally described with a patch of tiny pores on the shield positioned near the base of the V-rib. However, the patch of pores, not visible with LM, is not unique to *M. areolata*. The morphology of the long bristles and to a lesser extent, their position on the cell, may serve as the best diagnostic character for separating *M. alpina* from *M. areolata*. Both species have unilaterally serrated bristles, however, those of *M. areolata* are slightly thinner and end in a long needle-like tip.

Cells of *Mallomonas elongata* are clearly much larger than those of *M. alpina* or *M. areolata*. The height above the base plate and curvature of the V-rib and submarginal ribs are much greater in *M. elongata* than in any of the other six taxa discussed in this section; this feature is clearly evident with EM and, with practice, also discerned with LM. Two additional features of scales of *M. elongata* make them readily identifiable with EM. First, the distal part of the base plate has significantly fewer pores than the posterior end. Second, the dome, submarginal ribs and V-rib are often lined with a series of large pores aligned in rows. The long bristles of *M. elongata* are sometimes confused with those of *M. portae-ferreae*, however, the former are more strongly ribbed and each serration consists of several smaller teeth.

Cells of *Mallomonas alpina* can easily be confused with *M. tonsurata*, especially if the presence of a secondary layer is not clearly evident. For such specimens bristle morphology becomes critical.

In summary, the separation of species within the series Alpinae, Tonsuratae and Portaferreanae is often difficult, especially if whole cells are not available. Identifications are best made after considering a group of characters, especially bristle morphology.

Series Corconticae Asmund & Kristiansen 1986

Mallomonas corcontica (Kalina) Péterfi & Momeu

Synonym: *Mallomonas leboimei* var. *corcontica* Kalina

Cells are oblong-ovoid or elliptical and covered with long, curved bristles (Péterfi and Momeu, 1976a; Asmund and Kristiansen, 1986; Nicholls, 1987a). Scales possess domes and are arranged on the cell in the same manner as described for *M. striata*.

Scales are elongate-obovoid with pronounced lateral incurvings and possess a posterior rim encircling one-half of the perimeter and an acutely angled V-rib with long, straight sides that extend to the margin of the scale (Fig. 3.27E). Distinct anterior submarginal ribs are lacking; however, the anterior flange is marked with a series of ribs positioned roughly parallel to the margin (Figs. 3.27E-F). The shield, dome and posterior flange are also ornamented with ribs.

Unlike scales of *M. asmundiae*, *M. cratis* and *M. striata*, the ribbing on scales of *M. corcontica*, especially on the anterior end, is irregularly arranged. The shield consists of 9 to 12 equally spaced, slightly curved, parallel ribs, some of which are inter-connected with cross struts (Fig. 3.27E). Several smaller apical scales with fewer ribs have been observed (Fig. 3.27F). About 20 to 25 ribs, some branched, radiate from the V-rib onto the posterior flange. The ribs on the dome are irregularly arranged, some being continuous with those on the anterior flanges (Figs. 3.27E-F). An additional series of short struts is present; these radiate from the posterior border of the dome onto the shield and form right angles with the transverse shield ribs (Figs. 3.27E-F).

Bristles are ribbed along the entire length of the shaft, curved, and unilaterally serrated along their convex margin. The distal tip is either pointed or lance-shaped; on the latter bristle type, the ridge along the concave margin is expanded just below the tip and folded over to produce the lance-like appearance (Fig. 3.27E).

Series Mallomonas

Mallomonas acaroides var. *acaroides* Perty emend. Iwanoff

The morphologies of scales and bristles, and their arrangement on the cell are similar to those described for var. *muskokana*. The latter variety is described in greater detail because many more specimens were observed.

Four major features best serve to separate var. *acaroides* from var. *muskokana*. First, all scales on var. *acaroides* possess domes. Second, the helmet bristles of var. *acaroides* lack the subapical tooth on the side opposite of the helmet opening (Fig. 3.52A). Third, the domes of var. *acaroides* are rarely, if ever, ornamented with ribs and papillae (Figs. 3.52A-B). Fourth, the arms of the V-rib on scales of var. *acaroides* extend to the margin of the scale and, as such, are separate structures from the anterior submarginal ribs (Figs. 3.52A-B). Although this latter feature has not been previously described, it is consistent with published micrographs (e.g. Wee, 1982; Asmund and Kristiansen, 1986; Nicholls, 1987b).

Except for *Mallomonas acaroides* var. *inermis* Fott and var. *muskokana*, other previously described varieties have been shown to be within the accepted range of variation for var. *acaroides* as originally emendated by Ivanoff (Asmund and Kristiansen, 1986). Like var. *muskokana*, cells of var. *inermis* also have domeless scales in their posterior region. However, var. *inermis* lacks helmet bristles and has scales that are only one-half the size of those of var. *muskokana* (Nicholls, 1987). In addition, the structure of the shield and the juncture between the anterior submarginal ribs and the V-rib for var. *inermis* is similar to that of var. *acaroides*.

Mallomonas acaroides Perty emend. Ivanov var. *muskokana* Nicholls

Cells of *M. acaroides* var. *muskokana* are broadly ovoid to elliptical and covered with bristles over all but the posterior most region. Scales are orientated in spiral rows with their longitudinal axes roughly perpendicular to the longitudinal axis of the cell. Each scale is overlapped by the scale positioned behind it in the same spiral row and by the scales in the spiral row above it. Thus, on an intact cell usually only the anterior two-thirds

→

Fig. 3.43. Mallomonas acaroides var. *muskokana*. Fig. A. Remains of the siliceous cell covering of one cell. Note the length and morphology of the bristles. Portions of cells of *M. pugio* and *M. canina* are also observed. Scale bar is 10 μm. Fig. B. Domed body and domeless posterior scales. Except for the series of short struts lining the inner margin of the V-rib, the shields are unornamented. Note the continuation of the arms of the V-rib with the anterior submarginal ribs. Scale bar is 5 μm. Fig. C. Attachment of bristles to scales. Note the recurved hooks along the proximal portion of the bristle shaft and the ribs lining the front border of the dome. Scale bar is 2 μm. Fig. D. Posterior portion of a cell depicting the domeless body scales. Scale bar is 2 μm. Fig. E. Apical scales with large, highly sculptured domes and slightly winged anterior submarginal ribs (arrow). Scale bar is 1 μm. Fig. F. High magnification of a dome with ridges and papillae. Scale bar is 0.5 μm.

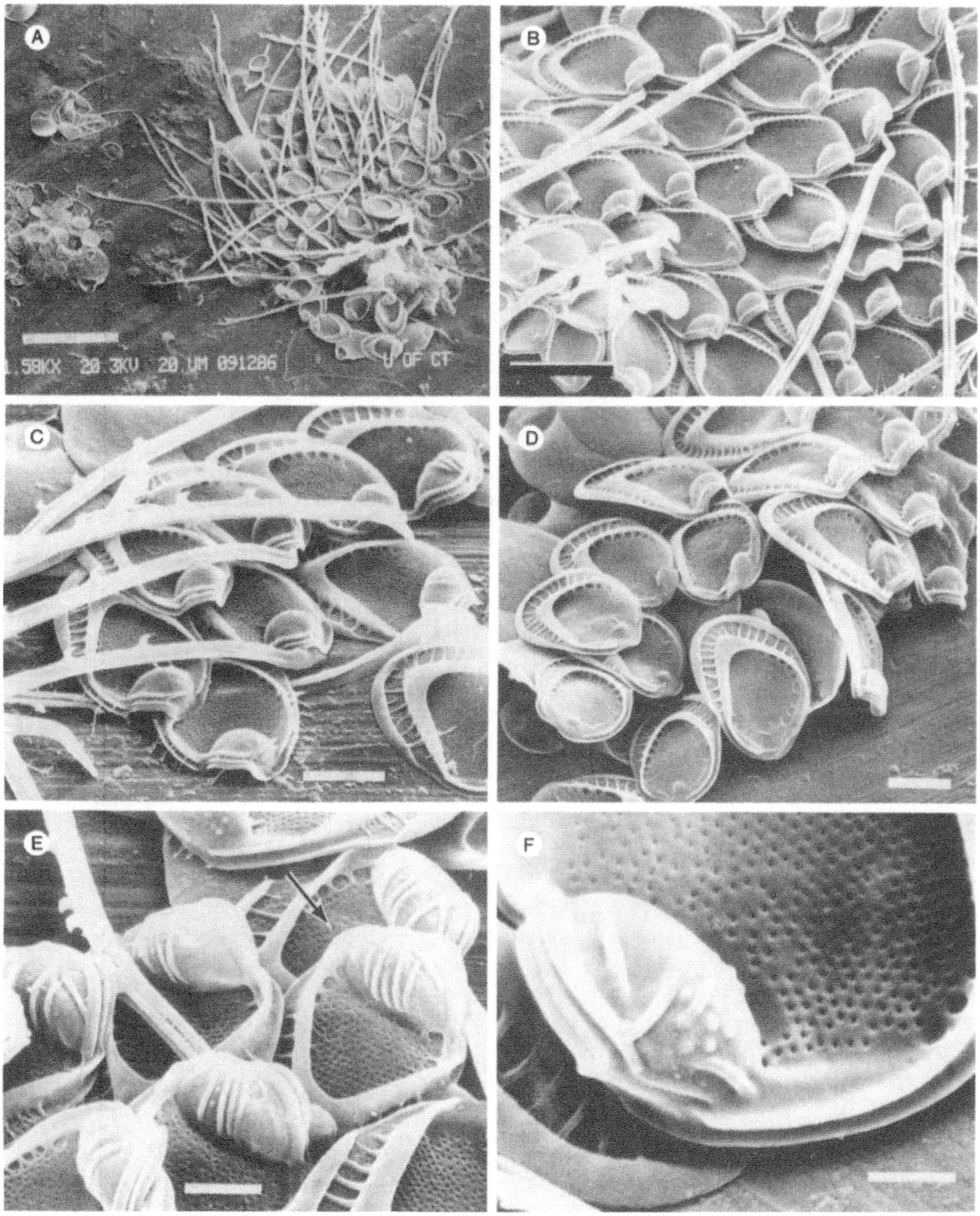
A
1.50KX 20.3KV 20.UM 091286
U OF CT
B
C
D
E
F

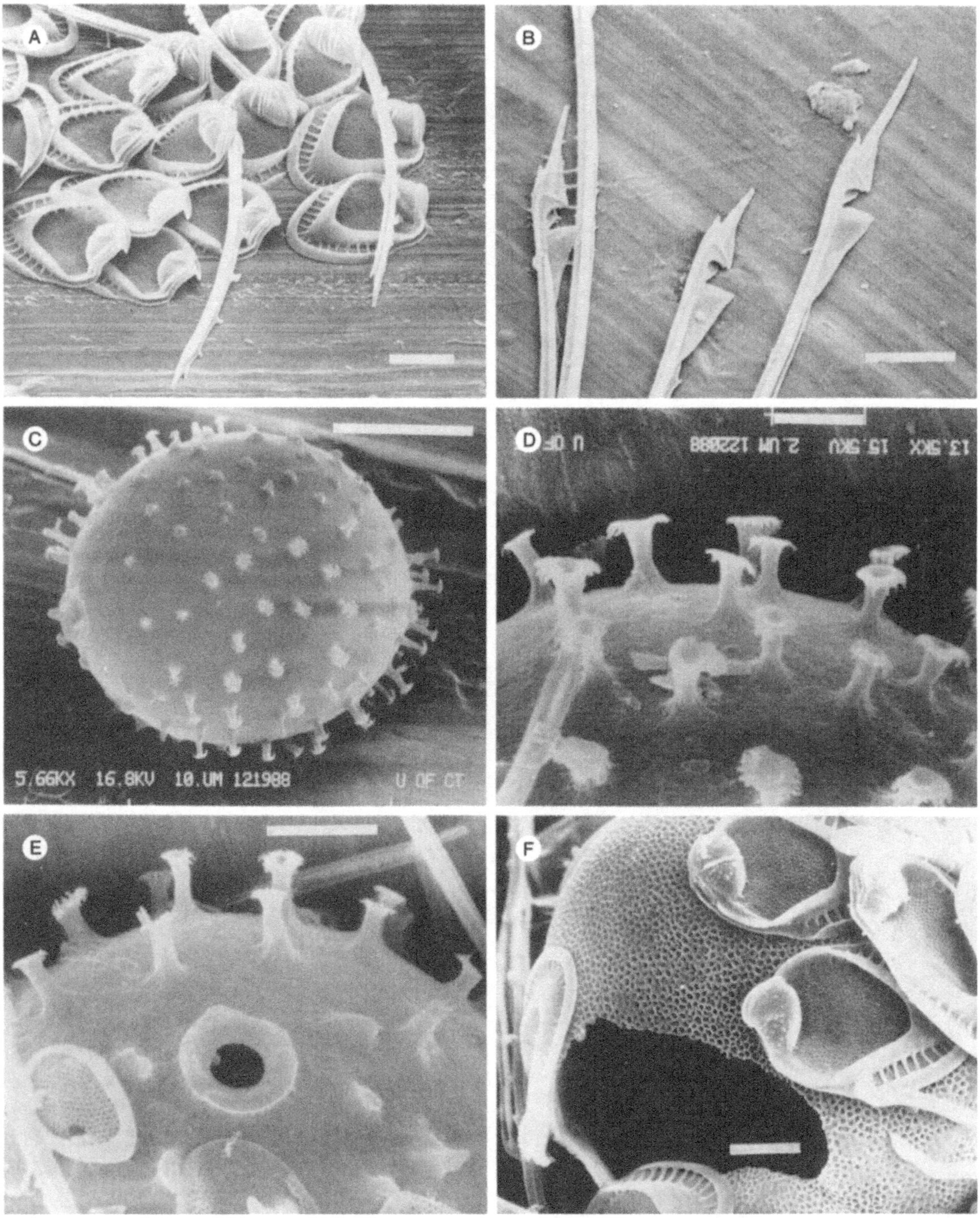
A
B
C
5,66KX 16.8KV 10.UM 121988 U OF CT
D
E
F

of the shield and dome (if present) are exposed (Fig. 3.43B).

Domed (Fig. 3.43C) and domeless (Figs. 3.43B,D) scales are present, the latter being restricted to the posterior end of the scale (Fig. 3.43D). Domed scales are broadly oval with slight lateral incurvings and have a base plate with evenly spaced pores, a posterior rim encircling one-half of the perimeter, a prominent V-rib with hood and a rather large sculptured dome (Figs. 3.43E-F; 3.44A). The distal ends of the arms of the V-rib curve and become continuous with the anterior submarginal ribs (Figs. 3.43C,F; 3.44A). On some scales the anterior submarginal ribs may be slightly winged (Fig. 3.43E). The right anterior submarginal rib continues around the perimeter of the dome and terminates above the point of emergence of the bristle (Figs. 3.43C; 3.44A). On domeless scales the submarginal ribs terminate close to a small swelling which is presumably a rudimentary dome-like structure (Figs. 3.43B,D).

Although a small degree of secondary ribbing has been reported on the shield (Nicholls, 1987b), they are generally not found (Figs. 3.43B-D; 3.44A). A series of short struts lines the interior side of the V-rib and on some scales may extend along the anterior submarginal ribs (Figs. 3.43B-D); the struts are somewhat vertically arranged forming large pore-like openings (Figs. 3.43B,D; 3.44A). A second series of equally spaced struts radiate at right angles from the V-rib onto the posterior flange.

Domes are largest on the apical most scales (Fig. 3.43E), and are usually marked with a series of ribs and papillae (Figs. 3.43F; 3.44A) that are commonly restricted to the left side (Fig. 3.43F). Some scales, usually in the posterior region, may have smooth domes (Fig. 3.43B). The front margin of the dome is always marked with at least two ribs, one of which is a continuation of the anterior submarginal rib (Figs. 3.43C,F).

Two types of bristles can be found on any given cell (Nicholls, 1987b; Siver, 1989b). Both types of bristles are strongly ribbed. One is slightly shorter, unilaterally serrated, terminates in a pointed apex, and has an additional subapical tooth positioned on the ridge opposite that with the serration (Fig. 3.44A). Some of the proximal teeth may be recurved (Fig. 3.44A). The second type is longer and with a helmet (Fig. 3.44B). The shaft of each helmet bristle may be smooth or serrated, however, a single subapical tooth is always present on the side of the helmet opposite of the cleft (Fig. 3.44B). All bristles emerge from a shallow U-shaped opening located to the right of center of the dome (Figs. 3.43C; 3.44A). Although cells with both helmet and serrated bristles may be found, most have only helmet bristles; cells with only serrated bristles have not yet been observed (Siver, 1989b). *M. acaroides* var. *muskokana* has been shown to be a warm water taxon (Siver, 1989b; see below). It is possible that bristle production is similar to that of *M. crassisquama* where helmet bristles were found to be formed under warm water conditions (Siver and Skogstad, 1988).

The structure of the stomatocyst (Figs. 3.44C-F) was shown to be similar to that of *M. acaroides* var. *acaroides* (Siver, 1991). The stomatocyst is slightly ovate in shape, covered with evenly spaced, ornamented spines and has a short cylindrical collar with a variable apex (Siver, 1991). The pore is surrounded by an internal annulus. Spines have a thickened base, a cylindrical middle section and a flattened apex with a ring of finger-like projections.

Mallomonas crassisquama (Asmund) Fott

Synonyms: *M. acaroides var. crassisquama* Asmund

Cells are ovoid, generally widest at the midsection,

←

Fig. 3.44. Mallomonas acaroides var. *muskokana*. Fig. A. Group of body scales with serrated bristles lacking helmets. Note the rather large and sculptured domes, the continuation of the right submarginal rib around the perimeter of the dome and the series of short struts lining the inner margin of the V-rib. Also, note the recurved proximal teeth on the bristles. Scale bar is 2 μm. Fig. B. Distal portions of three helmet bristles each with a single subapical tooth on the side of the shaft opposite of the cleft. Scale bar is 2 μm. Figs. C-F. Stomatocyst. Fig. C. Stomatocyst illustrating the distribution of the splayed spines and the morphology of the collar. Scale bar is 5 μm. Fig. D. Close-up of the surface of a stomatocyst. Scale bar is 2 μm. Fig. E. Close-up of the pore-collar complex. Scale bar is 2 μm. Fig. F. Immature stomatocyst with a 'honeycomb-like' surface structure. Scale bar is 2 μm.

A
B
C
D
E
F
U OF CT
4.73KX 16.8KV 10.UM 121988
U OF CT
5.90KX 18.2KV 2.UM 070984

and are largely covered with bristles, except for the posterior most region (Figs. 3.45A-C). The anterior is slightly wider on some cells.

Four types of scales, anterior domed scales, domed body scales, domeless body scales and posterior spined scales, are arranged along the length of the cell (Figs. 3.45A-B). Apical scales are smaller, asymmetric in shape, have a wing-like extension on the left side of the dome and form a ring around the flagellar opening (Fig. 3.46C). Apical scales are positioned in the ring such that: 1) each is overlapped by the scale to its left in the ring and by the scale positioned behind it in the same spiral row; 2) the wing-like extension on the left side of the dome faces forward (Fig. 3.45B). The domes of body scales are slightly asymmetric due to the positioning of the U-shaped bristle opening; otherwise, like domeless body scales, domed body scales are symmetric (Figs. 3.45E; 3.46A-B).

Scales are imbricated in a spiral arrangement with their longitudinal axes perpendicular to the longitudinal axis of the cell (Figs. 3.45A-C). The right posterior flange of each scale is overlapped by the scale positioned behind it in the same spiral row, and its left posterior flange by scales in the spiral row above it (Figs. 3.45B-C). Thus, only the shield, dome and part of the V-rib are exposed on intact cells (Fig. 3.45C). The spines on the asymmetric posterior scales project from the right-hand side (Fig. 3.46D), such that they are positioned parallel to the longitudinal axis of the cell (Figs. 3.45A-B).

Scales consist of a perforated base plate, a posterior rim that encircles one-half of the perimeter of the scale, an acutely angled V-rib with arms that extend close to or to the margin and, if present, a prominent dome (Figs. 3.46A-B,F). Base plate pores are lacking in extreme peripheral areas (Siver and Skogstad, 1988). The anterior submarginal ribs originate just below the distal ends of the arms of the V-rib and extend, parallel to the margin, to the sides of the dome on domed scales (Figs. 3.46A-B), meet and fuse on domeless scales (Figs. 3.45A,F), or become extended to form the spine on posterior scales (Fig. 3.46D) (Siver and Skogstad, 1988). Although the anterior submarginal ribs fuse with the arms of the V-rib, they are usually not a continuation of the latter and, as such, represent separate structures (Figs. 3.46A-B, F). On heavily silicified scales this distinction may be masked.

The shield is ornamented with a secondary layer of ribs that form a reticulum (Figs. 3.45E-F; 3.46 A-B). The degree of reticulation varies greatly between populations, cells from the same population and even on scales from the same cell (Siver and Skogstad, 1988). Some scales may have a weak network of reticulation such that the pores of the base plate are still readily seen (Figs. 3.45E; 3.46A). Most scales have a pattern of well defined ribs that fuse to form a secondary layer of larger pores (Fig. 3.46B); these scales have a window in the angle of the V-rib. Shields with little reticulation are most commonly found on apical domed and posterior spined scales. A series of struts radiate at right angles from the V-rib onto the posterior flange; the struts may fork or be connected with perpendicular ribs. The anterior flanges are narrow and unornamented (Figs. 3.46A-B). The degree to which the posterior rim and V-rib canopy varies; on highly silicified scales the canopies may obscure the posterior flange and window.

Domes are usually smooth (Figs. 3.45E; 3.46A) or sculptured with raised papillae (Figs. 3.45F; 3.46B); more rarely they are marked with parallel ridges. Domes that are either smooth or papillate may be found on the same cell (Siver and Skogstad, 1988). Bristles emerge from under the dome via a U-shaped opening, the perimeter of which may be lined with a small row of teeth or a short lateral wing (Siver and Skogstad, 1988).

←

Fig. 3.45. Mallomonas crassisquama. Figs. A-C. Whole cell depicting the orientation and arrangement of domed scales, domeless scales, posterior spined scales and bristles (Fig. C). Scale bars are 5 μm, 10 μm, and 5 μm, respectively. Fig. D. Surface detail of the stomatocyst of *M. crassisquama* var. *papillosa* showing the arrangement of the spines. Note the finger-like projections radiating from the flattened top of each spine. Scale bar is 2 μm. Fig. E. Body scales each with a relatively weakly formed secondary reticulum of ribs. Note the extension of the arms of the V-rib to the margin of the scale; the anterior submarginal ribs are attached to, but not continuous with, the arms of the V-rib. Scale bar is 2 μm. Fig. F. A group of heavily silicified domed and domeless body scales. The distal ends of the domeless scales and the domes are ornamented with papillae. Note the U-shaped bristle openings. Scale bar is 5 μm.

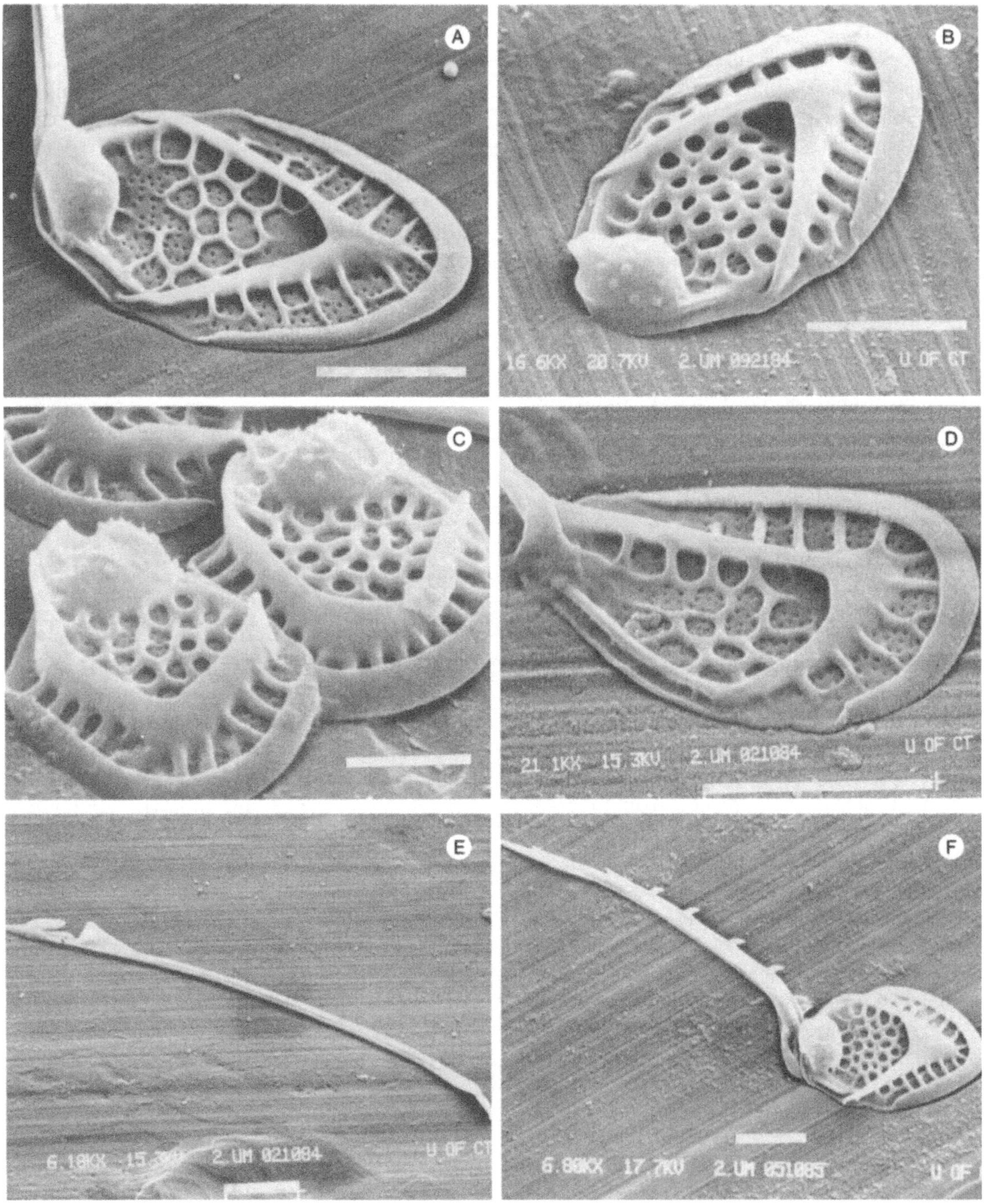

Fig. 3.46. Mallomonas crassisquama. Figs. A-B. Body scales with weakly (Fig. A) and heavily (Fig. B) silicified secondary ribs. Note the patch of small pores at the base of the V-rib on the scale in Fig. A. Scale bars are 2 μm. Fig. C. Asymmetric apical scales with a wing-like extension on the left side of the dome. Scale bar is 2 μm. Fig. D. Asymmetric posterior scale with a small spine. Scale bar is 2 μm. Fig. E. Helmet bristle lacking serrations on the shaft. Scale bar is 2 μm. Fig. F. Scale with serrated bristle. Note the recurved hooks on the proximal end. Scale bar is 2 μm.

Cells have either helmet bristles (Fig. 3.46E), serrated bristles (Fig. 3.46F), or both (Asmund and Kristiansen, 1986; Siver and Skogstad, 1988). All bristles are ribbed and have a distal tip that is slightly recurved and tapered to a point. The distal teeth of the serration project forward, however, the proximal teeth may form recurved hooks (Fig. 3.46F). Helmet bristles are generally longer than serrated bristles and have shafts that are smooth or serrated. Bristle production in *M. crassisquama* has been shown to be temperature dependent where cells form serrated or helmet bristles under cold or warm conditions, respectively (Siver and Skogstad, 1988).

M. crassisquama var. *papillosa* Siver & Skogstad differs from the type in possessing a single row of papillae deposited along both anterior submarginal ribs and often along the inner side of the V-rib (Figs. 3.45D; 3.48A). One cell of var.*papillosa* with a cyst was observed where the surface detail could be discerned (Fig. 3.45D). The surface of the cyst consists of equally-spaced spines roughly 1 μm long. Each spine is slightly tapered to a flattened top from which a ring of small finger-like projections radiate (Fig. 3.45D). A similar surface structure has been described for *M. crassisquama* var. *crassisquama* (Gretz et al., 1979; Asmund and Kristiansen, 1986) as well as for the related taxa *Mallomonas acaroides* var. *acaroides* (Cronberg, 1980; Asmund and Kristiansen, 1986), *M. acaroides* var. *muskokana* (Siver, 1991) and *M. pseudocoronata* (Smol, 1984).

Mallomonas duerrschmidtiae Siver, Hamer & Kling

Cells are elliptical to obovate in shape, with bristles covering the anterior one-half to two-thirds of the cell and a posterior group of scales with long spines (Figs. 3.48C-D; 3.49A-C).

The cell coat consists of four types of scales arranged from the anterior to posterior of the cell in the following order: domed anterior, domed body, domeless body and posterior spined (Figs. 3.48C-D; 3.49A-B). Body scales are arranged in spiral rows with their longitudinal axes perpendicular to the longitudinal axis of the cell (Figs. 3.48C-D). Each scale is overlapped by the scale positioned behind it in the same spiral row and by scales in the spiral row above it (Figs. 3.48C-D; 3.50A-C). The apical scales, which form the

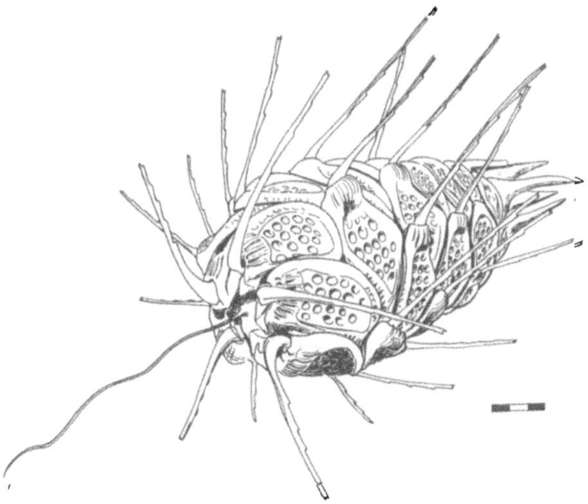

Fig. 3.47. Three-dimensional reconstruction of a whole cell of *Mallomonas duerrschmidtiae* depicting the distribution of scales and bristles. Scales are large and positioned in spiral rows such that their longitudinal axes are perpendicular to the longitudinal axis of the cell. Note the distribution of bristles and the posterior spined scales.

anterior most ring of scales, are highly asymmetric due to the position of the dome on the left side of the scale and a large serrated wing projecting from the left side of the dome (Fig. 3.48D); the wings collectively form a ring around the flagellar opening.

Body scales are large, usually highly silicified and curved such that the concave (ventral) surface conforms to the shape of the cell (Figs. 3.48F; 3.49D). The domes of body scales have a small serrated lip or wing on the right side of the dome lining the U-shaped bristle opening, giving the scale a slight asymmetry (Figs. 3.48B,F; 3.49E). Like the anterior domed scales, posterior spined scales are strongly asymmetrical due to the continuation of the submarginal ribs to form the long spine. The spine is orientated parallel to the longitudinal axis of the cell (Fig. 3.49B).

Body scales are elliptical and consist of a perforated base plate, posterior rim and V-rib. The posterior rim and V-rib are strongly hooded, forming extensive canopies (Figs. 3.48B,F). The arms of the V-rib curve and are continuous with the anterior submarginal ribs; the latter either terminate at the sides of the dome (Fig. 3.48B), meet and fuse in the domeless scales (Fig. 3.49A), or become extended and form the spines on the

Fig. 3.48A. Mallomonas crassisquama var. *papillosa.* Body scales on an intact cell. Note the papillae on the dome and along the anterior submarginal ribs, and the inverted U-shaped bristle openings. Scale bar is 2 μm. Figs. B-F. *Mallomonas duerrschmidtiae.* Fig. B. Body scale. Note the papillae, ribs and serrated lip of the dome. Scale bar is 1 μm. Figs. C-D. Whole cells denoting the orientation and pattern of overlap of the scales. Note the large, serrated, forward projecting wing on the apical most scales (arrow) of the cell in Fig. D. Scale bars are 5 μm. Fig. E. Domed scales, domeless scales and bristles. Scale bar is 5 μm. Fig. F. Domed body scales. Note the curved nature of the scales. Scale bar is 2 μm.

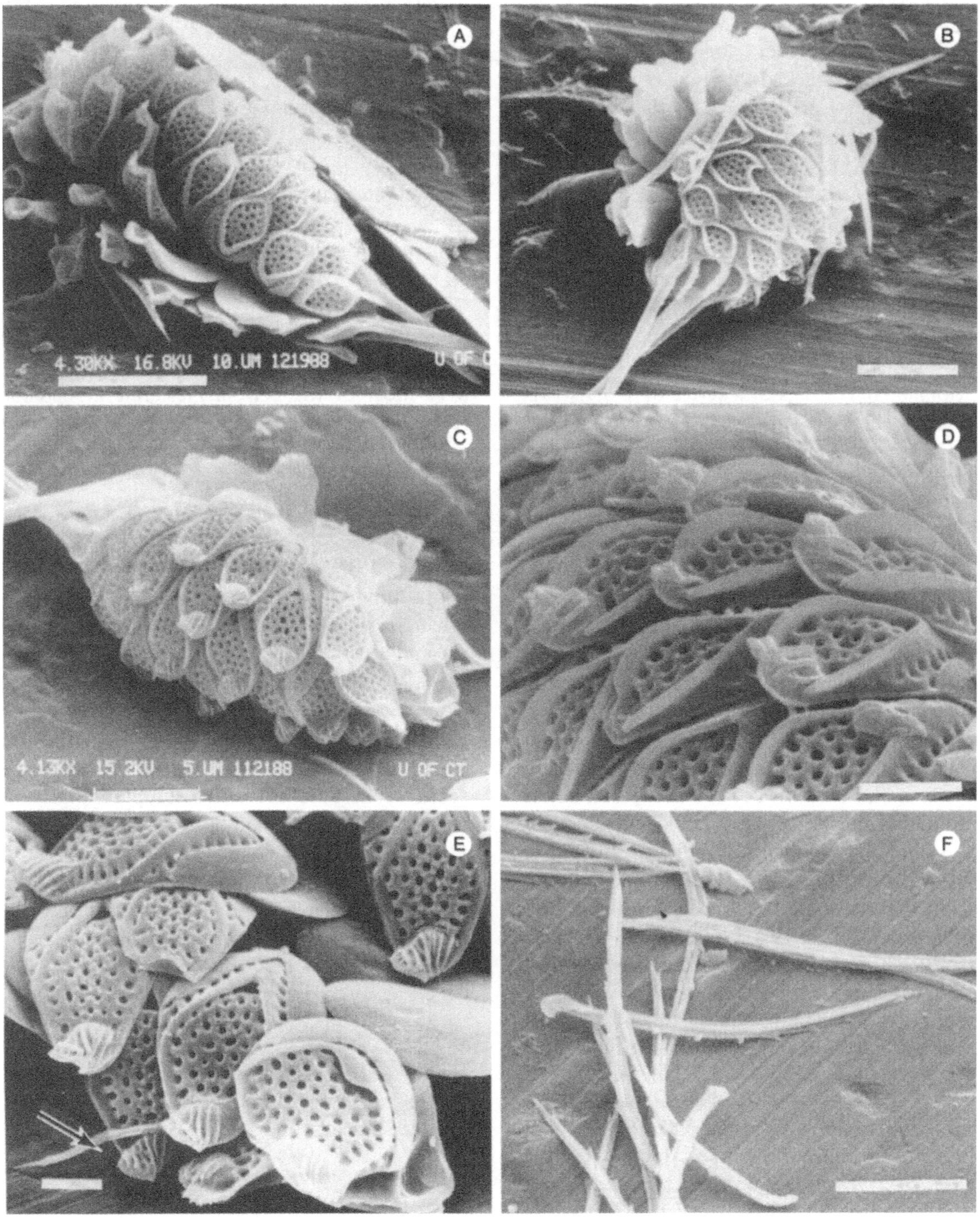

Fig. 3.49. Mallomonas duerrschmidtiae. Figs. A-C. Whole cells depicting the shape, arrangement of domed and domeless scales and the presence of long posterior spines. Scale bars are 10 μm, 5 μm and 5 μm, respectively. Fig. D. High magnification of body scales arranged in spiral rows on an intact cell. Note the winged anterior submarginal ribs. Scale bar is 2 μm. Fig. E. Body scales. Note the small, ribbed dome with serrated lip (arrow), the continuation of the arms of the V-rib with the anterior submarginal ribs and the curved nature of the scales. Scale bar is 2 μm. Fig. F. Serrated bristles. Scale bar is 5 μm.

posterior scales (Fig. 3.49B). The anterior submarginal ribs are often extended above the plane of the scale, forming short wings (Fig. 3.49D). Domes are relatively small, have a shallow bristle cavity and are most often marked with two to four parallel ribs positioned on the left side of the dome and orientated parallel to the longitudinal axis of the cell (Figs. 3.48B,D,F). The right anterior submarginal rib often continues across the dome and terminates above the bristle opening (Fig. 3.48B). Domes may also be ornamented with raised papillae. The shield consists of a secondary layer of interconnecting ribs forming large pores (Fig. 3.48B). The thicker the secondary ribs, the smaller are the diameters of the pores (Fig. 3.48F). The posterior flange consists of a series of short struts that connect the V-rib with the posterior rim (Figs. 3.48B,C,F); a series of circular pores alternates with the struts. On some scales the canopy of the posterior rim extends to the V-rib, completely obscuring the posterior flange (Fig. 3.48F). Anterior flanges are very narrow and unornamented.

The spines are long, thick and taper to a sharp point. Bristles are short, ribbed and unilaterally serrated (Fig. 3.49F). On some bristles the proximal teeth are recurved.

Many characteristics of *M. duerrschmidtiae* are similar to those of *M. crassisquama*, and the former has often been undoubtedly mistaken for the latter, especially if the identifications were based solely on isolated scales (Siver et al., 1990). The types, positions and arrangements of scales for both taxa are very similar. In addition, the overall appearances of the scales are similar. The following features serve to separate the two taxa (see Siver et al., 1990). First, the arms of the V-rib of *M. duerrschmidtiae* curve and become continuous with the anterior submarginal ribs. On scales of *M. crassisquama* the junction of the anterior submarginal rib and the V-rib is more angular and the two sets of structures are not continuous. With practice this feature can be observed using LM. Second, *M. duerrschmidtiae* lacks helmet bristles. Third, the scales of *M. duerrschmidtiae* tend to be larger with more heavily silicified features and a greater degree of canopying. Fourth, the domes of scales of *M. duerrschmidtiae* are smaller and marked primarily with parallel ridges, while those of *M. crassisquama* are larger and either smooth or marked with papillae. Fifth, the spines are significantly longer on cells of *M. duerrschmidtiae*. Sixth, there is a tendency for the anterior submarginal ribs of scales from *M. duerrschmidtiae* to form wings.

Sectio Pseudocoronatae Asmund & Kristiansen 1986

Series Pseudocoronatae Asmund & Kristiansen 1986

Mallomonas pseudocoronata Prescott

Cells are ovoid to elliptical and have a very irregular outline due to the morphology of the scales (Figs. 3.50; 3.51A-B). Bristles can be found over most of the cell surface and the posterior region consists of a group of scales with long, stout spines projecting parallel to the longitudinal axis of the cell (Figs. 3.50; 3.51A-B).

Apical domed scales, domed body scales and posterior spined scales are found along the length of the cell (Figs. 3.51A-B). All scales have a base plate with minute pores that are irregularly spaced, a wide posterior rim, a prominent and angled V-rib and forward projecting anterior wings (Figs. 3.51C,E,F; 3.52E-F). The wing-like structures originate along the anterior flanges and project upward at a 45° to 60° angle from the plane of the base plate. The size of the wings can vary between scales from different cells (compare Figs. 3.51A and 3.51B) as well as on the same cell (Fig. 3.51A). On body scales the wings fuse anterior to the dome, along a median axis running parallel to the lon-

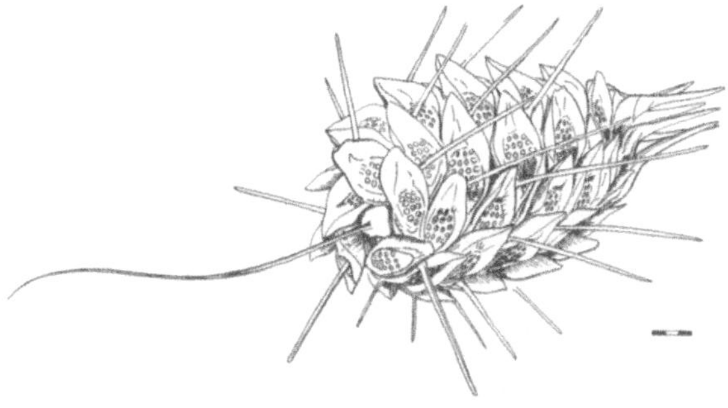

Fig. 3.50. Three-dimensional reconstruction of a whole cell of *Mallomonas pseudocoronata* depicting the distribution of scales and bristles. Scales are large and positioned in spiral rows such that their longitudinal axes are perpendicular to the longitudinal axis of the cell. Note the anterior wings of the scales, the distribution of bristles and the posterior spined scales.

Fig. 3.51. Mallomonas pseudocoronata. Figs. A-B. Whole cells illustrating the arrangement and pattern of overlap of scales and the posterior spines. Scale bars are 5 μm. Fig. C. Highly silicified body scale with a well developed anterior wing. Scale bar is 2 μm. Fig. D. Posterior scale with a long and ribbed spine. Note the lack of shield reticulation. Scale bar is 2 μm. Figs. E-F. Body scales with (Fig. E) and without (Fig. F) shield reticulation. Note that the anterior submarginal ribs (arrows) are separate structures from the anterior wings. Scale bars are 5 μm and 1 μm, respectively.

Fig. 3.52A-B. Mallomonas acaroides var. *acaroides*. Fig. A. Body scales and portions of bristles. The arms of the V-rib extend to the margin of the scale. The anterior submarginal ribs originate below the distal ends of the arms of the V-rib, forming a distinct angle. Note the unornamented domes and the distal tip of a helmet bristle. Scale bar is 2 μm. Fig. B. Body scale. Note the poorly developed posterior flange ribs, the continuation of the anterior submarginal ribs around the dome and the small diameters of the base plate pores. Scale bar is 2 μm. Figs. C-F. *Mallomonas pseudocoronata*. Figs. C-D. Aberrant scales with both wings and spines. Scale bars are 2 μm. Fig. E. Anterior portion of a cell. Note that the apical scales have a small wing along the right anterior submarginal rib only (arrow). The wings originate along the anterior submarginal ribs and are not simply extensions of the latter structures. Scale bar is 5 μm. Fig. F. Highly silicified body scale with a large wing. Scale bar is 5 μm.

gitudinal axis of the scale. On anterior dome scales the wings are smaller and may not meet and fuse, whereas on posterior scales they merge and form the spine (Fig. 3.51D). Aberrant scales with spines can be found (Figs. 3.52 C-D).

The arms of the V-rib curve and become continuous with the anterior submarginal ribs; the latter are orientated parallel to the margins of the scale and terminate at the dome, as described for *M. duerrschmidtiae* (Figs. 3.51C,E,F). Although the wings have been described as being extensions (Asmund and Kristiansen, 1986) or 'outgrowths' (Nicholls, 1987c) of the submarginal ribs, they are actually separate structures that originate from the base plate along the anterior flanges (Figs. 3.51E; 3.52E). The distinction between the anterior submarginal ribs and the base of the wings is less obvious on heavily silicified scales (Fig. 3.52F). The shield is most often marked with a dense reticulation of secondary ribs forming a series of large pores (Figs. 3.51 C,E; 3.52 E-F); however, cells can possess scales with little or no ribbing (Fig. 3.51F) (Nicholls, 1987c). Shield ribbing is commonly lacking on posterior spined scales (Fig. 3.51D). A series of struts radiates at right angles from the V-rib onto the posterior flange (Figs. 3.51C,E,F); the struts may be partially (Fig. 3.51E) or mostly (Fig. 3.52F) covered by the posterior rim. Domes are small, unornamented and rounded (Fig. 3.51C) or triangular (Fig. 3.51E) in shape. Because of the large size of the body scales (Table 4.2) they tend to be curved in order to conform to the shape of the cell (Fig. 3.51C).

The symmetrical body scales are arranged in spiral rows with their longitudinal axes perpendicular to the longitudinal axis of the cell. Each scale is overlapped by the scale positioned behind it in the same spiral row and by the scales located in the spiral row above it (Figs. 3.50; 3.51A-B). An anterior ring of domed scales surround the flagellar opening and a posterior group of spined scales form the posterior of the cell (Figs. 3.50; 3.51A-B). As described for both *M. crassisquama* and *M. duerrschmidtiae*, the anterior domed and posterior spined scales are strongly asymmetric. Each domed scale of the apical most ring, representing the first scale of a spiral row, is overlapped by the scale positioned behind it in the same spiral row and by the apical domed scale to its left (Figs. 3.50; 3.51B; 3.52E). Because body scales are primarily overlapped in the posterior flange and V-rib regions, the anterior wings do not interfere with the proper positioning of scales on the cell. However, the wings of apical domed scales, especially the left hand one, are usually smaller, presumably to accomodate for the overlapping of the scale within the ring to the right (Fig. 3.52E).

Bristles are short, slightly curved, ribbed along most of the shaft and tapered to a slightly rounded apex (Figs. 3.50; 3.52E). Bristles are easily disarticulated from the cell (Nicholls, 1987c).

Mallomonas pseudocoronata shares several features with, and is most closely related to, *M. duerrschmidtiae*. The overall cell morphology, arrangement of scales on the cell and scale symmetry are identical for both taxa. Both have similar types of bristles, lack helmet bristles and possess posterior scales with long, stout spines. Except for the large anterior wings, scale design is essentially identical. In both taxa the arms of the V-rib curve and are continuous with the anterior submarginal ribs. The short wings of *M. duerrschmidtiae* are true extensions of the anterior submarginal ribs, whereas on *M. pseudocoronata* they originate as separate structures. The two taxa also differ in their environmental requirements (see Chapter V).

Sectio Annulatae Asmund & Kristiansen 1986

Mallomonas annulata (Bradley) Harris

Synonyms: *M. papillosa* Harris & Bradley f. *annulata* Harris n.nud.
M. papillosa var. *annulata* Bradley

Cells are ellipsoidal with broadly rounded ends and bristles originating from both poles and scattered over the cell (Fig. 3.53A). Scales are either dome-bearing or domeless. Domed scales are found at each end of the cell as well as scattered over the surface of the cell. The majority of body scales are, however, domeless (Fig. 3.53A). Scales are in spiral rows, aligned with their longitudinal axes perpendicular to that of the cell and overlapped such that each scale is covered, in part, by the scale positioned behind it within the same spiral row and by the scale(s) above it (Fig. 3.53A). On the average, only 35% of the scale surface area remains exposed on an intact cell (see Chapter IV).

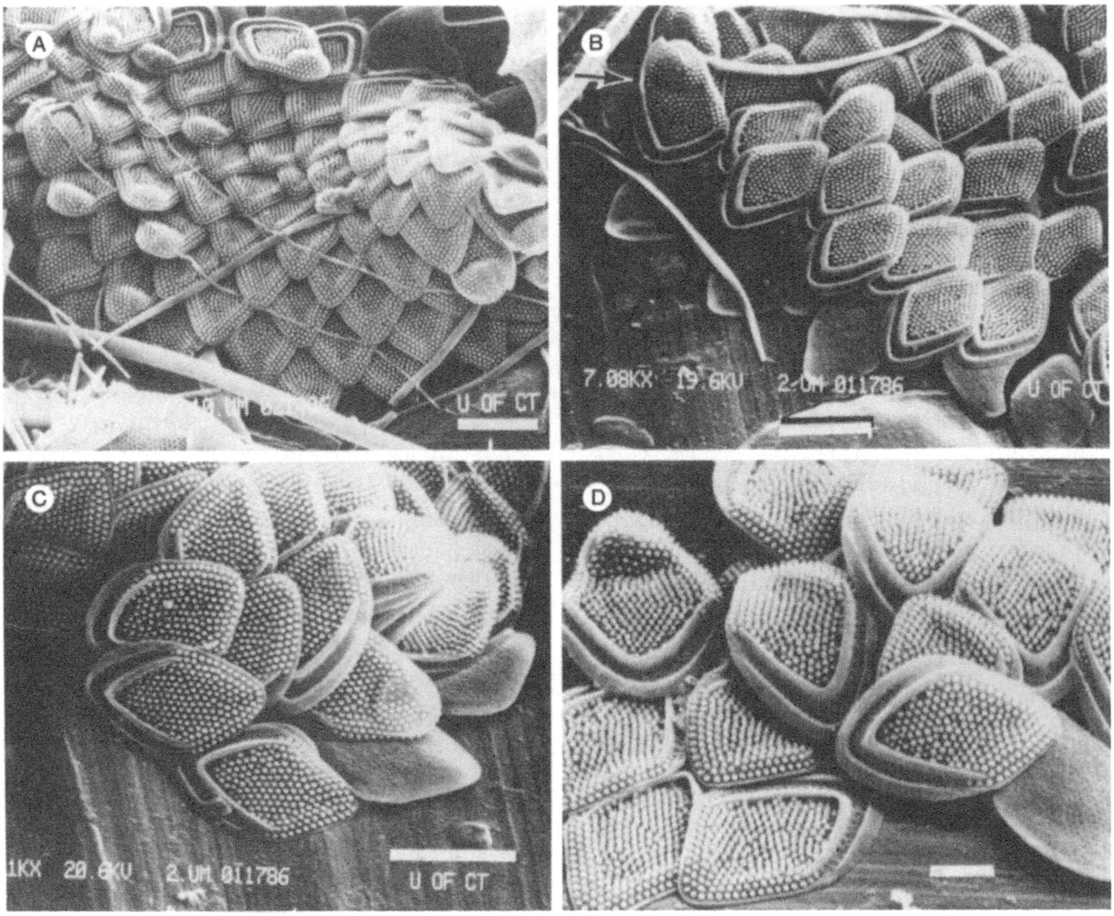

Fig. 3.53. Mallomonas annulata. Fig. A. Whole cell showing the arrangement, orientation and degree of overlap of the scales. Note the scattered positions of domed scales. Scale bar is 2 μm. Fig. B. Anterior end of a cell showing the morphology of scales and bristles. Note the extension of the left submarginal rib along the border of the dome (arrow) on the apical scales. Scale bar is 2 μm. Figs. C-D. Domed and domeless scales. Note the distribution of papillae on the shield, anterior flanges and domes, and the unornamented posterior flange. Scale bars are 2 μm and 1 μm, respectively.

Body scales are rhomboidal, symmetrical in outline and consist of a perforated base plate, a proximal rim and a V-rib with straight arms extending to the middle of the cell (Figs. 3.53B-D). The posterior flange is smooth and unornamented except for the presence of base plate pores (Fig. 3.53D). Although the areas corresponding to the anterior submarginal ribs are slightly raised relative to the shield, in SEM micrographs such ribs appear to be lacking on body scales (Fig. 3.53D). The shield, anterior flanges and dome are marked with closely spaced papillae (Figs. 3.53C-D). Occasionally in scattered locations on the shield, especially along the inner margin of the V-rib, papillae are lacking (Figs. 3.53B,D). The dome on body scales is large, prominent and symmetrically positioned (Fig. 3.53D).

Apical scales are more circular in outline, asymmetric, domed and form a ring around the proximal end of the flagellum (Fig. 3.53B). The left arm of the V-rib is extended distally along the side of the dome into a short wing giving the scale a distinct asymmetry. The V-rib is also more obtusely-angled than on the body scales. Posterior scales are only slightly smaller in size than body scales.

Based primarily on TEM, Asmund and Kristiansen (1986) discussed three types of ornamentation found on the shield: 1) regularly spaced papillae, but lacking in the angle of the V-rib; 2) patches of continuous secondary layer with pits; and 3) reticulate pattern of polygonal meshes, each of which encloses a group of pores. However, as viewed with SEM, the shield ornamentation on all specimens observed in this study never varied from that described above; the scales were always covered with closely spaced but separate papillae that were never fused into a reticulum.

Bristles are slightly curved, smooth and terminate in a blunt tip (Fig. 3.53B).

Isolated scales could be confused with those of *M. guttata*, however, the latter have circular pits lacking papillae and much smaller and structurally different domes. In addition, *M. guttata* lacks domeless scales. *M. annulata* also bears some resemblance to *M. papillosa*, *M. paxillata* and *M. rasilis*. The latter three taxa differ from *M. annulata* in possessing only domed scales and in having scales with much smaller domes. Other characteristics, such as the shape of the scale, morphology of the anterior submarginal ribs (if present), presence/absence of lateral incurvings and bristle structure also serve to distinguish between these taxa.

Mallomonas cf. *pillula*

A few isolated scales were found that were believed to belong to *M. pillula* (Fig. 3.54E), although insufficient material was observed to be able to verify the identification.

Sectio Torquatae Momeu & Péterfi 1979

Series Pumilae Momeu & Péterfi 1979

Mallomonas clavus Bradley

Synonym: *M. allantoides* Harris

The specimens referred to as *M. clavus* in this work (Figs. 3.54A-C) were difficult to identify and separate from *M. pumilio* (Harris and Bradley, 1957) Asmund, Cronberg and Dürrschmidt and *M. eoa* Takahashi. All three of these taxa share similarities in cell and scale features, especially concerning the morphology of the scales. According to Asmund and Kristiansen (1986) specimens in the Series Eoae, which contains *M. eoa*, have scales with well defined, smooth-edged, regularly spaced meshes comprising the reticulum of the shield. The meshes are most often (e.g. for *M. eoa*) circular in design. Taxa, including *M. pumilio* and *M. clavus*, within the Series Pumilae have scales with less defined and irregularly shaped meshes. The meshes of scales from *M. pumilio* var. *munda* Asmund, Cronberg and Dürrschmidt and *M. clavus* are more angular in outline and fairly evenly spaced (Asmund and Kristiansen, 1986). Based on the structure of the shield reticulum, I feel that the specimens are best placed in the Series Pumilae.

The distinction between *M. clavus* and *M. pumilio*, especially based on scale structure, is less clear. In his original description, Bradley (1964) stated, 'In fact, the scale structure (based on TEM) of *M. clavus* is identical with that of *M. pumilio* so that it might be confused with this species if isolated scales only were found in the electron microscope'. Asmund and Kristiansen (1986) also noted the similarity between scales of *M. pumilio* var. *munda* and *M. clavus*. Bradley (1964) described *M. clavus* as having larger cells with considerably larger posterior spines than *M. pumilio*. Cells described in this work ranged up to 25 μm in length and had a mean length of 20.4 μm, which more appropriately fit those of *M. clavus*. The posterior spines, ranging in length from 3.0 μm to 6.8 μm, were considerably longer than those reported for *M. pumilio* (Asmund, Cronberg and Dürrschmidt, 1982). In addition, the size of scales best reflected measurements reported for *M. clavus*, which has slightly larger scales than *M. pumilio*. Overall, although specimens did lack spines with narrow wing-like extensions (Asmund and Kristiansen, 1986), they best reflect *M. clavus*.

Cells are rather small and ellipsoidal, with a collar consisting of about four to five scales and a slightly tapered posterior end with spines (Fig. 3.54A). Scales are arranged in spiral rows with collar and body scales aligned with their longitudinal axes parallel and perpendicular, respectively, with the longitudinal axis of the cell. Each scale is overlapped by the one positioned behind it in the same row and by the scales positioned in the row above it (Fig. 3.54A). As is true with all species in the Sectio Torquatae, the collar scales form a ring around the flagellar pore and, as such, represent the distal most scales of the spiral rows.

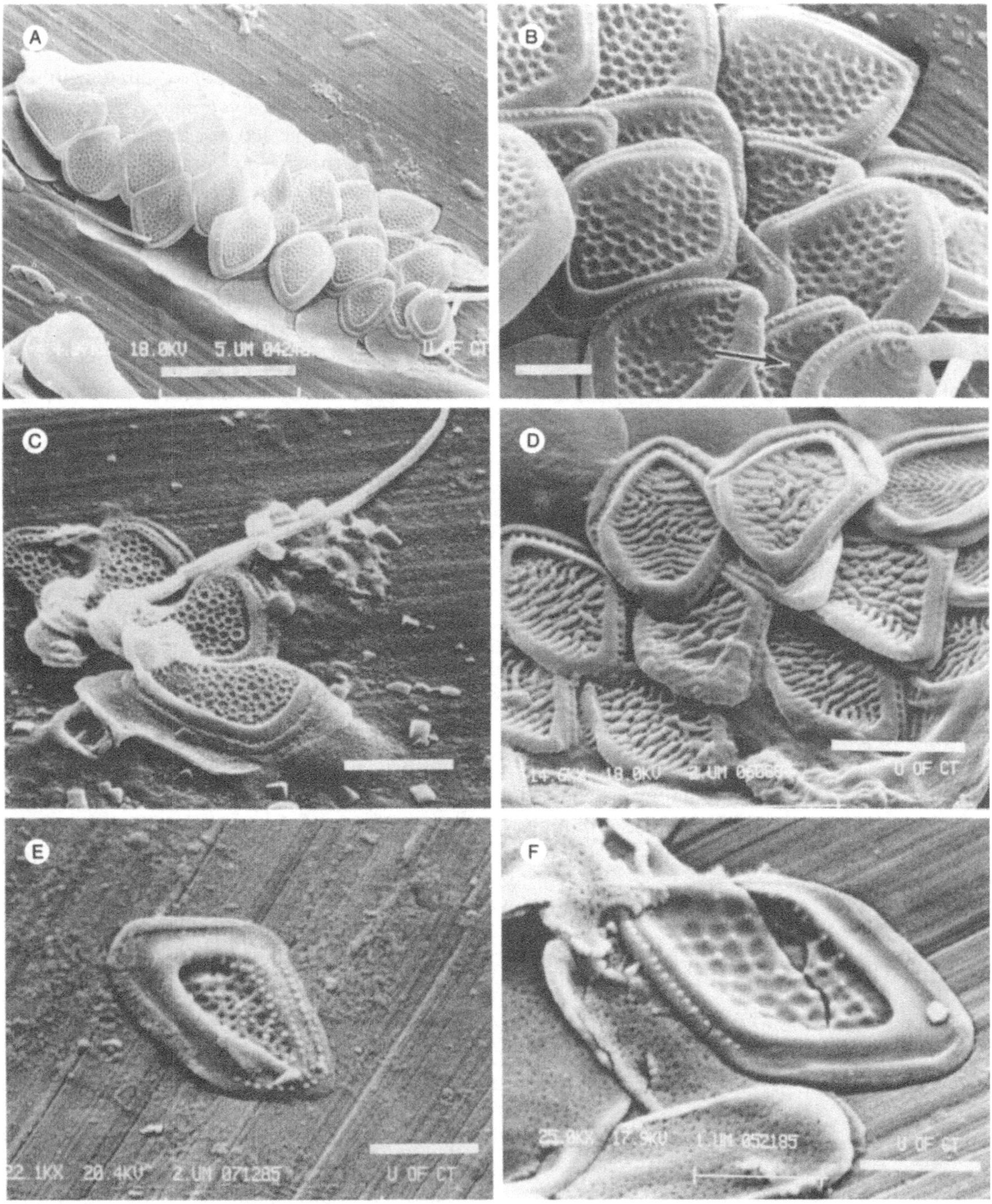

Fig. 3.54A-C. Mallomonas clavus. Fig. A. Whole cell depicting the arrangement and pattern of overlap of the scales, the collar region and the posterior spined scales. Scale bar is 5 μm. Fig. B. Close-up of the posterior region of the cell illustrated in Fig. A. Note the pattern of meshes on the shield. On several scales the right anterior submarginal rib is slightly thicker than the left rib. The scale with the posterior spine (arrow) has an unpatterned shield. Scale bar is 1 μm. Fig. C. Collar scales and a single bristle. Scale bar is 2 μm. Fig. D. *Mallomonas schwemmlei.* Body scales. The right anterior submarginal rib is thicker and extends further over the flange when compared to the left one. Also, note the varying height along the length of each transverse shield rib. Scale bar is 2 μm. Fig. E. *Mallomonas* cf. *pillula.* Isolated body scale. Scale bar is 1 μm. Fig. 6. *Mallomonas* sp. 2. Isolated body scale from an unidentified species from the Sectio Torquatae, probably belonging to the Series Pumilae.

Body scales are rhomboidal and possess a thin proximal rim and a V-rib with arms that are fused and continuous with the anterior submarginal ribs (Figs. 3.54A-B). The anterior submarginal ribs are rounded and often relatively indistinct. The height of the V-rib and the degree to which it forms a canopy over the shield are minimal (Fig. 3.54B). The shield consists of a reticulum of more or less evenly spaced and angular shaped meshes (Fig. 3.54B). Each mesh encloses from two to four and most often three pores (Fig. 3.54B). Each anterior flange is narrow and ornamented with a uniseriate row of papillae. The posterior flange is smooth and unornamented. The rear most scales are smaller, more circular in outline and possess a single spine ranging in length from 3 μm to 6.8 μm; these scales often lack shield reticulation and base plate pores (Figs. 3.54A-B). Several specimens had scales, positioned intermediate between the rear most spined scales and body scales, with a slightly more pronounced right anterior submarginal rim, similar to what is reported for *M. alata* Asmund, Cronberg and Dürrschmidt.

Collar scales are asymmetrical with a broadly rounded proximal end and a narrow dome bearing distal end (Figs. 3.54A,C). The proximal rim extends the length of the dorsal margin and terminates along the left lateral side of the dome. The dome is small with a short pointed peak (Figs. 3.54A,C). Shield reticulation is similar to that of the body scales. Bristles are approximately one-half or less the length of the cell, smooth, slightly bent and tapered. Most often, the bristles extend backwards and conform to the shape of the cell (Fig. 3.54A).

Mallomonas sp. 2

A few scattered scales were found that are believed to belong to a species from the Series Pumilae of the Sectio Torquatae (Fig. 3.54F). The shield is marked with a reticulum of smooth edged polygonal meshes each of which appears to be covered by an additional thin layer. Such a covering over the meshes is different from scales of other taxa within this series (e.g. *M. clavus*).

Series Mangoferae Asmund & Kristiansen 1986

Mallomonas mangofera Harris & Bradley

Cells are typically small and either obovate with a maximum diameter near the posterior end (Fig. 3.55A), or ovate-ellipsoidal with a maximum width near the middle of the cell. Each cell has a group of posterior scales with short spines and a collar consisting of approximately five bristle-bearing scales. Bristles range from one-half to 1 · the length of the cell (Fig. 3.55A). Scales are arranged in spiral rows and each scale is overlapped by the scale positioned behind it in the same row and by the scale(s) in the row above it (Fig. 3.55A). Body and collar scales are orientated with their longitudinal axes perpendicular and parallel, respectively, to the longitudinal axis of the cell.

Body scales are rhomboidal, domeless and have a proximal rim that encircles about one-half of the perimeter of the scale and a V-rib with straight arms that extend to the mid-section of the scale. The distal ends of each V-rib arm usually appear to be furcate, giving the appearance of a 'double' V-rib structure (note the exposed V-rib on posterior scales of cell in Fig. 3.55A and also Fig. 11 of Dürrschmidt, 1983a). The surface of the scale is ornamented with rather large and densely spaced papillae (Fig. 3.55A). Based on analysis of TEM micrographs, the papillae are solid. The posterior flange in unornamented.

There seems to be some question concerning the presence and morphology of anterior submarginal ribs. Asmund and Kristiansen (1986), Dürrschmidt (1983a) as well as Harris and Bradley (1960) in their original description all described scales with anterior submarginal ribs. However, Dürrschmidt (1983a) stated that they were 'incomplete' on some scales. In TEM images (e.g. Fig. 7 of Dürrschmidt, 1983a and Fig. 65F of Asmund and Kristiansen, 1986) the regions corresponding to those areas where the anterior submarginal ribs ought to be are darkened, suggesting more electron dense zones. However, in the same TEM micrographs sections of the 'anterior submarginal ribs' appear less electron dense and suggest the lack of such ribs. Scales in several SEM micrographs from Dürrschmidt (1983a) (e.g. Fig. 6) clearly indicate slightly raised regions corresponding to anterior submarginal ribs, but, these structures are less

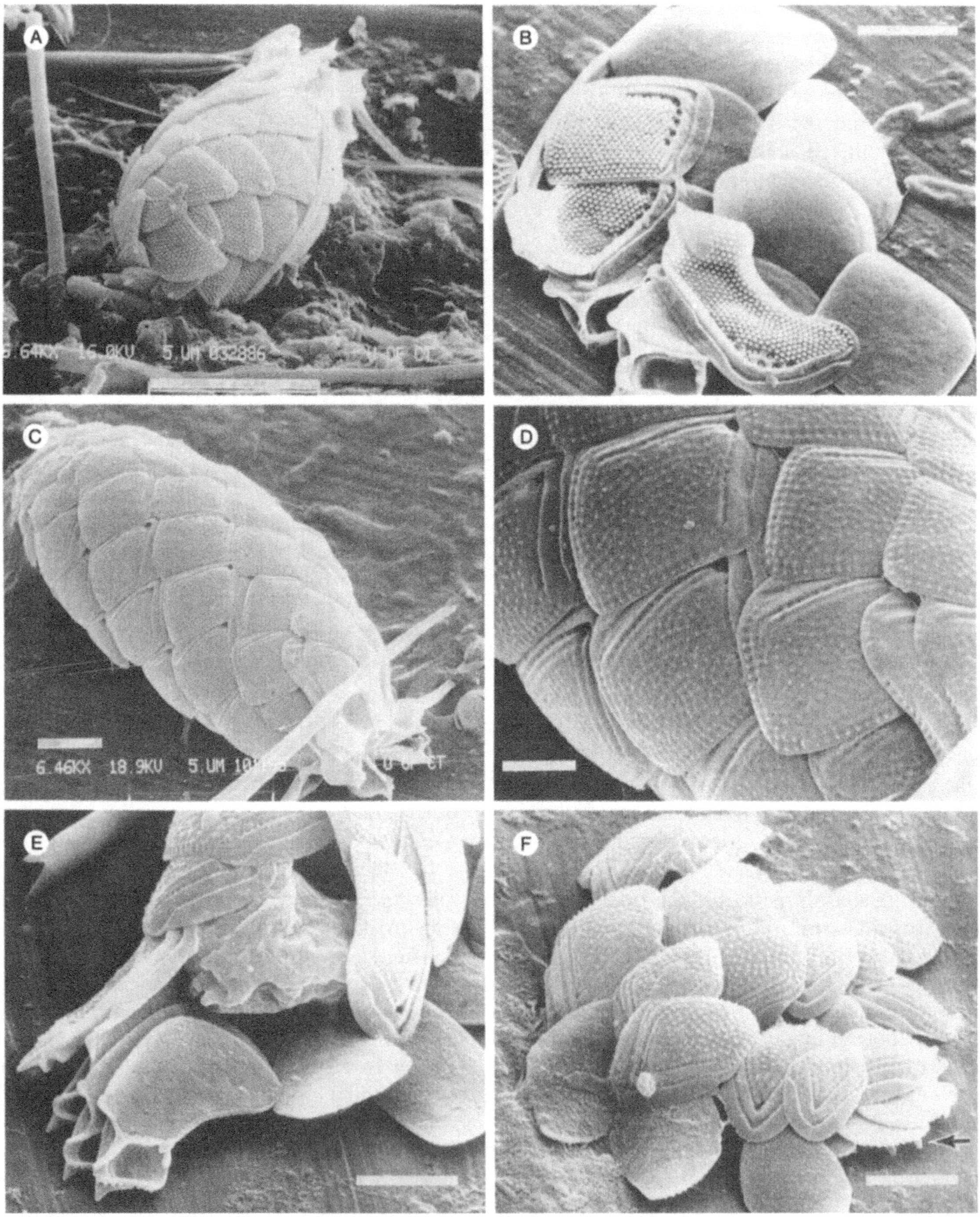
A
B
C
6.46KX 18.9KV 5.UM
D
E
F

evident in others micrographs (e.g. Figs. 11 and 18; the latter of f. *foveata*).

All of the scales that I have observed with SEM appear to lack well defined anterior submarginal ribs (Figs. 3.55A-B). I believe that scales of *M. mangofera* often, but not always, have slightly raised regions of the scale that correspond to and are equivalent to anterior submarginal ribs, but lack actual solid siliceous structures. As such, I believe that the apparently solid ribs corresponding to electron dense areas depicted in TEM micrographs are the result of a change in slope along the surface of the scale caused by the raised nature of the scale. Such an explanation, which accounts for the presence of incomplete ribs, is analogous to the situation described in Chapter I concerning the 'false impression' of a rib often depicted along the posterior border of the dome in some TEM images. Likewise, I do not believe that adjacent papillae are fused to form struts on the anterior flanges, as could be interpreted in some TEM micrographs (e.g. Fig. 65G of Asmund and Kristiansen, 1986). Because of the change in slope and angle between adjacent papillae, they become 'projected' into 2-space as a single rib. That is, when adjacent papillae along the slope are projected into 2-space (e.g. a TEM image) they can no longer be resolved as individual papillae.

Posterior scales, and often a number of the body scales, have a single enlarged spike-like papillae on their distal end (Fig. 3.55A). Collar scales are asymmetrical with a broad proximal end and a much narrower distal end (Fig. 3.55B). The proximal rim and left arm of the V-rib extend the length of the dorsal margin and terminate at the base of the dome. The dome is quite prominent with a well developed cavity and a forward projecting triangular shaped tooth (Fig. 3.55B). The shield and at least part of the dome are patterned with papillae as are the body scales. There is a region of the shield directly behind the dome that is sunken in nature and has less developed papillae (Fig. 3.55B). Based on TEM analysis this region, as well as the part of the dome lacking papillae, is thinner. The flange is smooth and unornamented.

Besides *M. mangofera* f. *mangofera*, there are two additional forms, f. *foveata* Dürrschmidt and f. *gracilis* Dürrschmidt, and one variety, var. *sulcata* Dürrschmidt, all of which vary only slightly from the type (Dürrschmidt, 1983a). Forma *foveata* differs from the type primarily due to the presence of a series of large circular pits each of which surrounds a single 'sunken' papillae (Fig. 3.55B). Each sunken papilla has a single pore that penetrates the base plate. The circular pits are aligned along the inner margin of the V-rib, often forming a V-shaped structure on the body scales (Fig. 3.55B). According to Dürrschmidt (1983a) scales of f. *gracilis* have smaller papillae that may fuse to form a continuous layer and an oval pit in the angle of the V-rib. Scales of var. *sulcata* differ from those of the type in having a denser, more complete secondary layer with small papillae and a prominent groove aligned somewhat parallel to the submarginal rib (Dürrschmidt, 1983a; Asmund and Kristiansen, 1986).

M. mangofera could be confused with the closely related species *M. favosa* (Nicholls, 1984). See the discussion under *M. favosa* for further details.

Mallomonas favosa Nicholls

In my opinion, *M. favosa*, a rare taxon reported only from Canada (Nicholls, 1984), Malaysia and Australia (Dürrschmidt and Croome, 1985), is most closely related to *M. mangofera* and *M. alveolata* Dürrschmidt. Because *M. favosa* is so rare, few micrographs are available for compara-

←

Fig. 3.55. Mallomonas mangofera f. *mangofera*. Whole cell showing the arrangement and pattern of overlap of the collar, body and posterior spined scales. Note the curved bristle conforming to the shape of the cell. Scale bar is 5 μm. Fig. B. *Mallomonas mangofera* f. *foreata*. Collar and body scales showing the nature and distribution of the large circular pits. On the collar scale, note the 'sunken' region with smaller papillae on the distal portion of the shield behind the dome. Scale bar is 2 μm. Figs. C-F. *Mallomonas favosa*. Fig. C. Whole cell illustrating the arrangement of scales in spiral rows. Scale bar is 2 μm. Fig. D. Close-up of the cell in Fig. C. Note the small surface papillae, the height of the shield relative to the V-rib and the series of shallow pores or depressions lining the inner margin of the V-rib. Scale bar is 1 μm. Fig. E. Apical region of a cell showing the outline of the collar scales and the cavity of the dome. Scale bar is 2 μm. Fig. F. Body and posterior scales. Note the thick nature of the scales, the wide proximal rim, V-rib and the pores along the inner surface of the V-rib. Arrow depicts the small spines on the posterior most scales. Scale bar is 2 μm.

tive study. The original description by Nicholls (1984) included only TEM micrographs and did not mention the presence of papillae (Asmund and Kristiansen, 1986). As a result, a match between the SEM micrographs in this work and the specimens of *M. favosa* illustrated in the original description by Nicholls could not be made. However, Dürrschmidt and Croome (1985) provided SEM micrographs of *M. favosa* f. *gemina* Dürrschmidt and Croome that appear to match closely those represented in this work. As a result, I feel that the populations of cells illustrated in this work (Figs. 3.55C-F) represent *M. favosa* and perhaps more appropriately f. *gemina*.

Cells are small, ovoid to elliptical with a broadly rounded posterior end and a rather wide collar (Fig. 3.55C). The arrangement, alignment and pattern of overlap of the scales are as described for *M. mangofera* and *M. dickii*.

Body scales are thick, rhomboidal and have both a proximal rim and V-rib that extend one-half the length of the scale (Figs. 3.55D,F). Although scales are described in TEM images as having anterior submarginal ribs, in SEM micrographs such ribs appear to be lacking, or at best weakly developed and not readily discernable from the shield (Fig. 3.55D). The shield consists of an internal reticulum of pentagonal meshes (observed in TEM images, but not readily seen with SEM) covered with a thin surface layer from which project small papillae (observed with SEM). As a result of the secondary structure the height of the shield is similar to that of the V-rib (Fig. 3.55D). The outline of the internal reticulum of the shield is barely visible in SEM micrographs (e.g. Fig. 3.55D) giving the scale a somewhat bumpy appearance. A circular shaped window is present at the base of the V-rib. The circular hole and pore described by Nicholls (1984) are probably positioned within the window. A series of shallow pores or depressions lines the inner surface of both arms of the V-rib resulting in an additional V-shaped structure that delineates the raised shield from the V-rib. It is very likely that these shallow pores are equivalent to the pores lining the V-rib observed in TEM images of f. *gemina* (e.g. Fig. 24 of Dürrschmidt and Croome, 1985). The height of the shield slopes back down to the base plate along the distal margin (Fig. 3.55D), which may, in part, result in the 'false impression' that anterior submarginal ribs and struts on the anterior flanges are present (see discussion under *M. mangofera*). A uniseriate row of papillae lines the distal margin of the scale. Posterior flanges are narrow and covered by the canopy of the proximal rim (Fig. 3.55F).

Posterior scales are smaller, with a more rounded outline than body scales and possess a short (<0.5 μm) spine (arrow, Fig. 3.55F). Many features of the collar scales of *M. favosa* are similar to those of *M. mangofera*. Collar scales from both taxa are similar in their shape and symmetry, in the structure and position of the dome and in having a large 'sunken' region behind the dome (Figs. 3.55C,E). The 'sunken' zone corresponds to the less electron dense areas depicted in TEM images (e.g. Fig. 23 of Dürrschmidt and Croome, 1985; Fig. 22 of Nicholls, 1984). Bristles, restricted to the collar scales, are slightly curved, smooth, tapered to a sharp point and approximately one-half the length of the cell.

The major difficulty in identifying specimens of *M. favosa* is due to the very different images obtained from using SEM vs. TEM. Unlike most species of *Mallomonas*, different conclusions could be made concerning the structure of the scales as viewed with SEM as opposed to TEM. On one hand, a TEM image depicts a scale with a thick proximal border, a continuous submarginal rib surrounding a shield consisting of a meshwork of polygonal meshes and anterior flange with struts. With SEM scales are seen to be thick, possess a well defined V-rib, lack anterior submarginal ribs and flange struts and have a shield with an apparently continuous secondary layer and ornamented with papillae. Similar differences in SEM and TEM images of scales are also seen in *M. bronchartiana* and *M. lychenensis*, both of which also possess surface papillae.

As stated above, *M. favosa* is very similar to *M. mangofera*. Both taxa have similar cell and scale morphologies, bristles, spike-like posterior spines and thin 'sunken' zones on their collar scales. Scales from both taxa also either lack or have weakly developed anterior submarginal ribs. Lastly, both species have sub-specific forms with pits and/or depressions lining the inner margin of the V-rib. The primary difference between the two species lies in the ornamentation of the scales. In *M. mangofera* the scales are covered with closely spaced papillae. The internal reticulum on the scales of *M. favosa* may actually represent partially fused papillae whose apices protrude above the thin continuous secondary covering.

Differences between *M. favosa* and *M. alveolata* or *M. pumilio* Harris and Bradley are described by Asmund and Kristiansen (1986) and Nicholls (1984), respectively.

Series Doignonianae Asmund & Kristiansen 1986

Mallomonas dickii Nicholls

Cells are small, elongate-ellipsoidal (Fig. 3.56) with a distinctively protruding collar (Fig. 3.57C), very short apical bristles (Fig. 3.57B) and posterior scales with very short spines (Fig. 3.57A). Scales are arranged in spiral rows with each row beginning with a collar scale and terminating with a posterior spined scale. Collar scales are orientated with their longitudinal axes parallel to the longitudinal axis of the cell and situated such that each one is overlapped by the collar scale to its left. Body scales are positioned at right angles to the longitudinal axis of the cell and each is overlapped by the scale situated behind it in the same spiral row and by the scales in the row above it.

Body and posterior scales are rhomboidal and consist of a narrow proximal rim, a well defined and continuous submarginal rim and a series of closely spaced, parallel ribs that transverse the shield (Figs. 3.57F; 3.59A). The arms of the V-rib are fused with the anterior submarginal ribs forming a continuous submarginal rib. However, the distal and proximal halfs of the submarginal rib are morphologically distinct (Fig. 3.57F). The distal half of the submarginal rib is broad, essentially terminates at the perimeter of the scale and is ornamented with a uniseriate row of pores and 3 to 5 rows of small papillae (Figs. 3.57F; 3.59A). Generally, a single row of papillae are found below the row of pores. The row of pores does not penetrate the base plate and, as such, represents a series of pores in the secondary structure of the scale only (Fig. 3.57F). The papillae are not observed with TEM (see Nicholls, 1982; 1988c; Asmund and Kristiansen, 1986). The proximal half of the submarginal rib (i.e. the V-rib) is smooth and unornamented.

As observed by Nicholls (1982), base plate pores appear to be lacking or are so small as to not be easily resolved. However, a single rather large pore is found on the undersurface of the distal end of each scale; this pore is located directly below the spines on each posterior scale (Figs. 3.57D; 3.59B). The number of transverse shield ribs ranges from 8 to 24; body scales have a significantly larger number of ribs. The ribs situated near the mid-section of the shield are sometimes branched and a circular pore is often formed where each rib fuses with the submarginal rib. The height of a rib is more or less constant across its length, however, slightly wavy ribs, similar to those found on *M. schwemmlei*, are occasionally seen (Fig. 3.59A). Posterior scales are smaller and possess a single small spine ranging in size from 0.2 μm to 0.8 μm (Figs. 3.57A,D; 3.59B).

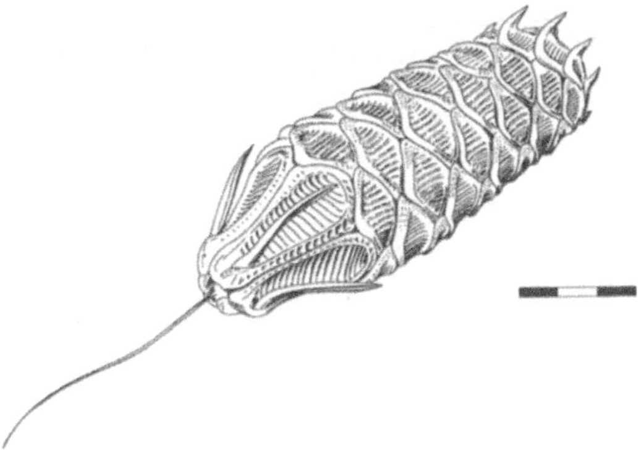

Fig. 3.56. Three-dimensional reconstruction of a whole cell of *Mallomonas dickii* depicting the collar and the distribution of scales and bristles. Body scales are positioned in spiral rows such that their longitudinal axes are perpendicular to the longitudinal axis of the cell. Note the anterior collar scales with very short bristles and the short posterior spines.

Collar scales are asymmetrical, tripartite with a broadly rounded proximal end and a more narrow distal end (Figs. 3.57B,C,E,F). The dome is small, rounded and distally extended into a sharp triangular point (Figs. 3.57B,E,F). The proximal rim extends the length of the dorsal margin and terminates near the dome. There is a series of pores along the inner ventral side of the submarginal rib positioned where the transverse ribs adjoin; a single row of pores is also found along the outer ventral margin (Fig. 3.57C). The anterior half of the shield is patterned as are the body scales; however, on the proximal end of the shield the transverse ribs are still parallel but obliquely positioned (Figs. 3.57A-C).

Fig. 3.57. Mallomonas dickii. Figs. A-B. Whole cells showing the arrangement of scales, protruding collar (Fig. A), short apical bristles (Fig. B) and small posterior spines. Scale bars are 2 μm. Fig. C. Close-ups of the collar illustrating the transition between collar and body scales. Scale bar is 2 μm. Fig. D. Posterior region of a cell. Note the progressively smaller scales, the small spines on the rear most scales and the large pore (arrow) on the undersurface of the distal end of the scale. Scale bar is 2 μm. Fig. E. Whole, but collapsed, cell. Note the morphology of the collar and body scales and the small apical bristles. Scale bar is 5 μm. Fig. F. Anterior end of a cell illustrating the morphological details of both body and collar scales. Scale bar is 2 μm.

Bristles are remarkedly short, with a mean length of 1.9 μm, are slightly curved and tapered to a point (Figs. 3.56; 3.57B). Each bristle emerges on the right side of the dome from a shallow U-shaped opening, lies close to the cell proper and projects backwards, extending less than the length of the collar (Fig. 3.57B).

Cells and scales of *M. dickii* could be potentially confused with those of *M. doignonii* and *M. schwemmlei*. Scale features of *M. dickii* are similar to those of *M. doignonii*, however, cells, scales, spines and bristles of the former are significantly smaller than those of the latter (see Chapter 4; Nicholls, 1982; Nicholls, 1988c). For example, the surface area of scales and the lengths of cells, spines and bristles of *M. dickii* are one-third or less those of *M. doignonii*. As such, morphometric features are useful in separating *M. dickii* from *M. doignonii*. In addition, the collar of *M. dickii* is usually much more pronounced from the rest of the cell than that of *M. doignonii* (compare Figs. 3.57A and 3.59C). Cells and spines of *M. schwemmlei* are closer in size to those of *M. dickii*, but, the former has scales and bristles that are much larger and similar in size to those of *M. doignonii*. Scale features also differ between *M. dickii* and *M. schwemmlei*.

Mallomonas doignonii var. *tenuicostis* Asmund and Cronberg

Cells are ellipsoidal or elongate-ellipsoidal with a protruding collar and long posterior spines (Figs. 3.58; 3.59C). Scales are arranged, orientated and overlapped in the same manner as described for *M. dickii*. Scales are arranged in spiral rows with collar and body scales being orientated parallel and perpendicular, respectively, to the longitudinal axis of the cell (Fig. 3.59C). Each scale is overlapped by the scale positioned behind it in the same row and by the scales in the row above it.

Body scales are rhomboidal and consist of a perforated base plate, a narrow proximal rim, a well defined submarginal rib and a series of parallel transverse shield ribs (Fig. 3.59D). The fusion of the arms of the V-rib to the anterior submarginal ribs to form the continuous submarginal rib is as described for *M. dickii*. A series of short parallel struts radiate from the submarginal rib onto each of the narrow anterior flanges; the struts are broad and completely mask the flanges (Fig. 3.59D). A uniseriate series of pores alternates with the struts. The anterior half of the submarginal rib is ornamented with one to three rows of papillae; the papillae within each row are aligned with the series of short struts. The proximal half of the submarginal rib (i.e. the V-rib) is narrow and unornamented.

Unlike *M. dickii*, base plate pores of *M. doignonii* var. *tenuicostis* are present and easily resolved with SEM, especially on the undersurface of the scale (Fig. 3.59C). As seen in scales of *M. dickii* a single slightly larger pore is located on the distal most end of the scale (Fig. 3.59C). The shield usually has 18 or more transverse ribs. The distal and proximal most ribs of the shield are slightly bowed forward and backward, respectively; the middle ribs are more or less straight (Fig. 3.59D). Ribs may be branched, usually closer to where they attach to the inner portion of the submarginal rib. The posterior flange is unornamented.

The rear most scales each have a single backward projecting spine ranging in size from 0.7 μm (small spike) to 11.3 μm (long, tapered spine) (Figs. 3.59C-D). Collar scales are elongated with a morphology similar to those of *M. dickii*, have a small dome with a forward projecting triangular shaped point and are patterned as are the body scales. Bristles, restricted to collar scales, are slightly bent, needle-like and range in size from 9.3 μm to 11.3 μm (see Chapter IV).

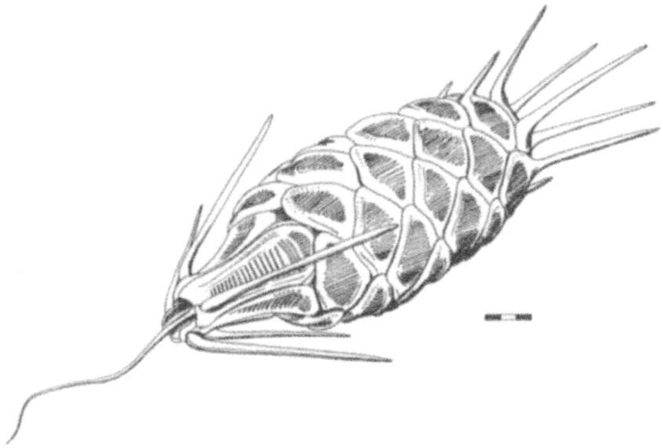

Fig. 3.58. Three-dimensional reconstruction of a whole cell of *Mallomonas doignonii* depicting the collar and the distribution of scales and bristles. Body scales are positioned in spiral rows such that their longitudinal axes are perpendicular to the longitudinal axis of the cell. Note the anterior collar scales, bristles and long posterior spines.

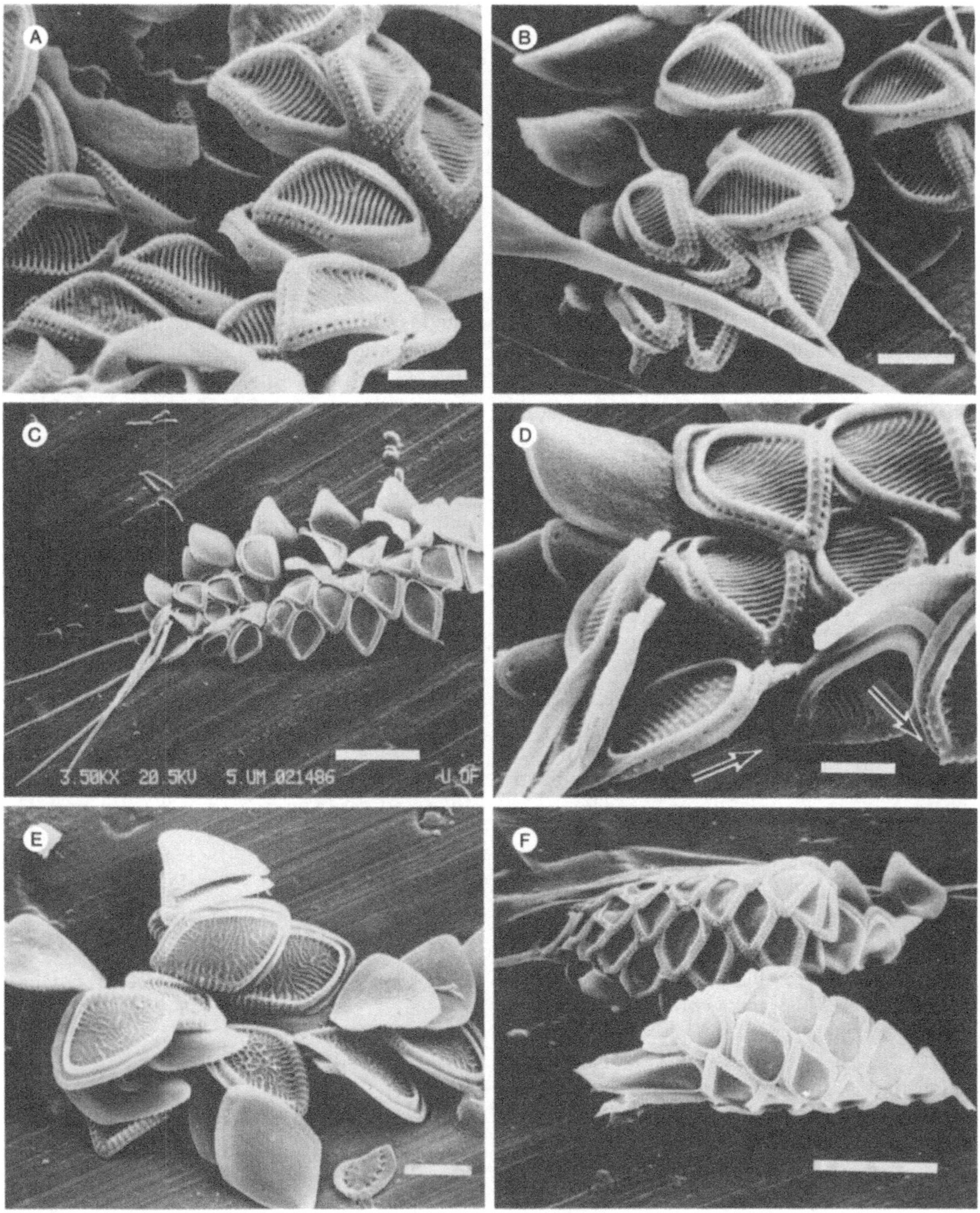
A
B
C
3.50KX 20.5KV 5.UM 021486 U.OF
D
E
F

There are two other closely related varieties of *M. doignonii*, var. *doignonii* Bourrelly emend. Nicholls and var. *robusticostis* Nicholls, that differ primarily on the basis of the number and density of transverse ribs and the size of spines and bristles (Asmund and Cronberg, 1979; Nicholls, 1988c). *M. doignonii* var. *tenuicostis* is separated from var. *doignonii* on the basis of a greater number of transverse ribs and larger spines (Asmund and Cronberg, 1979). Variety *robusticostis* has an even greater number of thick transverse ribs and also differs from the other two varieties in having shorter cells, spines and bristles (Nicholls, 1988c). The small spike-like spines noted for *M. dickii* (Nicholls, 1982; 1988c) and var. *robusticostis* (Nicholls, 1988c) are documented here for var. *tenuicostis* (Fig. 3.59D). A comparison of *M. doignonii* with *M. dickii* is given under the discussion of the latter and by Nicholls (1988c).

Mallomonas torquata f. *simplex* Nicholls

Cells are elongate-ellipsoidal with a collar, short apical bristles and relatively long posterior spines (Figs. 3.60; 3.61A-C). Scales are arranged, orientated and overlapped in the same manner as described for *M. dickii*. The pattern of overlap and the role of the V-rib in the spacing of body scales described in Chapter II are nicely observed in Fig. 3.61D.

Body scales are rhomboidal and consist of a perforated base plate, a narrow proximal rim, a well developed continuous submarginal rib and a relatively unpatterned shield and posterior flange (Figs. 3.61E-F). The base plate pores are very small, distributed over most of the scale and are best observed at high magnification (Fig. 3.61F). The distal half of the submarginal rib is much broader and ornamented relative to the proximal (V-rib) end (Figs. 3.61D-F). Short ribs or struts radiate from the distal half of the submarginal rib onto both anterior flanges and the shield (Figs. 3.61D-F). Short uniseriate rows of papillae extend across the broad submarginal rib between the two groups of short struts (Figs. 3.61D-F). Each row consists of 5 to 12 papillae and is not observed in TEM images. The spaces between the adjoining struts along the outer margin of the submarginal rib (Fig. 3.61D) and often along the inner margin (Fig. 3.61F) appear as a row of pores alternating with the struts. Other than the short struts radiating from the submarginal rib, additional secondary structures are lacking on the shield (Fig. 3.61D).

The collar consists of approximately five scales, each of which is asymmetric, tripartite with a broad proximal end and a small triangular shaped dome (Figs. 3.59F; 3.61F). The proximal rim extends the length of the dorsal margin and terminates near the base of the dome. The distal end of the dome is extended forward into a sharp apex (Fig. 3.61F). The dome is often perforated with pores (Fig. 3.61F). The shield and submarginal rib are patterned like the body scales.

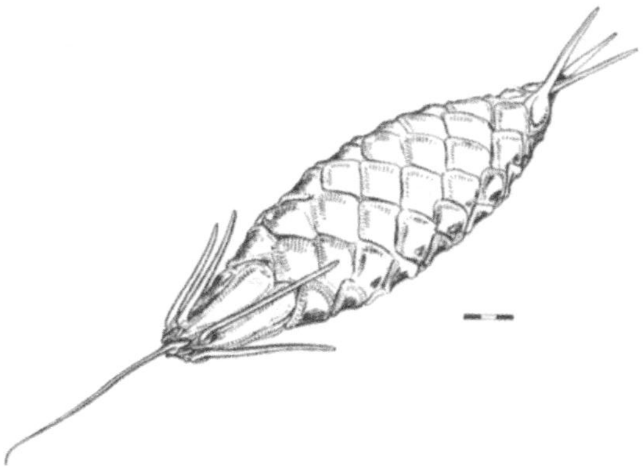

Fig. 3.60. Three-dimensional reconstruction of a whole cell of *Mallomonas torquata* f. *simplex* depicting the collar and the distribution of scales and bristles. Body scales are positioned in spiral rows such that their longitudinal axes are perpendicular to the longitudinal axis of the cell. Note the anterior collar scales, bristles and long posterior spines.

←

Fig. 3.59A-B. Mallomonas dickii. Fig. A. Body scales. Note the arrangement of shield ribs, the ornamentation of the anterior submarginal ribs and the curved nature of the scales. Scale bar is 1 μm. Fig. B. Posterior region of a cell. Scale bar is 1 μm. Figs. C-D. *Mallomonas doignonii* var. *tenuicostis*. Fig. C. Whole cell depicting orientation and arrangement of scales, collar, and the long posterior spines. Scale bar is 5 μm. Fig. D. Close up of the posterior region of the cell in Fig. C. Note the morphology of the scales, the presence of base plate pores and a single large distal pore on the undersurface of the scale, and the small spike-like spines (arrows). Scale bar is 1 μm. Fig. E. *Mallomonas torquata* f. *torquata*. Body scales showing the markings of the shield and anterior submarginal ribs. Scale bar is 2 μm. Fig. F. *Mallomonas torquata* f. *simplex*. Whole cell sheared in half along a spiral row. Note the unpatterned shields and the sharp apex on each collar scale. Scale bar is 5 μm.

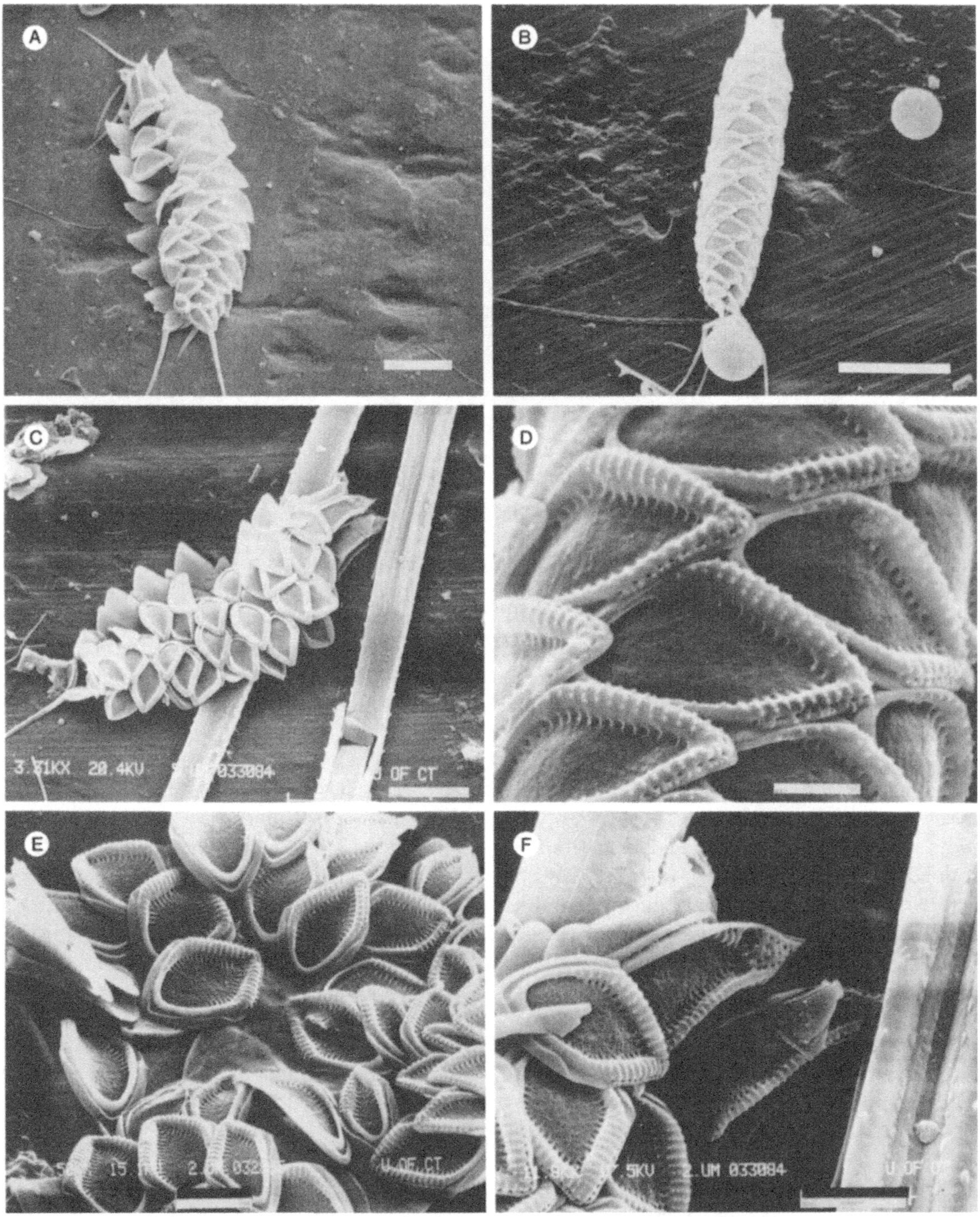

Fig. 3.61. Mallomonas torquata f. *simplex*. Figs. A-C. Whole cells depicting the collar, arrangement of body scales, short apical bristles (Fig. A) and the relatively long posterior spines. Scale bars are 5 μm for Figs. A and C, and 10 μm for Fig. B. Fig. D. Close-ups of an intact cell showing the precise arrangement of body scales. Details of the morphology of the shield and anterior submarginal ribs are also depicted. Scale bar is 1 μm. Fig. E. Body scales. Note the relatively unpatterned shield, the broad and ornamented anterior submarginal ribs and the narrow posterior flange. Scale bar is 2 μm. Fig. F. Close-up of collar scales illustrating the relatively small, triangular shaped and perforated domes. Scale bar is 2 μm.

The rear scales are smaller, have more rounded outlines and each possesses a single spine ranging in length from 3.4 μm to 12.8 μm (Figs. 3.61A-C). Smaller rudimentary spines may also be found (Fig. 3.59F). Bristles, restricted to the collar scales, are only slightly longer than the collar, smooth, slightly bent and tapered (Fig. 3.61A).

M. torquata f. *simplex* differs from *M. torquata* var. *torquata* Asmund and Cronberg in the degree of ribbing on the scales. In addition to the series of short struts radiating from the submarginal rib, scales of var. *torquata* also have a reticulation of ribs on the shield (Fig. 3.59E). All other characteristics are similar. There is also a similarity in the structure of body scales between *M. torquata* f. *simplex* and *M. phasma* Harris and Bradley (Nicholls, 1988c). However, cells of *M. phasma* have scales that are pitted (Harris and Bradley, 1960; Asmund and Kristiansen, 1986) and differently shaped collar scales (Nicholls, 1988c).

A last point is noteworthy concerning the distribution and arrangement of papillae on the submarginal ribs of all taxa described in this book belonging to the Series Doignonianae. This feature, observed only with SEM, appears to be a relatively stable character that could serve as an additional taxonomic tool for distinguishing different taxa.

Mallomonas schwemmlei Glenk emend. Glenk & Fott

Synonyms: *M. coronifera* Matvienko
M. coronifera Matvienko var. *schwemmlei* (Glenk) Schmid
M. pumilio Harris & Bradley var. *schwemmlei* (Glenk) Harris

Cells are broadly ovoid or elongate-ovoid with a relatively short collar and posterior scales with short spines (Figs. 3.62A-C). Scales are arranged in spiral rows. The apical most scale of each spiral row (i.e. the collar scale) is orientated with its longitudinal axis more or less parallel to that of the cell, while the body scales of each spiral row are situated such that their longitudinal axes are perpendicular to the longitudinal axis of the cell (Figs. 3.62A,C). Each scale is overlapped by the scale situated behind it in the same spiral row and by the scale(s) positioned in the spiral row above it (Figs. 3.62A,C). The scales of the collar are aligned in a ring such that each one is overlapped by the scale to its left.

Body scales are rhombic and consist of a perforated base plate, a narrow proximal rim and a continuous submarginal rib (Figs. 3.62B,F). The arms of the V-rib are continuous with the anterior submarginal ribs which in turn fuse at the distal tip forming one continuous structure (Fig. 3.62F). The anterior submarginal ribs are morphologically different; the rib on the right-hand side is thicker, more raised and extended over the anterior flange (Fig. 3.62F). The anterior flanges are narrow and marked with a series of closely spaced, short ribs that radiate from the submarginal rib; the spaces between the anterior flange ribs appear as an alternating series of pores (Figs. 3.62D,F). The alternating pattern of rib and pore is hidden in both SEM and TEM images along the right flange due to the overhanging of the submarginal rib. Like the anterior flanges, the narrow posterior flange may also, but not always (Fig. 3.62F), be marked with a series of short struts. A row of base plate pores is also present in the posterior flange. The shield consists of a wavy fingerprint-like pattern of ribs that are positioned on the base plate between the pores; base plate pores can be seen between the ribs (Fig. 3.62F). A characteristic feature of the scales is that the height of each rib varies along its length, alternating between lower and higher portions (Fig. 3.62F).

The asymmetric collar scales, usually five to six in number, are rather broad relative to their length (Figs. 3.62B-D). The proximal rim extends along the dorsal edge almost to the distal end of the scale and, as in the body scales, the right side of the submarginal rib is thicker and extends over the ventral edge (Fig. 3.62D). The markings of the flanges and shield are similar to those on body scales. Domes are relatively small.

The right anterior submarginal rib of the smaller posterior scales is extended into a very short spine (Figs. 3.62A-C); the latter are usually not visible with light microscopy. Although rare, several posterior scales were observed where the distal end of the submarginal rib formed a structure that resembled a dome (Fig. 3.62B).

Bristles are short, 1.5 to 2 times the length of the collar scales, curved to conform to the shape of the cell and needle-like (Figs. 3.62A-B). Bristles are rolled up such that a longitudinal slit extends

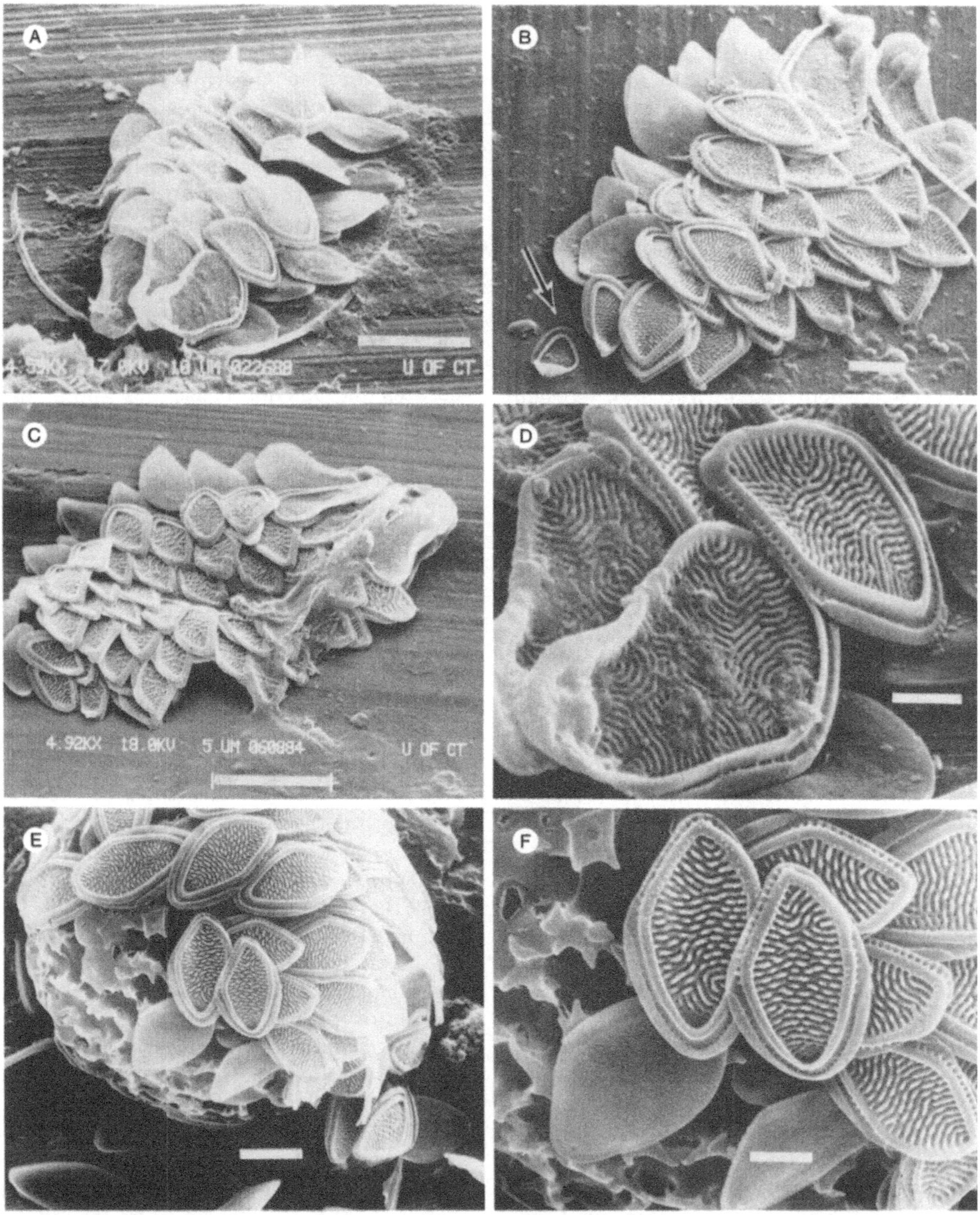

Fig. 3.62. Mallomonas schwemmlei. Figs. A-C. Whole cells showing the nature and arrangement of scales and bristles. Note the fingerprint-like pattern of ribs and the slightly thicker and more extended right anterior submarginal rib; the latter is most evident on the posterior end of the cell (Fig. C) where it often becomes extended into a short spine (Fig. B). Also, note the rear scale where the submarginal rib formed a 'dome-like' structure (arrow, Fig. B). Scale bars are 5 μm, 2 μm and 5 μm, respectively. Fig. D. Collar and body scales. Note the pattern of ribs on the shield, the varying height along the length of each rib and the ornamentation of the anterior flanges. Scale bar is 1 μm. Figs. E-F. Cell with stomatocyst. Surface consists of closely spaced, thick, star-like spines. Scale bars are 2 μm and 1 μm, respectively.

along the length of the shaft.

The walls of the stomatocyst of *M. schwemmlei* consist of closely spaced, thick star-like spines (Figs. 3.62E-F). Such an ornamentation is in contrast to the smooth appearance described, using LM, by Glenk (1956). The star-like spines are similar to those of the closely related taxon, *M. torquata* (Asmund and Cronberg, 1979).

CHAPTER IV

Morphology: Quantitative Features of the Siliceous Cell Coat

Over 10,000 images from more than 2,000 SEM micrographs were traced using a digitizer in order to determine the morphometric dimensions of scale, bristle and cell characters for individual taxa. In addition, numerous measurements of cells were made directly from LM and SEM preparations. Results are presented here for 36 taxa for which sufficient data were available. Although results for each taxon are given, the discussion will focus primarily on relationships between organisms. The main objective is to present a comparison of the quantitative characteristics of scales, bristles and cells which in turn offer an additional tool for discriminating between taxa.

Body Scales

A 40-fold difference in surface area was found between the smallest and largest body scales measured for all species (Table 4.1). The area of body scales varied from a low of 1.4 μm^2 (*M. akrokomos*) to a high of 39.3 μm^2 (*M. caudata*), or 58.7 μm^2 if the wing of *M. pseudocoronata* was included (Figs. 4.1-4.2). Commonly for a given taxon, the ratio between the area of the largest scale to that of the smallest scale was close to 2; twenty-six of thirty-six species had ratios between 1.5 and 2.5 (Table 4.1). An eight-fold difference in length was found between the smallest and largest body scales (Table 4.1); scales ranged in length from 1.6 μm (*M. canina*) to 12.1 μm (*M. pseudocoronata*) (Figs. 4.2-4.3). The ratio in length between the largest and smallest scales for twenty-five of thirty-six species was between 1.5 and 2.0 and reached a maximum of 2.2 for *M. akrokomos*. Similarly, an eleven-fold difference in the widths of body scales was found between species (Table 4.1); scales ranged in width from 0.6 μm (*M. akrokomos*) to 6.6 μm (*M. caudata*).

Body scales of *M. canina*, *M. akrokomos* and *M. dickii* were consistently the smallest in terms of mean area, length and width, whereas the largest scales belonged to *M. lychenensis*, *M. elongata*, *M. caudata* and *M. pseudocoronata* (Figs. 4.2, 4.4, 4.5). The mean area, length and width of scales of the former three taxa were less than 3 μm^2, 2.6 μm and 1.7 μm, respectively. The latter four taxa had mean scale area, length and width values greater than 19.5 μm^2, 6.5 μm and 3.9 μm, respectively

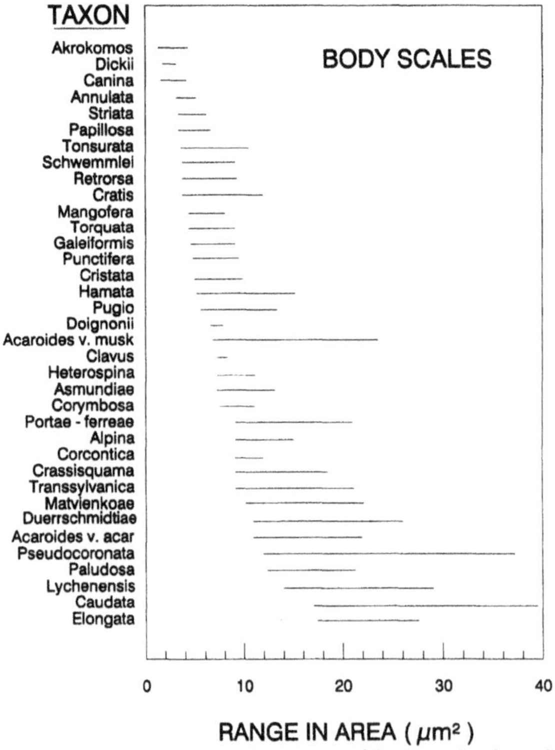

Fig. 4.1. Ranges in the surface areas of body scales from 36 taxa of *Mallomonas*. Values were obtained from scanning electron micrographs using an image analyzer.

Table 4.1. Ranges in the length, width and surface area of body scales from 36 taxa of *Mallomonas*. The ratio between the maximum and minimum values is also given. Ranges are given twice for *M. pseudocoronata*, once including the wing and once without. Taxa are arranged according to their minimum scale lengths.

	Length			Width			Area		
Taxon	Min	Max	Ratio	Min	Max	Ratio	Min	Max	Ratio
Canina	1.6	2.8	1.7	1.2	2.2	1.8	1.7	4.3	2.5
Akrokomos	1.7	3.8	2.2	0.6	2.0	3.3	1.4	4.5	3.2
Dickii	1.9	2.6	1.4	1.2	1.8	1.5	1.7	3.0	1.8
Conspersa	2.3	4.2	1.8	1.4	2.4	1.7	4.4	6.0	1.4
Retrorsa	2.6	3.9	1.5	1.7	3.1	1.8	4.0	9.1	2.3
Striata	2.6	3.9	1.5	1.6	2.3	1.4	3.5	6.7	1.9
Punctifera	2.6	4.2	1.6	2.0	2.9	1.4	5.0	9.2	1.8
Annulata	2.7	3.6	1.3	1.7	2.4	1.4	3.2	5.5	1.7
Tonsurata	2.7	4.4	1.6	1.8	3.2	1.8	3.5	10.5	3.0
Schwemmlei	2.8	4.5	1.6	1.7	2.8	1.6	3.9	8.6	2.2
Torquata f. simplex	2.9	4.8	1.7	1.8	3.3	1.8	4.4	9.0	2.0
Pugio	2.9	5.1	1.8	2.0	3.6	1.8	5.8	13.3	2.3
Cratis	2.9	5.4	1.9	1.7	2.8	1.6	4.0	11.9	3.0
Galeiformis	3.0	4.1	1.4	2.0	2.9	1.4	4.8	8.7	1.8
Papillosa	3.0	4.4	1.5	1.3	2.4	1.8	3.5	6.8	1.9
Magnofera	3.1	4.4	1.4	2.0	3.3	1.6	4.3	7.9	1.8
Cristata	3.1	4.8	1.5	2.1	2.7	1.3	5.3	9.7	1.8
Hamata	3.1	5.9	1.9	1.6	3.9	2.4	5.6	14.6	2.6
Heterospina	3.2	5.4	1.7	2.6	3.0	1.2	6.8	10.9	1.6
Doignonii	3.4	4.3	1.3	2.1	3.6	1.7	6.4	7.7	1.2
Clavus	3.6	3.9	1.1	2.5	3.0	1.2	6.7	8.0	1.2
Transsylvanica	3.8	5.4	1.4	2.7	4.0	1.5	9.2	21.0	2.3
Crassisquama	3.8	6.7	1.8	2.4	4.4	1.8	9.2	18.4	2.0
Acaroides v. muskokana	3.9	7.6	1.9	2.0	4.4	2.2	6.5	23.7	3.6
Corcontica	4.0	5.1	1.3	2.7	3.1	1.1	9.2	11.8	1.3
Portae-ferreae	4.0	7.5	1.9	2.9	4.1	1.4	8.8	20.5	2.3
Paludosa	4.1	7.3	1.8	2.8	3.7	1.3	12.3	21.4	1.7
Corymbosa	4.2	4.9	1.2	2.5	3.1	1.2	7.8	10.9	1.4
Asmundiae	4.3	5.5	1.3	2.2	3.1	1.4	7.0	12.7	1.8
Matvienkoae	4.3	6.7	1.6	2.8	4.7	1.7	9.8	22.1	2.3
Acaroides v. acaroides	4.3	6.7	1.6	3.1	4.4	1.4	10.8	22.0	2.0
Alpina	4.5	4.9	1.1	2.1	2.8	1.3	9.1	14.7	1.6
Lychenensis	4.7	8.1	1.7	3.3	4.3	1.3	14.4	28.7	2.0
Duerrschmidtiae	4.7	8.6	1.8	2.9	4.2	1.4	10.7	25.6	2.4
Caudata	5.1	8.7	1.7	3.0	6.6	2.2	16.7	39.3	2.4
Elongata	5.6	8.3	1.5	3.5	5.0	1.4	17.0	27.4	1.6
Pseudocoronata –wing	5.9	9.3	1.6	3.7	5.6	1.5	11.9	36.7	3.1
Pseudocoronata +wing	6.4	14.1	2.2	3.7	5.6	1.5	22.6	58.7	2.6

(Table 4.2). Two-thirds of the taxa had scales with average area, length and width measurements less than 12 μm^2, 5 μm and 3 μm, respectively.

Two characteristics, the mean L:D ratio and the mean form factor, were used to compare shapes of body scales between species. The mean L:D, the ratio of the mean scale length: width, ranged from 1.3 to 2.0 (Table 4.2). The form factor is another dimensionless ratio that compares the area to the perimeter using the following relationship:

$$\text{Form factor} = (4\ \pi\ A) / P^2$$

where A = area of and P = perimeter of the enclosed area. The ratio is normalized so that the form factor of a circle = 1 and of a line = 0. Since the overall outline of a scale is much closer to that of a circle than a line, the mean values are clustered closer to 1. Mean form factor scores ranged from a low of 0.83 for both *M. paludosa* and *M. duerrschmidtiae* to a high of 0.97 for *M. canina* (Table 4.2).

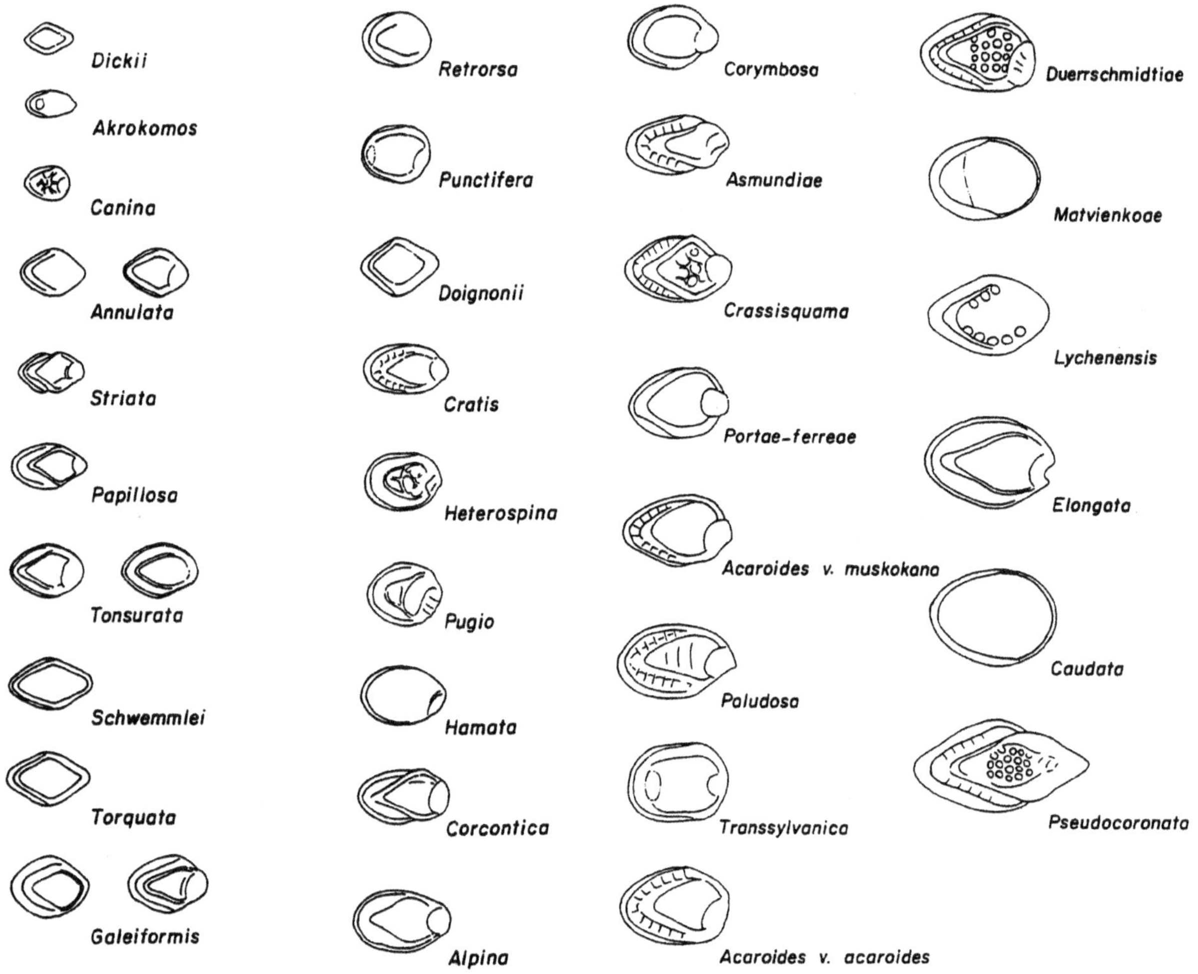

Fig. 4.2. Comparative drawings from 33 taxa of *Mallomonas* depicting the relative differences in the sizes of body scales. The mean length, width and surface area of scales from each taxa were used to construct the drawings. Domed and domeless scales are shown for *M. annulata*, *M. tonsurata* and *M. galeiformis*.

For most species there was a significant linear relationship between the area and length of body scales (Tables 4.3, 4.4). A linear regression model, based on scale data from all taxa yielded a highly significant relationship ($p < 0.001$) with R^2 and SE values of 0.90 and 0.52, respectively (Fig. 4.6). A similar relationship, with R^2 and SE values of 0.93 and 0.36, was found between the mean areas and lengths of scales from 36 taxa (Table 4.3; Fig. 4.7). Adding the mean form factor as a second independent variable to the model yielded an even lower SE of 0.28 (Table 4.3). Thus, the mean area of a scale can be accurately estimated from the mean length, or vice versa.

The linear relationship between the area and length of body scales was also significant for many, but not all, individual species (Table 4.4). The R^2 values for individual taxa ranged from an insignificant and low of 0.1 for *M. pugio* and *M. caudata* to a significant and high of 0.89 for *M. cristata*. In general, those taxa with more rounded scales (i.e. those with form factors closer to 1) had less significant relationships (Table 4.4).

Although neither the length, width or area could be used to accurately predict the form factor of a body scale, there was a significant relationship ($p < 0.05$; $R^2 = 0.54$) between the L:D ratio and the form factor (Table 4.3; Fig. 4.8).

The mean percentage of the total surface area of a body scale covered by the dome ranged from a low of 5.8% for *M. pseudocoronata* to a high of over 20% for *M. pugio*, *M. corymbosa*, *M.*

Table 4.2. Mean morphometric values of body scales for species of *Mallomonas*. Taxa are arranged by increasing surface area. The '% domed' is the percentage of the total area of the scale covered by the dome. The '% exposed' is the percentage of the total surface area exposed on an intact cell. For those taxa with domed and domesless body scales data was combined. Values for *M. pseudocoronata* are given twice, once including the wing and once without.

Taxon	Area (μm^2)	Form factor	Length (μm)	Width (μm)	Ratio L:D	% Domed	% Exposed
Dickii	2.33	0.90	2.27	1.42	1.6		59
Akrokomos	2.74	0.85	2.58	1.26	2.0		59
Canina	3.02	0.97	2.29	1.74	1.3	12.7	64
Annulata	4.50	0.92	3.10	2.08	1.5	25.6	35
Striata	5.0	0.91	3.19	2.00	1.6	13.0	53
Papillosa	5.02	0.85	3.50	1.90	1.8	11.0	45
Conspersa	5.10	0.90	3.24	2.01	1.6	12.0	
Magnofera	5.86	0.88	3.59	2.50	1.4		42
Tonsurata	5.90	0.93	3.40	2.54	1.3	22.0	46
Schwemmlei	6.14	0.85	3.90	2.20	1.8		51
Torquata f. simplex	6.50	0.89	3.80	2.44	1.6		65
Galeiformis	6.60	0.91	3.52	2.44	1.4	24.0	45
Retrorsa	6.74	0.96	3.40	2.38	1.4		54
Punctifera	6.80	0.93	3.24	2.49	1.3		60
Doignonii	6.95	0.96	3.79	2.67	1.4		59
Clavus	7.10	0.96	3.71	2.77	1.3		64
Cratis	7.56	0.88	4.25	2.24	1.9	12.0	44
Cristata	7.74	0.90	4.11	2.45	1.7	16.0	49
Heterospina	8.25	0.91	3.90	2.77	1.4	10.4	63
Pugio	8.64	0.94	3.90	2.90	1.3	20.0	65
Hamata	9.29	0.95	4.26	2.62	1.6		49
Corcontica	10.23	0.90	4.52	2.89	1.6	17.0	
Alpina	10.50	0.93	4.70	2.50	1.9	29.0	45
Corymbosa	10.70	0.93	4.60	3.02	1.5	21.0	48
Asmundiae	11.83	0.84	5.10	2.88	1.8	17.3	47
Crassisquama	12.70	0.93	5.12	3.31	1.5	13.7	49
Portae-ferreae	13.47	0.92	5.24	3.44	1.5	17.8	50
Acaroides v. muskokana	13.51	0.89	5.27	3.23	1.6	13.9	50
Paludosa	14.90	0.83	5.84	3.20	1.8	12.0	50
Transsylvanica	15.20	0.96	4.70	3.60	1.3		65
Acaroides v. acaroides	15.75	0.91	5.69	3.66	1.6	13.5	54
Duerrschmidtiae	16.00	0.83	6.40	3.52	1.8	10.8	52
Matvienkoae	16.60	0.94	5.76	3.67	1.6		56
Lychenensis	19.50	0.91	6.51	3.86	1.7		59
Elongata	20.82	0.92	6.72	4.24	1.6	11.4	45
Pseudocoronata –wing	23.70	0.85	7.59	4.68	1.6	5.8	46
Caudata	27.00	0.96	6.79	4.80	1.4		48
Pseudocoronata +wing	34.0	0.80	9.52	4.68	2.0	5.8	46

tonsurata, *M. galeiformis*, *M. annulata* and *M. alpina* (Table 4.2). For the majority of species the dome covered between 10% and 18% of the scale (Fig. 4.9). In general, taxa with domed body scales restricted to the anterior half of the cell have larger domes.

For most species the range in the size of body scales on a given cell is similar to that found between cells of different populations. Commonly, body scales gradually decrease in size progressing from about the middle region of the cell to the posterior end and it is often difficult to decide where body scales end and posterior scales begin. In the data presented above scales on the lower one-fourth of the cell were considered as posterior scales.

Despite the overlap in scale dimensions for some species, it is clear that accurate measurements of

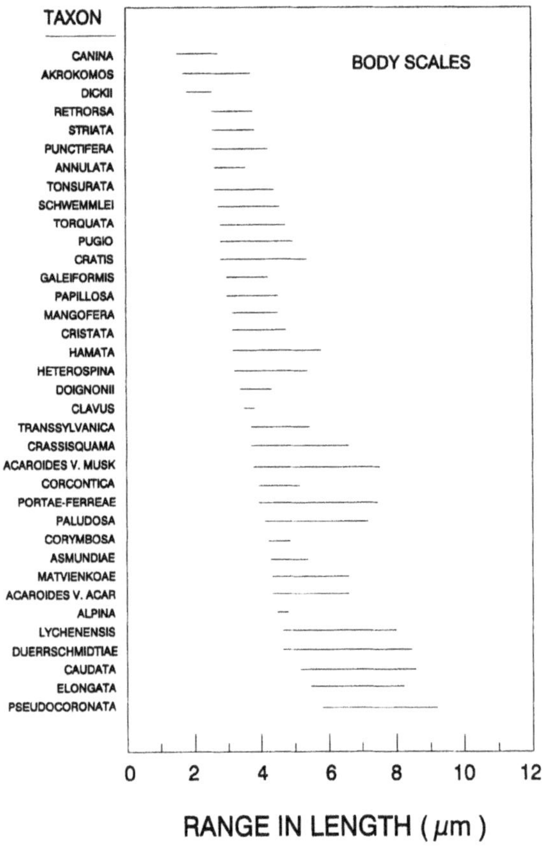

Fig. 4.3. Ranges in the lengths of body scales from 36 taxa of *Mallomonas*. Values were obtained from scanning electron micrographs using an image analyzer.

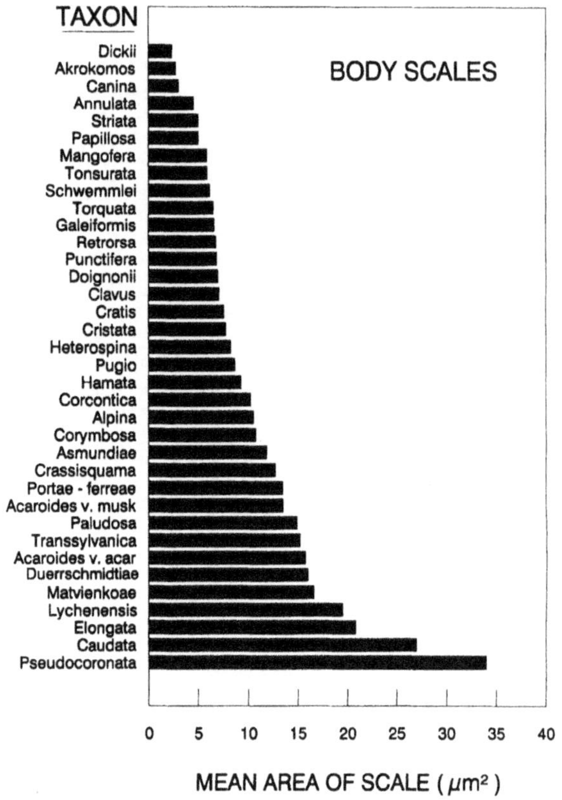

Fig. 4.4. Mean surface areas of body scales from 36 taxa of *Mallomonas*. Values were obtained from scanning electron micrographs using an image analyzer. The value for *M. pseudocoronata* includes the anterior wing.

Table 4.3. **Results of regression analyses depicting the relationships between various morphological characteristics of body scales. The mean values of body scales for the 36 taxa listed in Table 4.2 were used for each analysis. The independent variable(s), dependent variable, R^2, standard error (SE) of the estimate for Y and the regression model are listed for each analysis. Graphic illustrations of those relationships marked with an * are presented elsewhere in the text.**

	Independent Variable(s)	Dependent Variable	R^2	SE of est. Y	Regression Model
1.	Area*	Length	0.93	0.36	$Y=0.21(X)+2.22$
2.	Area (X_1) Form factor (X_2)	Length	0.96	0.28	$Y=0.21(X_1)-5.7(X_2)+7.4$
3.	Area	Width	0.94	0.21	$Y=0.13(X)+1.45$
4.	Length (X_1) Width(X_2)	Area	0.96	1.17	$Y=2.13(X_1)+3.8(X_2)-9.83$
5.	L:D*	Form factor	0.54	0.03	$Y=-0.16(X)+1.15$
6.	Length[a]	Form factor	0.03	0.04	$Y=-0.005(X)+0.93$
7.	Area[a]	Form factor	0	0.04	$Y=0(X)+0.91$

[a]Relationships are not significant ($p > 0.05$).

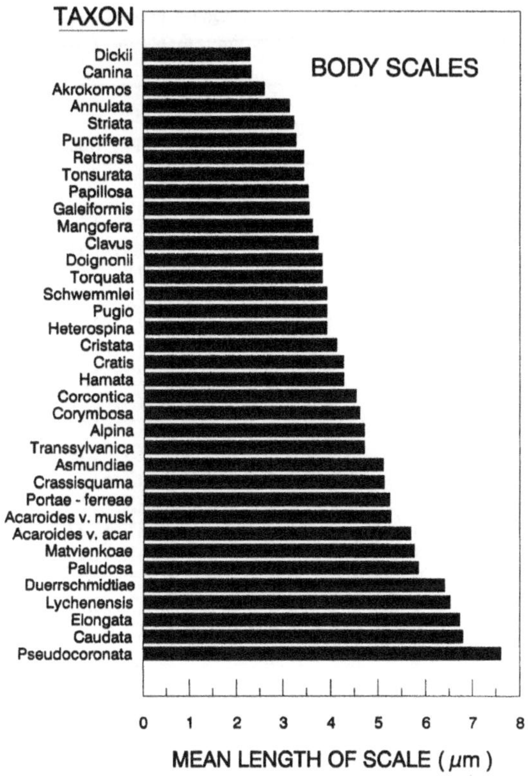

Fig. 4.5. Mean lengths of body scales from 36 taxa of *Mallomonas*. Values were obtained from scanning electron micrographs using an image analyzer. The value for *M. pseudocoronata* does not include the anterior wing.

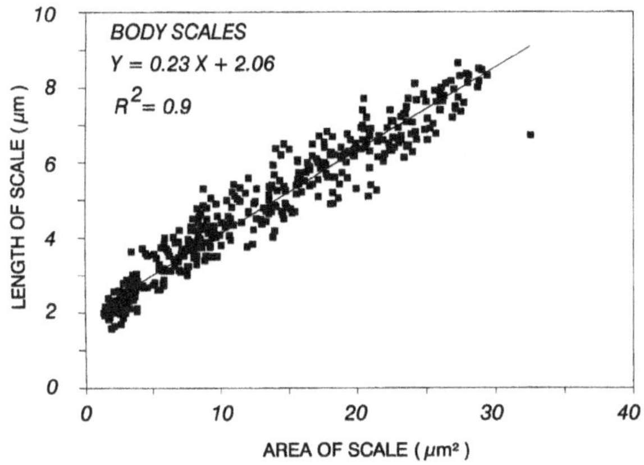

Fig. 4.6. Plot of the surface area (independent variable) versus the length (dependent variable) of body scales for all species of *Mallomonas*. The regression analysis is based on data from over 1000 scales. Only a random number of the actual data points are shown on the plot, including the farthest outliers. The regression analysis is significant at the $p < 0.001$ level.

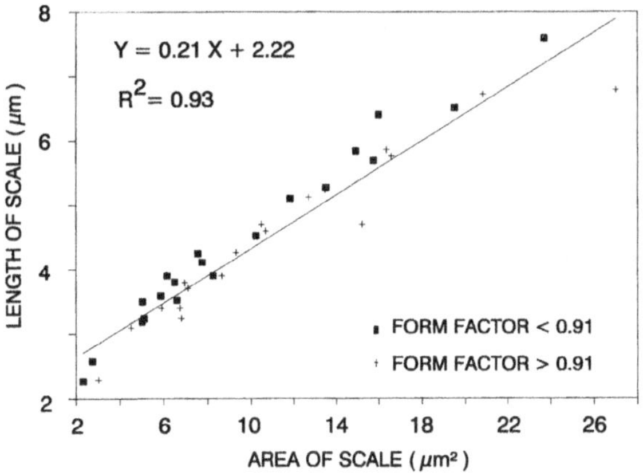

Fig. 4.7. Plot of the mean surface area (independent variable) versus the mean length (dependent variable) of body scales for 38 taxa of *Mallomonas*. Taxa are divided into two groups based on the mean form factor of the scale. The regression analysis is significant at the $p < 0.001$ level.

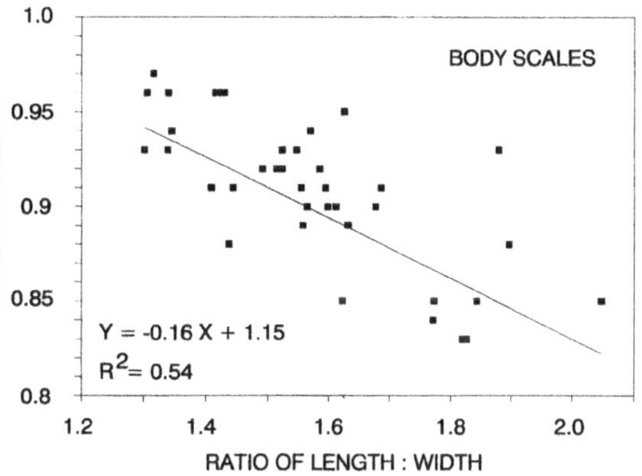

Fig. 4.8. Plot of the mean length:width ratio (independent variable) versus the form factor (dependent variable) of body scales for 38 taxa of *Mallomonas*. The regression analysis is significant at the $p < 0.05$ level.

body scales can be used to help discriminate between taxa, especially those with similarly shaped scales. For example, the morphology of scales is similar for *M. dickii* and *M. doignonii*, however, little overlap is found in the dimensions of their body scales. A similar situation was documented for scales of the related taxa *M. punctifera* and *M. transsylvanica* (Fig. 4.2). Likewise, scales of *M. tonsurata* are significantly smaller than those of the closely related taxa *M. corymbosa*, *M. portae-ferreae* and *M. elongata* (Fig. 4.10). On the other hand, the dimensions of body scales could not be

Table 4.4. **Results of regression analyses depicting the relationships between the mean surface area (independent variable) and mean length (dependent variable) of body scales from 32 taxa of *Mallomonas*. The R^2, standard error (SE) of the estimate for Y and the regression model are listed for each taxon. Individual taxa are listed in order of increasing R^2 values.**

Taxon	R^2	SE of est. Y	Regression Model
caudata	0.10	0.45	Y=0.04(X)+5.36
pugio	0.10	0.50	Y=0.13(X)+2.7
retrorsa	0.12	0.28	Y=0.08(X)+2.84
asmundiae	0.24	0.30	Y=0.11(X)+3.96
galeiformis	0.33	0.26	Y=0.17(x)+2.42
matvienkoae	0.35	0.58	Y=0.15(X)+3.2
hamata	0.47	0.47	Y=0.23(X)+2.21
papillosa	0.48	0.28	Y=0.30(X)+2.02
crassisquama	0.52	0.44	Y=0.19(X)+2.70
schwemmlei	0.52	0.34	Y=0.25(X)+2.33
canina	0.54	0.20	Y=0.38(X)+1.14
transsylvanica	0.57	0.46	Y=0.15(X)+2.46
acaroides v. acaroides	0.60	0.46	Y=0.19(X)+2.5
pseudocoronata	0.61	0.59	Y=0.16(X)+3.73
paludosa	0.61	0.73	Y=0.24(X)+2.29
elongata	0.64	0.41	Y=0.22(X)+2.2
torquata	0.65	0.24	Y=0.27(X)+2.07
alpina	0.71	0.09	Y+0.07(X)+4.0
tonsurata	0.71	0.26	Y=0.22(X)+2.14
heterospina	0.72	0.26	Y=0.40(X)+0.58
duerrschmidtiae	0.74	0.48	Y=0.24(X)+2.57
annulata	0.74	0.12	Y=0.37(X)+1.44
akrokomos	0.74	0.25	Y=0.50(X)+1.22
dickii	0.75	0.11	Y=0.59(X)+0.89
punctifera	0.77	0.20	Y=0.26(X)+1.41
striata	0.77	0.20	Y=0.35(X)+1.42
corymbosa	0.80	0.20	Y=0.25(X)+1.5
cratis	0.82	0.19	Y=0.32(X)+1.89
lychenensis	0.86	0.40	Y=0.22(X)+2.3
portae-ferreae	0.87	0.29	Y=0.26(X)+1.7
acaroides v. muskokana	0.88	0.39	Y=0.17(X)+3.0
cristata	0.89	0.25	Y=0.39(X)+1.07

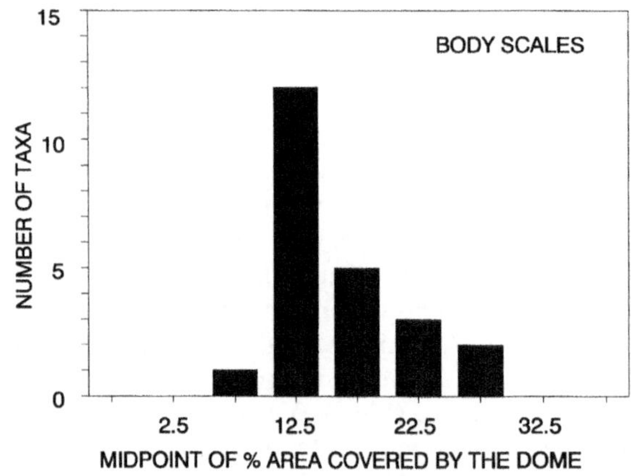

Fig. 4.9. Frequency distribution of the mean percentage of the total surface area covered by the dome for 23 species of *Mallomonas*.

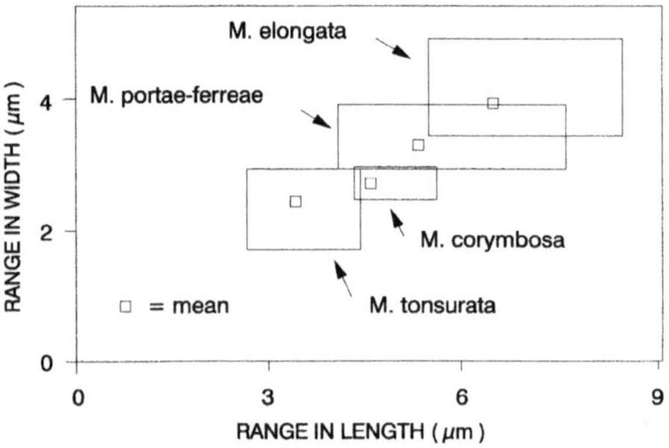

Fig. 4.10. Plot of the ranges in length versus width of body scales from four closely related species of *Mallomonas*, *M. tonsurata*, *M. corymbosa*, *M. portae-ferreae* and *M. elongata*. The ranges are represented by the larger rectangular boxes and the mean values by the smaller squares.

used to separate *M. tonsurata* from *M. galeiformis* (Tables 4.1-4.2). The comparative morphometric data for scales presented in this chapter could serve as an additional tool for discriminating between taxa.

Cells

A 15-fold difference in the lengths of cells (including spines) was observed. Cells ranged from a minimum of 6 μm for both *M. canina* and *M. dickii* to a maximum of 91 μm for *M. caudata*. The mean lengths of cells ranged from 9.4 μm for *M. canina* to 70.1 μm for *M. caudata* (Table 4.5). Cells of *M. caudata* were consistently larger than all other taxa (Fig. 4.11). Seventy-two % of the taxa had mean cell lengths less than 30 μm (Table 4.5; Fig. 4.11). The smallest cells were of *M. canina*, *M. dickii*, *M. mangofera*, *M. papillosa*, *M. punctifera* and *M. tonsurata*, all of which had mean lengths less than 13 μm (Table 4.5). Nine species had mean lengths above 30 μm; *M. pseudocoronata* and *M. torquata* fall into this category because of the presence of long posterior spines (Table 4.6).

The widths of cells ranged from a low of 2.9

Table 4.5. The mean dimensions of cells for species of *Mallomonas*. Taxa are ordered according to increasing length. The outlines of cells used to calculate the volumes and surface areas are shown in Figure 4.13. The volume and surface area were calculated using rotational geometry. See text for details. The number of scales per cell was estimated by dividing the surface area by the mean area of a scale that is exposed on an intact cell. The length and form factors of cells include the spines, if present.

Taxon	Form factor	Length (μm)	Width (μm)	L:D Ratio	Volume (μm^3)	Surface (μm^2)	No. of scales per cell
Canina	0.95	9.4	8.0	1.2	280	131	68
Mangofera	0.91	9.6	5.7	1.7	151	89	36
Dickii	0.57	11.1	3.2	3.5	92	69	50
Papillosa	0.88	11.7	6.8	1.7	342	152	67
Punctifera	0.78	12.2	5.6	2.2	204	109	27
Tonsurata	0.87	12.9	7.5	1.7	370	160	59
Striata	0.87	14.6	8.6	1.7	563	217	82
Heterospina	0.90	14.7	11.0	1.3	920	299	58
Pugio	0.95	15.0	10.0	1.5	833	276	49
Schwemmei	0.84	15.3	8.0	1.9	483	200	64
Galeiformis	0.84	16.0	8.0	2.3	476	190	64
Annulata	0.73	17.9	7.4	2.4	540	216	137
Clavus	0.62	20.4	6.0	3.4	354	168	37
Akrokomos	0.44	21.4	5.9	3.6	276	176	109
Asmundiae	0.98	22.0	16.0	1.4	2661	597	107
Lychenensis	0.86	22.1	11.9	1.9	1685	441	38
Transsylvanica	0.76	22.4	9.5	2.4	1102	342	35
Paludosa	0.71	23.3	9.4	2.5	1253	365	49
Crassisquama	0.73	24.3	10.9	2.2	1167	356	57
Duerrschmidtiae	0.53	26.6	10.1	2.6	902	296	36
Hamata	0.62	26.7	10.0	2.7	1477	421	92
Doignonii	0.39	27.2	5.9	4.6	430	229	56
Corymbosa	0.74	28.0	10.5	2.7	1749	469	91
Acaroides v. acaroides	0.91	29.0	15.0	1.9	3585	734	86
Acaroides v. muskokana	0.88	30.0	12.1	2.5	2682	616	91
Cratis	0.74	30.2	9.8	3.1	1907	499	150
Matvienkoae	0.80	31.0	17.0	1.8	4034	798	86
Portae-ferreae	0.72	31.6	12.0	2.6	2889	647	96
Pseudocoronata	0.55	32.4	13.3	2.4	2297	556	51
Torquata	0.38	33.1	6.6	5.0	729	285	68
Retrorsa	0.44	33.9	8.8	3.9	1163	430	118
Elongata	0.67	40.2	15.0	2.7	4990	932	100
Caudata	0.59	70.1	19.5	3.6	12228	2073	160

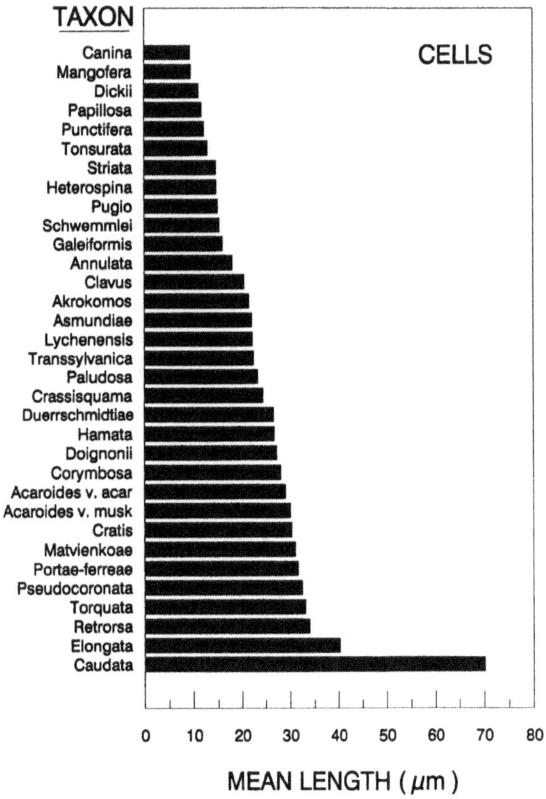

Fig. 4.11. Mean lengths of cells from 33 taxa of *Mallomonas*. Values were obtained from scanning electron micrographs using an image analyzer. The values include the length of spines, if present.

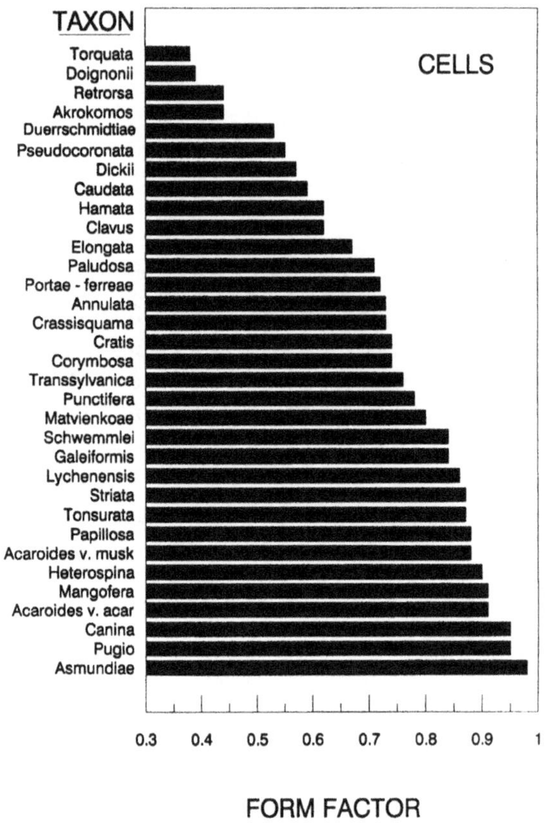

Fig. 4.12. Mean form factors of cells from 33 taxa of *Mallomonas*. Values were obtained from scanning electron micrographs using an image analyzer. Form factor measurements do not include any posterior spines. Species with smaller means have more elongated shapes, while those with values close to 1 have shapes similar to a sphere.

µm for *M. dickii* to a high of 25 µm for both *M. caudata* and *M. matvienkoae* (Table 4.5). The mean widths of cells ranged from 3.2 µm for *M. dickii* to 19.5 µm for *M. caudata*. About half of the species had cells with a mean width less than 10 µm (Table 4.5).

The ratio of the mean length to width (L:D) of a cell ranged from a low of 1.2 for *M. canina* to a high of 5.0 for *M. torquata* (Table 4.5). In general, taxa with the largest L:D values either had a long caudal tail (e.g. *M. retrorsa*, *M. akrokomos* and *M. caudata)*, or were taxonomically positioned in the Sectio Torquatae (e.g. *M. torquata*, *M. doignonii* and *M. dickii*). A much greater range in the form factor was observed between the shapes of cells than was found for body scales (Table 4.5; Fig. 4.12). The form factor ranged from a low of 0.38 for *M. torquata* to a high of 0.98 for *M. asmundiae*. Species with either elongated caudal tails, long spines or within the Sectio Torquatae had cell shapes that most closely resembled a line and, hence, had correspondingly low form factor values. Cells of *M. asmundiae*, *M. pugio*, *M. canina*, *M. acaroides* var. *acaroides*, *M. mangofera* and *M. heterospina* had values above 0.9; these taxa most closely approached a true spherical shape (Fig. 4.13). A strong negative correlation, with an R^2 and standard error of 0.83 and 0.07, respectively, was found between the mean L:D ratios and from factors of cells (Table 4.7; Fig. 4.14).

Most surveys of *Mallomonas* simply list length and width measurements of cells and give a qualitative statement describing the shape of the cell (e.g. Takahashi, 1978; Wee, 1982; Asmund and Kristiansen, 1986). In this work I have attempted to quantify the general outline of the cell (form

Table 4.6. Mean values and ranges in the lengths of bristles and spines for 35 taxa of *Mallomonas*. Taxa are arranged by increasing bristle length. For those taxa with two types of bristles, the names are noted in parentheses. A '<0.5' appears in the spine column for those taxa with spines less than 0.5 µm. All lengths are given in µm.

TAXON	BRISTLE						SPINE	
	Type I			Type II				
		Mean	Range		Mean	Range	Mean	Range
dickii		1.9	1.8-2.2				0.5	0.2-0.8
retrorsa		3.6	3.2-4.5					
canina	(helmet)	4.0	3.3-4.9	(needle)	4.5	3.9-5.7		
papillosa		4.5	3.0-6.4					
hamata	(helmet)	5.2	4.2-7.3	(body)	18.3	9.5-28.8		
pugio		5.7	4.0-7.7					
punctifera	(apical)	5.8	4.3-7.3	(body)	11.7	9.7-13.7		
heterospina	(helmet)	6.2	4.9-7.5	(needle)	7.6	6.6-8.1		
torquata		6.3	6.0-6.6				8.1	3.4-12.8
tonsurata	(apical)	6.7	5.0-8.5	(body)	12.6	7.7-17	<0.5	
akrokomos	(apical)	6.9	5.1-12.2	(body)	16.3	10.2-23.9		
mangofera		7.3	6.0-10.2				0.5	0.2-1.4
galeiformis	(apical)	7.4	6.8-8.2	(body)	12.1	8.6-14.3	1.9	0.2-3.5
striata		7.4	5.3-10.3					
annulata		7.4	6.5-8.0					
duerrschmidtiae		8.0	6.2-9.9				7.8	5.7-9.7
schwemmlei		8.0	7.0-9.2				<0.5	
clavus		8.8	8.6-9.1				4.4	3.0-6.8
pseudocoronata		9.4	7.8-12.9				9.3	7.7-11.5
transsylvanica	(apical)	10.2	6.4-13.8	(body)	19.2	13.6-25.4		
doignonii		10.2	9.3-11.3				7.1	0.7-11.3
cristata		10.4	6.1-12.5					
alpina	(apical)	10.7	8.1-13.0	(body)	18.2	12.1-21.5	<0.5	
lychenensis		12.1	10.2-14.5				1.6	1.1-2.0
acaroides v. musk.	(needle)	12.7	11.6-14.2	(helmet)	30.0	8.2-40.9		
cratis		13.1	8.9-18.1					
asmundiae		13.2	9.7-16.5					
acaroides v. acar.	(helmet)	14.4	12.3-16.6	(serrated)	32.0	31.0-32.5		
crassisquama	(serrated)	16.0	11.0-20.4	(helmet)	16.6	12.6-21.0		
matvienkoae		16.5	8.3-25.7					
elongata		19.0	13.9-24.3				<0.5	
paludosa		21.0	16.3-23.6					
portae-ferreae		32.0	17.5-49.3				<0.5	
caudata		34.7	21.5-51.85					
corymbosa		37.7	20.0-51.5				<0.5	

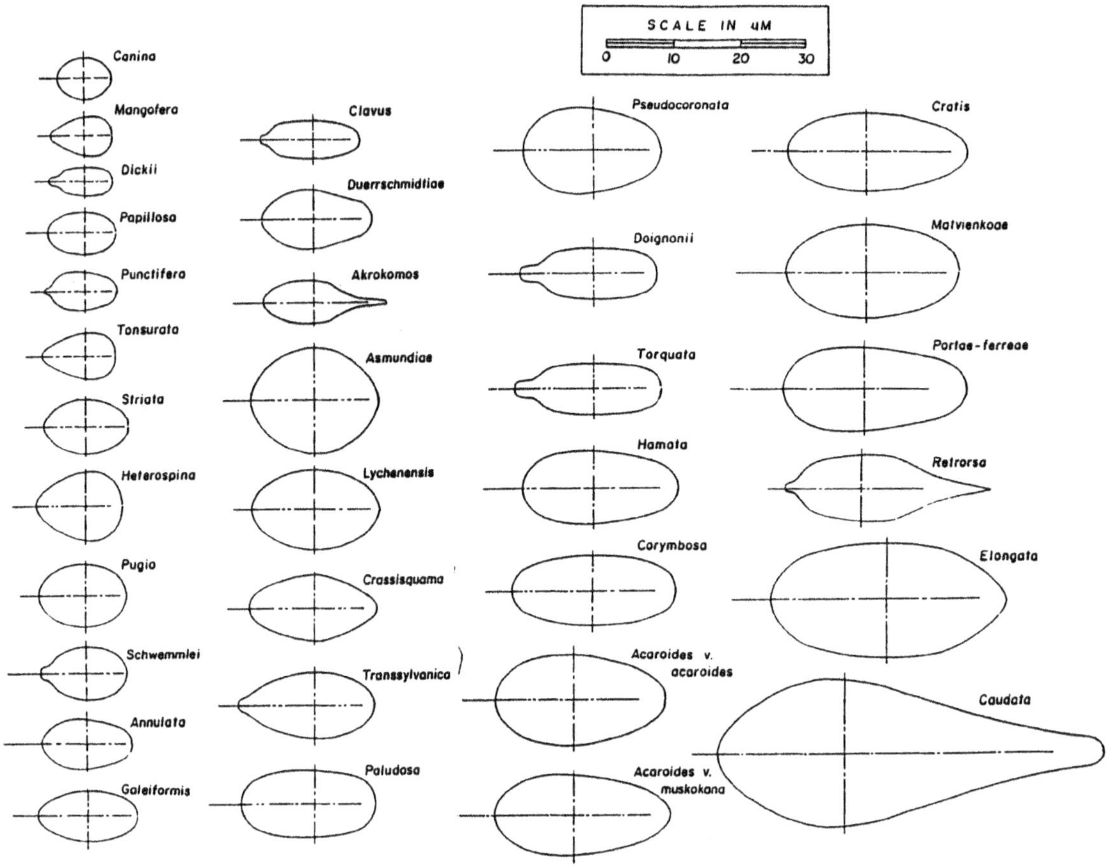

Fig. 4.13. Composite outlines of the two-dimensional shapes of species of *Mallomonas*. Each outline was reconstructed using the mean length, width and form factor for a given taxa (see Table 4.5).

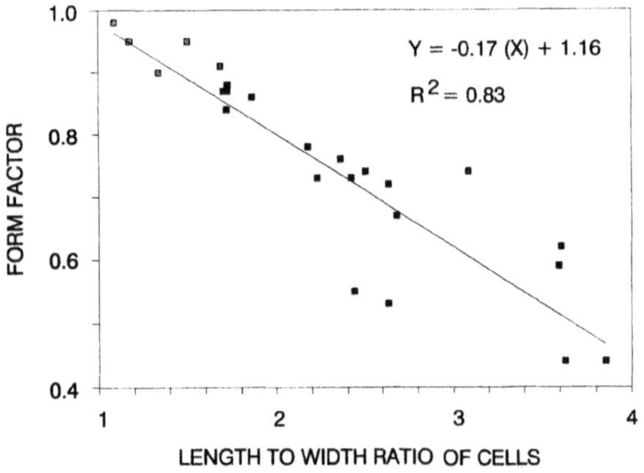

Fig. 4.14. Plot of the mean length:width ratio (independent variable) versus the form factor (dependent variable) for cells of *Mallomonas*. The regression analysis is significant at the $p < 0.05$ level.

factor) and to provide mean surface area and volume estimates for individual taxa (Table 4.5; Figs. 4.15-4.16). Such measurements are of particular interest not only for studies incorporating biovolume estimates of phytoplankton, but also because they can be used in conjunction with measurements of scale size and the degree to which scales are overlapped on the cell to reconstruct the number of cells in a population from counts of isolated scales (see below).

Usually, species of Chrysophyceae and Synurophyceae (including *Mallomonas*) with siliceous scales are underestimated in studies where phytoplankton are enumerated (Siver, 1990; Siver and Smol, in press), primarily because of poor preservation and/or disarticulation of the siliceous coat (Taylor, Wee and Wetzel, 1986). Recent studies incorporating scaled chrysophytes and synurophytes to infer current (Siver and Hamer, 1990) and

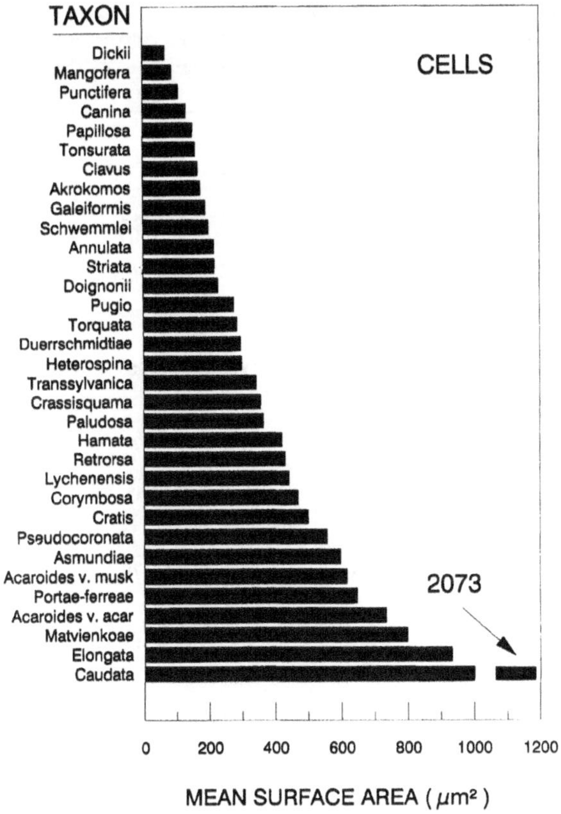

Fig. 4.15. Mean surface areas of cells from 33 taxa of *Mallomonas.* Values were obtained using rotational geometry on the two-dimensional outlines of each cell. See the text for details.

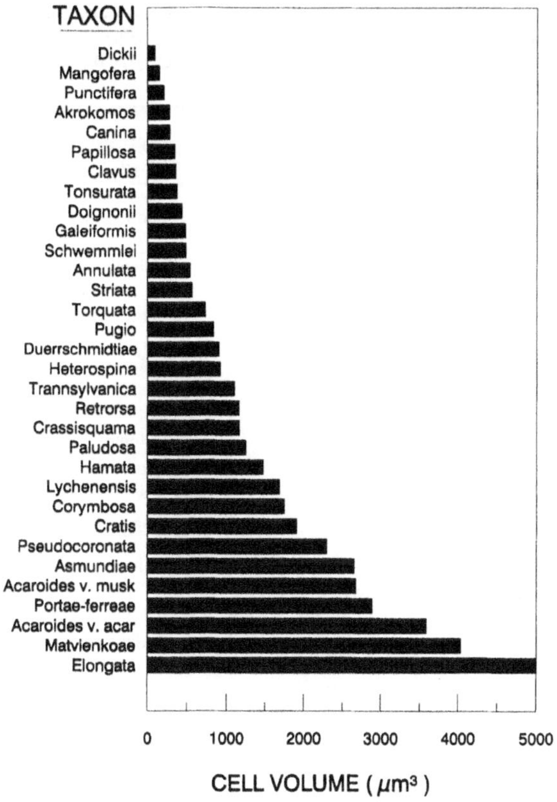

Fig. 4.16. Mean volumes of cells from 32 taxa of *Mallomonas.* Values were obtained using rotational geometry on the two-dimensional outlines of each cell. See the text for details. The value for *M. caudata*, 12,228 μm^3, was not included.

historical environments (e.g. Charles and Smol, 1988; Dixit et al., 1988) are based solely on the relative abundances of scales and not cells. Consequently, by using such inference models, the importance of species with greater or lesser numbers of scales per cell are overrepresented or underrepresented, respectively. Thus, once the mean numbers of scales per cell for each taxon are known, a theoretically more accurate estimate of the size of each population can be made and, in turn, used to develop more precise environmental inference models.

Estimating the number of scales per cell by direct enumeration of scales on whole cells was not always feasible since the cell coat often becomes disarticulated. It was also difficult to accurately estimate the percentage of the total number of scales that could actually be enumerated from micrographs of whole cells. In this study quantitative measurements of cells and scales were made with image analysis and then used to estimate the surface area and volume of cells and the number of scales per cell. The following steps were used to estimate the number of scales per cell for a given taxon:

1) The mean length, width and form factor of the cell and the mean surface area of the portion of a scale not overlapped on an intact scale coat were estimated from micrographs using image analysis.

2) From micrographs of whole cells, a 2-dimensional composite of the outline of the cell was reconstructed such that it conformed to the mean length, width and form factor estimated in step 1. The composite outlines of the cells are illustrated in Fig. 4.13.

3) The assumption was made that a cell was symmetric about the anterior to posterior axis (the longitudinal axis). Each composite outline was

Table 4.7. Results of regression analyses depicting the relationships between various morphological characteristics of cells. The mean values of the 33 taxa listed in Table 4.5 were used for the analyses. The independent variable(s), dependent variable, R^2, standard error (SE) of the estimate for Y and the regression model are listed for each analysis. Graphic illustrations of those relationships marked with an * are presented elsewhere. Variables are for features of cells unless otherwise stated.

	Independent Variable(s)	Dependent Variable	R^2	SE of est. Y	Regression Model
1.	Length	Surface area	0.83	155	Y=28.8(X)-276.4
2.	Length	Volume	0.77	1081	Y=170(X)-2386
3.	L:D*	Form factor	0.83	0.07	Y=-0.17(X)+1.16
4.	Form factor (X_1)* Length (X_2)	Volume	0.91	700	$Y=5441(X_1)+206(X_2)-7222$
5.	Surface area*	Volume	0.99	257	Y=6.06(X)-827
6.	Surface area* (<250 μm² only)	Volume	0.88	55	Y=2.7(X)-88.5
7.	Surface area* (>250 μm² only)	Volume	0.997	151	Y=6.44(X)-1170
8.	Form factor (X_1)* Length (X_2)	Surface area	0.95	82	$Y=852(X_1)+34.4(X_2)-1034$
9.	Surface area	Number scales per cell	0.35	27.9	Y=0.05(X)+52.8
10.	Scale length (X_1)	Number scales per cell	0.51	24.5	$Y=-13.4(X_1)+2.73(X_2)+69.6$
11.	Ratio of cell to scale surface area*	Number scales per cell	0.88	12	Y=2.08(X)-5.02

divided in half along the longitudinal axis. The circumference, surface area and center of area (centroid) were determined for each of the two-dimensional half-cell outlines.

4) Using rotational geometry, the circumference and surface area of the half-cell image was rotated about a circle with a radius equal to the centroid (distance from the longitudinal axis of the cell to the center of area) to obtain the surface area and volume of the cell, respectively. The following formulae apply:

$$\text{Surface area of cell} = C * [2 \pi r]$$

$$\text{Volume of cell} = A * [2 \pi r]$$

where C = circumference of one-half of the cell outline; A = area of one-half of the cell outline; r = distance along a perpendicular from the longitudinal axis of the cell to the center of area, or centroid, of the outline of one-half of the cell.

5) To determine the mean number of scales per cell, the surface area of the cell was divided by the mean area of a scale exposed on an intact cell.

A 30-fold difference in surface area was found between species of *Mallomonas* (Table 4.5). Only *M. dickii* and *M. mangofera* had mean cell surface areas less than 100 μm^2, while only one organism, *M. caudata*, had a surface area greater than 1000 μm^2 (Table 4.5; Fig. 4.15). Seventy-five percent of the taxa had mean surface areas less than 500 μm^2. A significant relationship ($p < 0.05$), with an R^2 of 0.83, was found between the mean length (independent variable) and the surface area (dependent variable) of a cell (Table 4.7). When the mean form factor was added to the model as a second independent variable, an R^2 of 0.95 was obtained (Fig. 4.17), indicating that the surface area of a cell can be accurately determined from measurements of cell length and form factor.

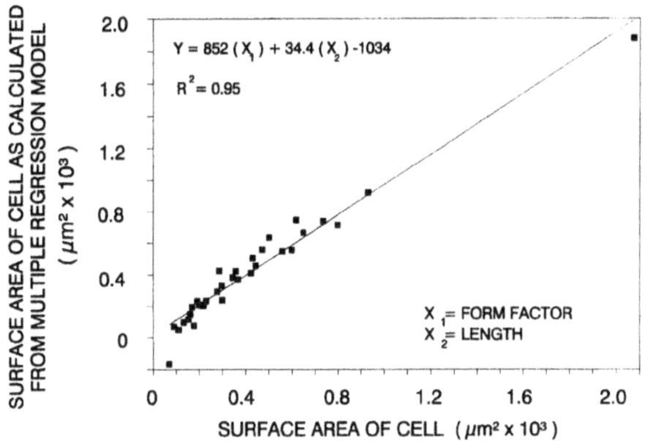

Fig. 4.17. The relationship between the measured mean surface area (independent variable) of cells and the predicted surface area obtained from a multiple regression model based on the form factor and length of the cell. A total of 33 taxa of *Mallomonas* are represented. The regression analysis is significant at the $p < 0.001$ level.

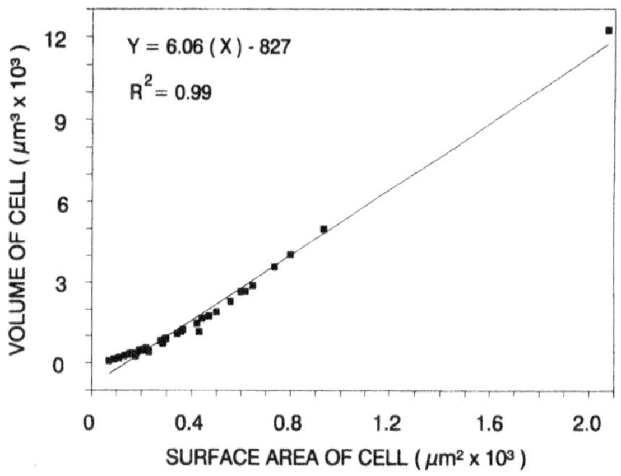

Fig. 4.18. The relationship between the mean surface area (independent variable) and volume (dependent variable) of cells from 33 species of *Mallomonas*. Although the regression model is significant at the $p < 0.001$ level, it underestimates the cell volume for smaller species. See also Figs. 4.19-4.20.

A 133-fold difference in the mean cell volumes of the 33 taxa was documented, with volume measurements ranging from a low of 92 μm³ for *M. dickii* to a high of 12,228 μm³ for *M. caudata*. Except for *M. caudata*, all taxa had a mean cell volume less than 5,000 μm³ (Table 4.5; Fig. 4.16). The four taxa with the largest mean surface areas, *M. caudata*, *M. elongata*, *M. matvienkoae* and *M. acaroides var. acaroides*, also had the largest cell volumes. Volume measurements were comparable to those cited for other similar sized phytoflagellates (e.g. Vollenweider, 1969; Wetzel, 1975). The cell volume calculated for *M. caudata* (12,228 μm³) was essentially identical to the 12,000 μm³ value reported in the literature (Wetzel, 1975).

As was determined for the surface area of a cell, the mean volume can also be accurately estimated ($R^2 = 0.91$) with a multiple regression model utilizing cell length and form factor as independent variables (Table 4.7). The mean surface area and volume of a cell are also highly correlated with an R^2 of 0.99 (Table 4.7; Fig. 4.18). Although this latter relationship is statistically very significant, it underestimates the volumes of small cells with surface areas less than 250 μm². When the taxa were divided into two groups, one with mean surface areas less than 250 μm² and one greater than 250 μm², and the analyses redone, significant models with R^2 values of 0.88 (Fig. 4.19) and 0.997 (Fig. 4.20), respectively, were found (Table 4.7).

The average percentage of a scale that is exposed (i.e. not overlapped by adjacent scales on an intact cell ranged from a low of 35% to a high of 65% (Table 4.2). Despite this apparently large range, over 70% of the taxa had a much narrower range between 45 and 60% (Fig. 4.21). The degree to which scales were overlapped on an intact cell varied between the actual type of scale. Cells with domeless body scales had a much wider range in the degree of overlap than species that possess

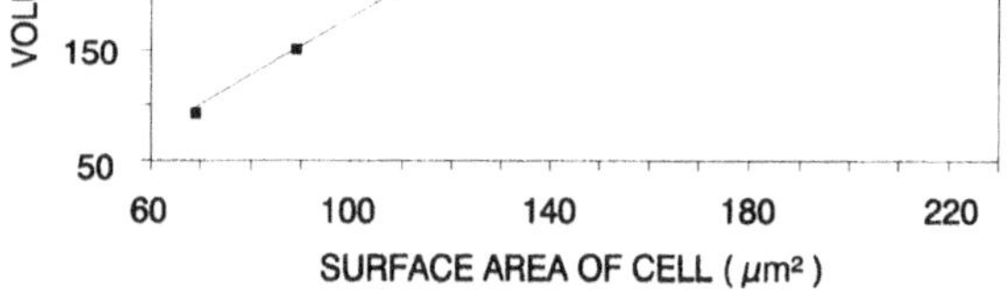

Fig. 4.19. The relationship between the mean surface area (independent variable) and volume (dependent variable) of cells from 13 species of *Mallomonas*. Only smaller species, each with a mean cell surface area less than 250 μm², were used in this analysis. See also Figs. 4.18 and 4.20.

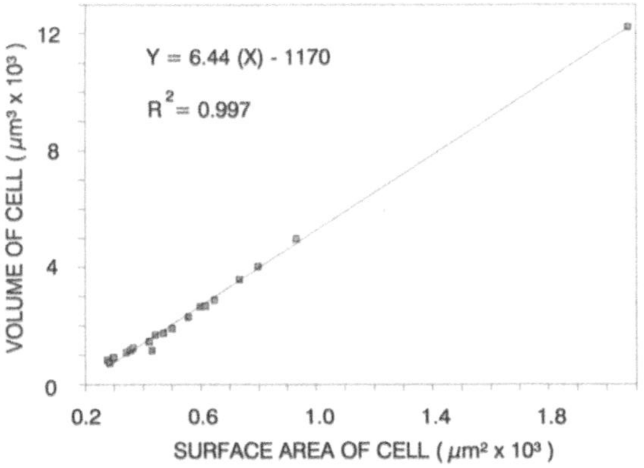

Fig. 4.20. The relationship between the mean surface area (independent variable) and volume (dependent variable) of cells from 20 larger species of *Mallomonas*. Each taxon has a mean cell surface area greater than 250 μm^2. The regression model is significant at the $p < 0.001$ level. See also Figs. 4.18 - 4.19.

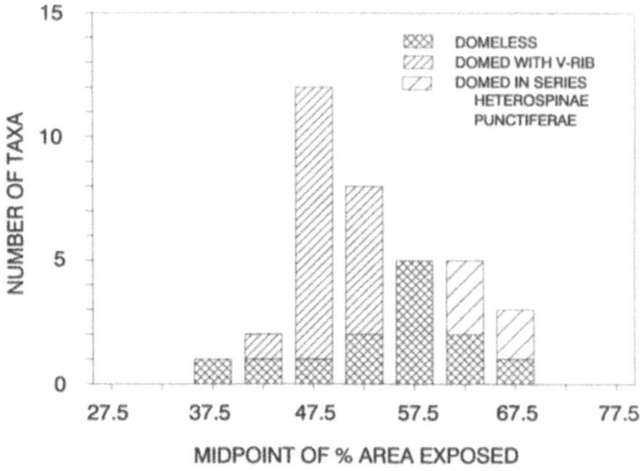

Fig. 4.21. Frequency distribution showing the mean percentage of surface area of a scale exposed on an intact cell for 36 species of *Mallomonas*. Species are divided into groups according to whether they have domeless or domed scales, or are within the Series Heterospinae or Punctiferae.

scales with domes and a well defined V-rib (Fig. 4.21). All species examined within the Sections Heterospinae and Punctiferae had relatively high scale exposure values of between 60 and 65%.

The mean number of scales per cell varied between a low of 27 for *M. punctifera* to a high of 160 for *M. caudata* (Table 4.5; Fig. 4.22). Thus, interpretations based on relative percentages of scales may overestimate or underestimate the actual number of cells by as much as a factor of six. Over 80% of the taxa had less than 100 scales per cell (Table 4.5). A significant but weak (R^2 = 0.35) relationship was found between the surface area and the mean number of scales per cell (Table 4.7). A slightly stronger relationship (R^2 = 0.51) was found when mean cell and scale lengths were used to estimate the number of scales per cell. The most significant model ($p < 0.001$; R^2 = 0.88) for estimating the number of scales per cell was obtained using the ratio between the surface area of the cell to that of the scale (Fig. 4.23).

Using the above regression models, a reasonably good estimate of the number of scales per cell can be made in a two-step process if the mean length and form factor of the cell and the surface area of the scales are known. First, the mean cell length and form factor are used to estimate the mean surface area of the cell (Fig. 4.17). Second, the mean cell surface area is used with the mean scale surface area to estimate the number of scales per

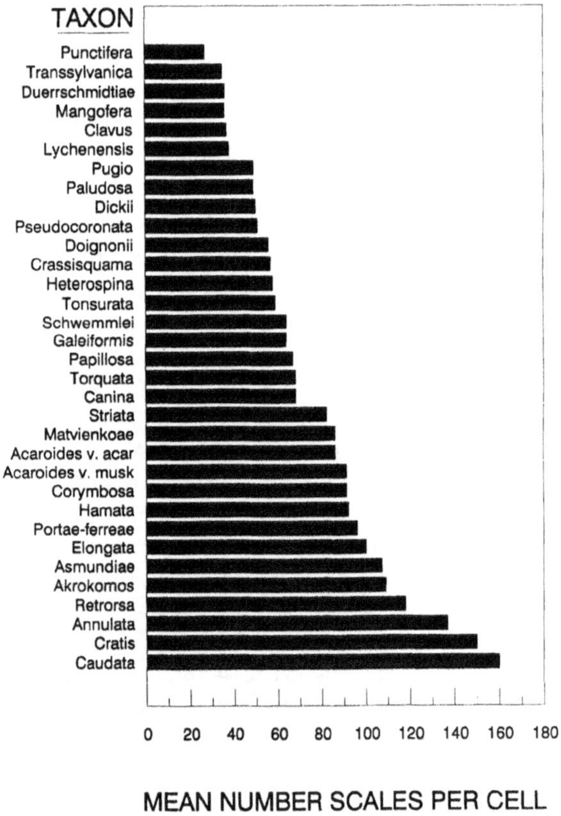

Fig. 4.22. Mean number of scales per cell for 33 taxa of *Mallomonas*. There is approximately a six-fold difference between the species with the least number of scales (*M. punctifera*) and the one with the most (*M. caudata*). See the text for details.

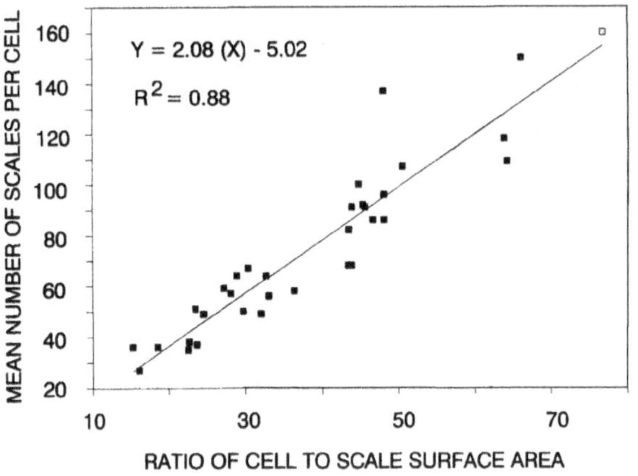

Fig. 4.23. The relationship between the ratio of the surface area of the cell:scale (independent variable) and the mean number of scales per cell (dependent variable) for 33 species of *Mallomonas*. The regression model, significant at the $p < 0.001$ level, could be used to estimate the number of scales per cell for rare species or ones that become easily disarticulated.

cell (Fig. 4.23). These relationships will be valuable for estimating the number of scales per cell for rare taxa as well as those that are easily disarticulated.

A summary diagram comparing the mean L:D ratios of scales and cells was prepared for 33 taxa (Fig. 4.24). Taxa in the lower left region of the diagram are those with spherical cells and scales with a more or less circular outline, while taxa in the upper right quadrant have elongated cells with elongated scales. Likewise, taxa in the upper left region have elongated cells with more or less circular scales while those in the lower right have spherical cells and elongated scales.

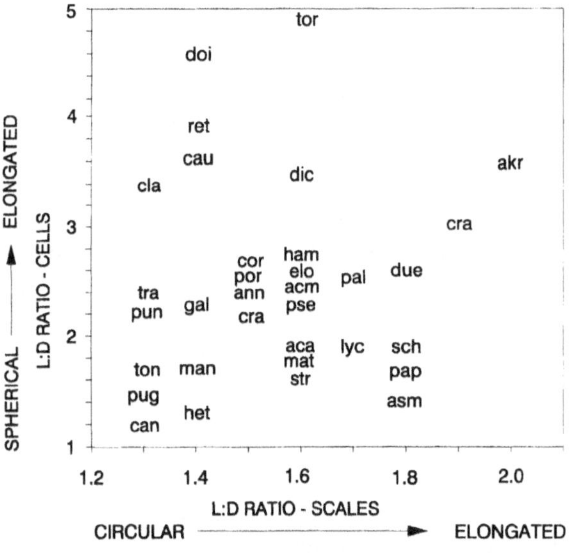

Fig. 4.24. The relationship between the shapes of both cells and scales for 33 species of *Mallomonas*. The ratios of the length to width of body scales are plotted against the ratios of the length to width of cells. Groups of taxa with similar shaped scales and cells can be identified. See the text for details.

KEY

can = canina	pal = paludosa
man = mangofera	cra = crassisquama
dic = dickii	due = duerrschmidtiae
pap = papillosa	ham = hamata
pun = punctifera	doi = doignonii
ton = tonsurata	cor = corymbosa
str = striata	aca = acaroides v. acaroides
het = heterospina	acm = acaroides v. muskokana
pug = pugio	cra = cratis
sch = schwemmlei	mat = matvienkoae
gal = galeiformis	por = portae-ferreae
ann = annulata	pse = pseudocoronata
cla = clavus	tor = torquata
akr = akrokomos	ret = retrorsa
asm = asmundiae	elo = elongata
lyc = lychenensis	cau = caudata
tra = transsylvanica	

Key to Fig. 4.24.

CHAPTER V

The Ecology of *Mallomonas*

Introduction

The primary goal of this chapter is to describe the ecological tolerances of all common species and sub-specific taxa of *Mallomonas*. Emphasis has been placed on the distributions of each taxon along pH, specific conductance, temperature, trophic and seasonal gradients. To accomplish this goal three sets of data totalling results from over seven hundred collections were used. Four hundred of the collections were taken from localities in southern New England from the State of Connecticut. Another one hundred of the collections were from waterbodies in the Adirondack Mountain region of New York State. The remainder of the collections were taken from the literature and represent waterbodies from different parts of the world (see Siver, 1989a). For the purposes of discussion the three data sets will each be considered as separate floras and referred to as the Connecticut, the Adirondack and the literature floras.

All taxa of *Mallomonas* from each of the Connecticut and Adirondack collections were identified with electron microscopy. Only literature studies where documentation of species was made with electron microscopy and where the presence-absence of the species observed were reported for each collection were used for quantitative purposes. The pH, specific conductance, temperature and total phosphorus levels were measured for each of the collections from Connecticut and the Adirondacks and used to construct frequency graphs for each species along the different environmental gradients and to calculate weighted mean and standard deviation values. Sets of distributional graphs are presented for each taxon found in a minimum of ten of the collections. The weighted means were based on the frequencies of each taxon along a specific gradient using the following equation:

$$\text{weighted mean (variable)} = \Sigma\, P_i\, (X_i) \,/\, \Sigma\, P_i$$

where P_i is the frequency of occurrence of the taxon in the ith interval for the given variable, and X_i is the midpoint of the ith interval. The intervals used for each variable were as depicted on the frequency graphs. Weighted mean pH scores were also calculated from the literature records. Except in a few instances for temperature, weighted means for the other variables could not be calculated from the literature records due to insufficient data. In addition to the literature references used in a quantitative manner, many others were used, in part, to describe the niche requirements of individual taxa. The literature records represent a review of both contemporary and paleolimnological studies and are presented in tabular form.

In assembling the collections comprising the Connecticut flora an attempt was made to equalize the number of samples for each month of the year. Except for the month of August (n=25), between 30 and 40 collections were made during each month. The Adirondack collections were evenly distributed between the spring, summer and autumn seasons; the collection of samples in the winter under the ice was not feasible. The percentages of the total number of collections made along gradients for each of the environmental variables are compared in Figure 5.1 for both the Connecticut and the Adirondack floras. Due to the lack of sampling during the winter in the Adirondacks no samples were taken below 6°C. In contrast, 32% of the collections from Connecticut represented water temperatures less than 6°C.

The distribution of collections made along pH, specific conductance and total phosphorus gra-

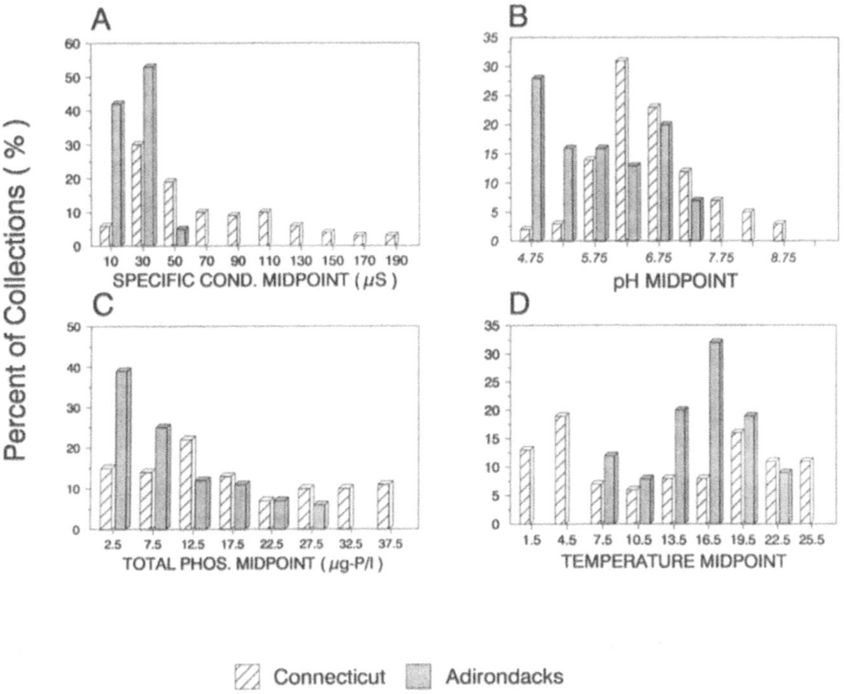

Fig. 5.1. The relative percent of the total number of collections made at different intervals along (A) temperature, (B) pH, (C) specific conductance and (D) total phosphorus gradients for Connecticut and the Adirondacks.

dients illustrates some of the basic differences in the water chemistry between both regions. The majority of collections from the Connecticut localities were made between pH 5.5 and 7.5. Only 5% and 15% of the collections were made below and above this pH range, respectively (Fig. 5.1). In contrast, 44% of the collections from the Adirondacks represented collections with a pH below 5.5 and none were made above pH 7.5, illustrating the more acidic nature of waterbodies in the Adirondacks as compared to Connecticut. The literature collections represent more alkaline conditions than either the Connecticut or Adirondack data sets. Although literature records represent pH values ranging from 4.0 to 9.5, over 27% of the collections had a pH above 7.5.

In general, waterbodies from Connecticut are higher in total dissolved salts than those from the Adirondacks (Fig. 5.1). A total of 78% of the collections from Connecticut had a specific conductance between 20 μS and 120 μS. Only 6% and 16% of the collections represented specific conductance values below and above this range, respectively. In contrast, all of the collections from the Adirondack localities had specific conductance below 60 μS.

A larger percentage of the collections from the Adirondacks were taken from oligotrophic waterbodies. There is a greater degree of development of the watersheds surrounding the lakes in Connecticut resulting, in part, in waterbodies with a higher trophic status. While 76% of the collections from the Adirondacks had total phosphorus concentrations below 15 μg-P/l, 50% of those from Connecticut represented concentrations above 15 μg-P/l (Fig. 5.1). In summary, waterbodies from Connecticut are higher in pH, specific conductance and total phosphorus levels.

Several points regarding the ecological summaries for each taxon are worth noting. First, information from both the Connecticut and Adirondack data sets were combined in order to prepare the frequency graphs and calculate weighted scores. This was decided upon after careful analysis of the data indicated that species had similar distributions along environmental gradients in both regions. For example, an acidic, cold water taxon in the Adirondacks was found under similar conditions in the Connecticut localities. Thus, by combining the data a better representation of each environmental gradient was achieved. Second, differences in the numbers of collections made within each interval for each variable are accounted for in the calculation of weighted means and relative frequencies. Third, the terminology introduced by Hustedt (1939) to describe the distributions of taxa along a pH gradient are referred to throughout this chapter. Hustedt's groupings include:

- acidobiontic = organism occurs only in acidic environments and has an optimum distribution at pH values below 5.5;
- acidophilic = organism is mostly distributed below pH 7;
- indifferent = organism is distributed around a pH of 7;
- alkaliphilic = organism is mostly distributed above pH 7;
- alkalibiontic = organism occurs only in alkaline environments above pH 7.

The phrase 'pH indifferent' is utilized in this chapter to describe organisms that are distributed about pH 7 as defined by Hustedt, but that are also less common at pH extremes. The term circumneutral is used in the same context.

The Genus *Mallomonas*

The number of taxa of *Mallomonas* found per collection along seasonal, pH, temperature, specific

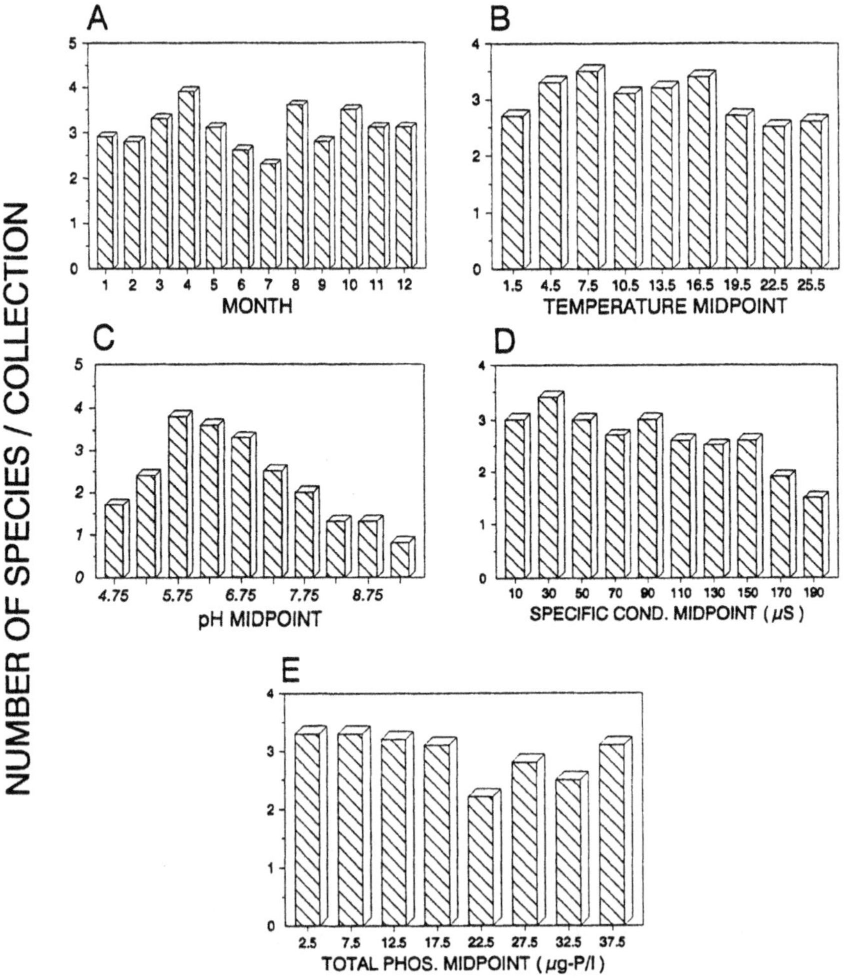

Fig. 5.2. The mean number of taxa of *Mallomonas* found per collection for each month (A) and different ranges of temperature (B), pH (C), specific conductance (D) and total phosphorus (E).

conductance and total phosphorus gradients are shown in Figure 5.2. The water temperature, total phosphorus content and month of the year had little effect on the number of *Mallomonas* taxa observed per collection. No significant differences in the number of taxa per collection were found between the nine categories of water temperature or eight categories of total phosphorus tested. In addition, there were no significant differences between the number of organisms found in collections between different months of the year. On the average, three taxa of *Mallomonas* are found in a given locality, regardless of the temperature, total phosphorus content or month sampled.

The pH and specific conductance of the locality had an effect on the number of taxa of *Mallomonas* found in a given collection. A mean close to four taxa per collection was observed between a pH of 5.5 and 6.5. As the pH raised or lowered from the 5.5 to 6.5 interval, the number of organisms found in a collection significantly decreased. Approximately two and one taxa were observed in collections below pH 5.5 or above 8, respectively. The water conductance had less of an effect on the number of organisms found per collection. A mean of three taxa per collection was recorded at specific conductance values below 180 μS. A significant decrease in the number of organisms found did not occur until conductance values rose above 180 μS.

In summary, the environmental variables pH and specific conductance had a greater influence in determining the number of taxa of *Mallomonas* found in a waterbody at a given point in time. In contrast, at the genus level, the month, water temperature and total phosphorus content played a lesser role. However, at the species or subspecific level, these generalizations do not normally hold. Individual species or subspecific taxa are commonly restricted along temperature, seasonal and total phosphorus gradients as well as those for pH and specific conductance. Thus, from the standpoint of utilizing the genus as an indicator taxon, or in the reconstruction of historical environments, it must be dealt with at the specific and sub-specific levels.

Autecology of Selected Taxa

Mallomonas acaroides var. *acaroides*

Mallomonas acaroides has recently been demonstrated to be an important component of the paleosediment record from lakes in the Adirondacks that have been adversely affected by acid deposition (Smol et al., 1984a; Charles and Smol, 1988; Christie and Smol, 1986; Siver, 1988c). The acidic nature of *Mallomonas acaroides* in Adirondack localities as reported by Smol and his co-workers (1984a) was in direct contrast to the literature records where this species was most often reported as being restricted to alkaline waters (Siver, 1989a). Siver (1989a; 1989b) realized that the conflicting reports were the result of comparing records from two morphologically distinct taxa, var. *acaroides* and var. *muskokana*. The two varieties were clearly separated along a pH gradient (Siver, 1989a; 1989b).

Although *M. acaroides* var. *acaroides* was quite common in a survey of the literature, it was relatively rare in Connecticut localities and lacking from the Adirondacks (Siver, 1988c; 1989a). *M. acaroides* var. *acaroides* was most common during late summer and autumn, although it was also present in spring and winter (Fig. 5.3A). It was primarily found under cooler conditions, from 2°C to 19°C, and had a weighted mean temperature of 13.3°C (Fig. 5.3B).

Based on a survey of over 200 collections from the literature, *M. acaroides* var. *acaroides* was found between 6°C and 19°C and to have a weighted mean temperature of 12.8°C (Table 5.1). Asmund (1959) and Kristiansen (1985; 1988) reported *M. acaroides* var. *acaroides* during the spring and fall at temperatures of 8°C to 19°C and 0.8°C to 14.2°C, respectively. In addition, this taxon has been reported at temperatures of 5.5°C to 17.6°C (Skogstad, 1982), 7°C to 19°C (Roijackers and Kessels, 1986; Wee and Gabel, 1989), 8°C to 10°C (Cronberg and Kristiansen, 1980) and 6.5°C to 11°C (Dürrschmidt, 1984), supporting the view that *M. acaroides* var. *acaroides* predominates at cool temperatures. Cronberg (1980) reported cyst formation in October.

It is clear from a review of the literature that *M. acaroides* var. *acaroides* is best described as an alkalobiontic organism (Table 5.1). In earlier com-

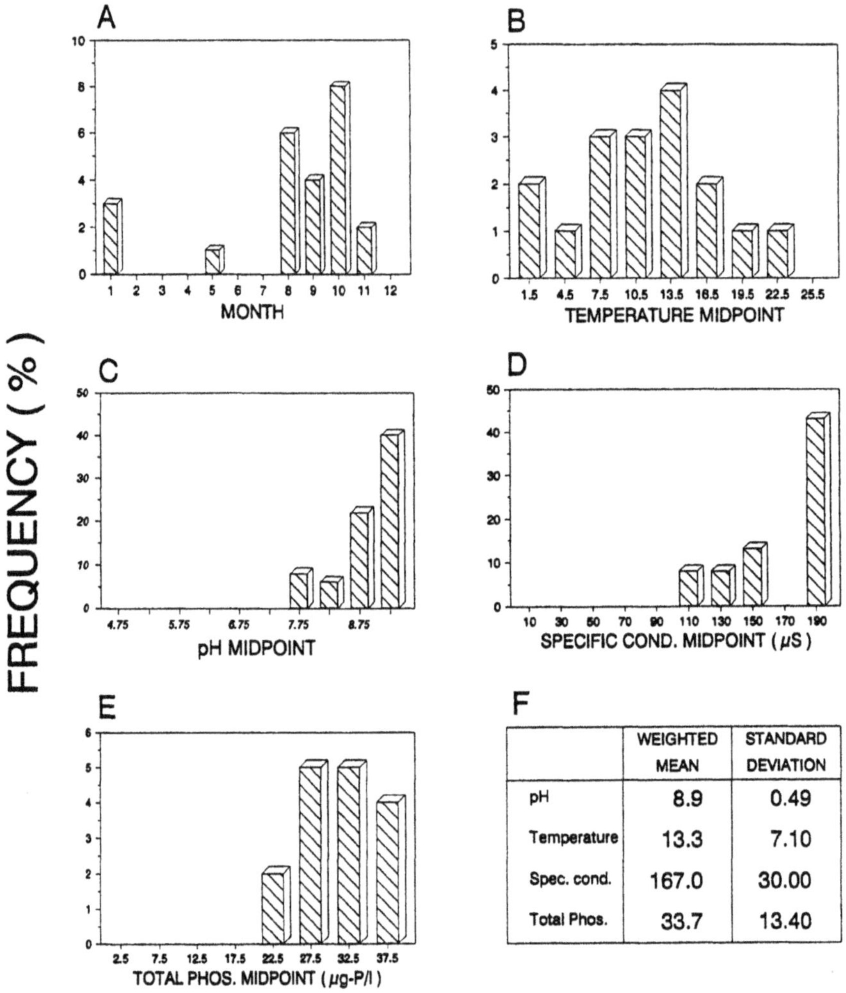

	WEIGHTED MEAN	STANDARD DEVIATION
pH	8.9	0.49
Temperature	13.3	7.10
Spec. cond.	167.0	30.00
Total Phos.	33.7	13.40

Fig. 5.3. The frequencies of occurrence for *Mallomonas acaroides* var. *acaroides* along seasonal (A), temperature (B), pH (C), specific conductance (D) and total phosphorus (E) gradients. Weighted mean and standard deviation values are also given (F).

munications Siver (1989a; 1989b) reported weighted mean pH values for var. *acaroides* of 8.1 and 8.4 from Connecticut localities and the literature survey, respectively. Based on a larger data set, *M. acaroides* var. *acaroides* was found between pH 7.9 and 9.5, had a maximum occurrence above pH 8.5 and a weighted mean pH of 8.9 (Fig. 5.3C), making this organism one of the most alkaline taxa of *Mallomonas*. The alkaline nature of *M. acaroides* var. *acaroides* is strongly supported by a review of the literature (Siver, 1989a) (Table 5.1). This taxon has been reported to have a maximum occurrence between pH 7 and 8 (Roijackers and Kessels, 1986), 8.5 and 9.0 (Siver, 1989b) and above 8 (Dürrschmidt, 1984; Kristiansen, 1985; 1988; Wee and Gabel, 1989). Asmund (1959), Skogstad (1982), Kristiansen (1986) and Nicholls (1987b) also found this organism to be alkaline in nature. Although Eloranta (1989) found *M. acaroides* var. *acaroides* between pH 6.1 and 7.2, it was alkaliphilous relative to the other taxa observed in his study lakes. Péterfi and Momeu (1976a) reported a population of *Mallomonas acaroides* var. *echinospora* near a pH of 6.0.

In Connecticut localities *M. acaroides* var. *acaroides* was restricted to waters of high specific conductance and had one of the highest weighted mean values of 167 µS (Fig. 5.3D). The waterbodies harboring this variety are best described as mesotrophic to eutrophic and the taxon also had a

Table 5.1. Literature records indicating the ecological preferences of *Mallomonas acaroides* v. *acaroides*. Records for this taxon were separated from those of v. *muskokana* based on electron micrographs contained within each reference.

Reference	Seasonality	Trophic Status	Temperature (°C)	pH and Related Factors	Other
1. Asmund, 1959	Spring and autumn	Mesotrophic to eutrophic	8 to 19	Alkaline	Bristle morphology may be related to water temperature. Denmark, Reported as v. *striatula*
2. Cronberg, 1980	September to October	Eutrophic	---	---	Formed cyrts in October. Lake Trummen, Sweden.
3. Cronberg and Kristiansen, 1980	---	Eutrophic	---	---	Sweden
4. Ito, 1988	Summer	---	Warm	---	Lake Biwa, Japan
5. Kristiansen, 1985	---	Eutrophic	Higher temperatures, but below 19.6	Alkaline, pH 8.2 - 9.1	Denmark, Lake Tystrup Sø
6. Kristiansen, 1988	Preference for winter/ spring and fall	Eutrophic	0 - 14	Alkaline	Denmark, Lake Tystrup Sø
7. Roijackers and Kessels, 1986	---	---	Wide temperature range	Maximum frequency between pH 7 and 8	Netherlands
8. Siver, 1989a	---	---	Rather wide temperature range	Alkaline, most common above pH 8	Rare in Connecticut, absent from Adirondack waterbodies.
9. Siver, 1989b	---	---	---	Alkalibiontic, WMpH 8.1	Connecticut waterbodies.
10. Wee and Gabel, 1989	---	---	---	Alkaline, pH 8.0 - 8.7	Iowa

WMpH = weighted mean pH.

high weighted mean total phosphorus value of 33.8 μg-P/l. The lack of *M. acaroides* var. *acaroides* in oligotrophic waters has been noted by Asmund (1959), Cronberg and Kristiansen (1980), Kristiansen (1985; 1988) and Dürrschmidt (1984).

Mallomonas acaroides var. *muskokana*

M. acaroides var. *muskokana*, a recently described taxon from Canada (Nicholls, 1987b), was much more common in Adirondack than Connecticut waterbodies, present in 47% and 12% of the samples, respectively. Siver (1988c) recorded it from 65% of all waterbodies surveyed in the Adirondacks. In both localities, this organism was primarily found during the late spring to early autumn period (Fig. 5.4A). It was reported from 31% of the collections made from July through September, but in only 1% of those from December through March. Because of its predominance during the warm months, var. *muskokana* had a relatively high weighted mean temperature of 17.5°C. It was present in 1%, 17% and 25% of the collections less than 6°C, 6°C to 15°C and above 15°C, respectively (Fig. 5.4B).

Like *M. duerrschmidtiae*, *M. acaroides* var. *muskokana* is restricted to acidic waterbodies low in specific conductance (Figs. 5.4C-D). Its maximum occurrence was below pH 5 where it was found in over 60% of the collections. It was found in

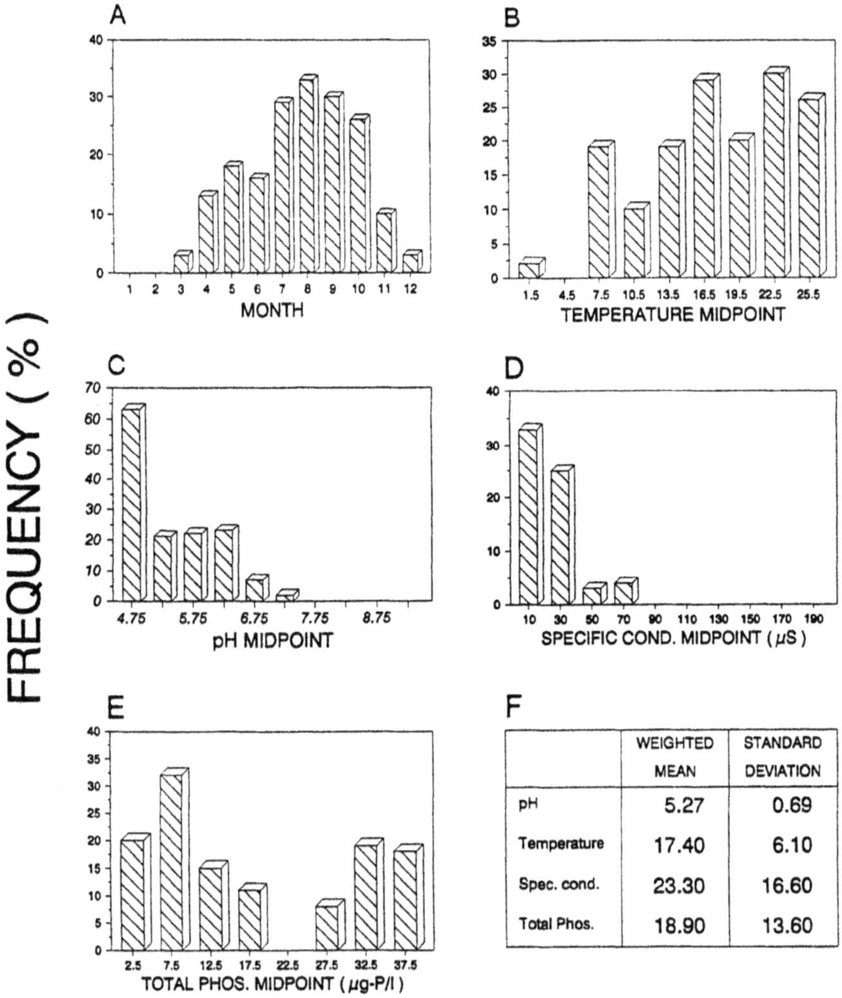

	WEIGHTED MEAN	STANDARD DEVIATION
pH	5.27	0.69
Temperature	17.40	6.10
Spec. cond.	23.30	16.60
Total Phos.	18.90	13.60

Fig. 5.4. The frequencies of occurrence for *Mallomonas acaroides* var. *muskokana* along seasonal (A), temperature (B), pH (C), specific conductance (D) and total phosphorus (E) gradients. Weighted mean and standard deviation values are also given (F).

Table 5.2. Literature records indicating the ecological preferences of *Mallomonas acaroides* v. *muskokana*.

Reference	Seasonality	Trophic Status	Temperature (°C)	pH and Related Factors	Other
1. *Charles and Smol, 1988	---	---	---	Acidic organism.	Adirondacks
2. *Christie and Smol, 1986	---	---	---	Increase correlated with decline in pH.	Adirondacks, Upper Wallface Lake.
3. *Dixit et al., 1988a	---	---	---	AWMpH 5.8	May be effected by high salt and/or metal levels. Sudbury region.
4. *Dixit et al., 1988b	---	---	---	Acidic, but decrease in abundance below pH 5.0.	---
5. Nicholls, 1987	---	Dilute, oligotrophic	---	Acidic softwater.	Lakes of Precambrian shield. Ontario.
6. Siver, 1988c	---	Oligotrophic	Warm water	Acidic; range pH 3.98 - 7.1	Common in bogs. Adirondacks
7. Siver, 1989b	Late spring through early fall	Oligotrophic	Warm water	Acidobiont, WMpH 5.3	Also common in humic stained bogs. Adirondacks and Connecticut.
8. *Smol, 1986	---	---	---	Maximum frequency in water with pH < 5.0.	Common in bogs.
9. *Smol et al., 1984a	---	---	---	Acidic organism.	Adirondacks

AWMpH = abundance weighted mean pH.
*Indicates a paleolimnological study.

only 23% of the samples between pH 5 and 6.5 and was not found above pH 7.2. As a result, var. *muskokana* is an acidobiontic taxon with a weighted mean pH of 5.37, similar to that reported by Siver (1989a,b). A total of 96% of all collections containing this organism had a specific conductance below 40 μS (Fig. 5.4D). Its preference for acidic softwater lakes most likely explains its widespread and common occurrence in the Adirondacks (Siver, 1989b).

Nicholls (1987b) reported var. *muskokana* as a common species in dilute softwater lakes of the Canadian shield and Charles and Smol (1988) found it to have a weighted mean pH of 5.25. Dixit et al. (1988a) also found var. *muskokana* to be an acidic taxon in lakes of the Sudbury region with a maximum occurrence between pH 5.5 and 6.5 and a weighted mean value of 5.8. They attributed its reduced occurrence in samples below pH 5.5 to the overall high specific conductance and metal concentrations of the lakes in this region. Increased percentages of *M. acaroides* var. *muskokana* in core samples have often been correlated with a decline in lakewater pH (Christie and Smol, 1986; Dixit et al., 1988b. See Table 5.2).

Although *M. acaroides* var. *muskokana* was found over a relatively large gradient of total phosphorus, it is an oligotrophic taxon with a maximum frequency of occurrence below 10 μg-P/l (Fig. 5.4E). In addition, Siver (1988c; 1989b)

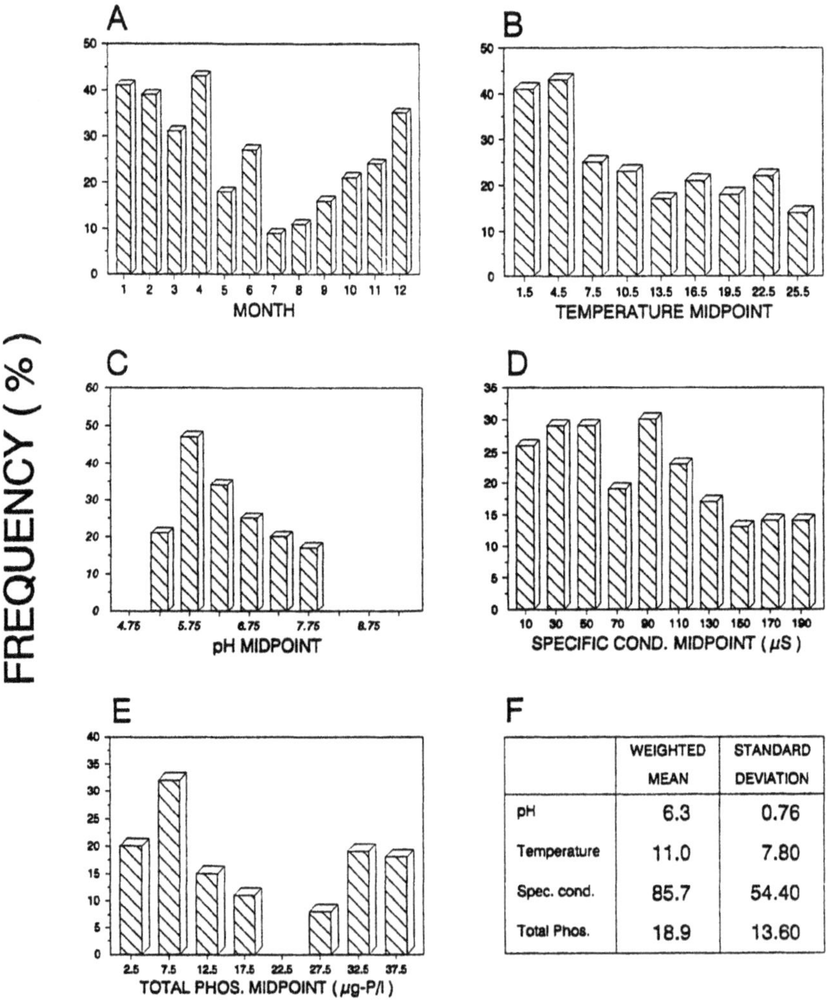

	WEIGHTED MEAN	STANDARD DEVIATION
pH	6.3	0.76
Temperature	11.0	7.80
Spec. cond.	85.7	54.40
Total Phos.	18.9	13.60

Fig. 5.5. The frequencies of occurrence for *Mallomonas akrokomos* along seasonal (A), temperature (B), pH (C), specific conductance (D) and total phosphorus (E) gradients. Weighted mean and standard deviation values are also given (F).

Table 5.3. Literature records indicating the ecological preferences of *Mallomonas akrokomos*.

Reference	Seasonality	Trophic Status	Temperature (°C)	pH and Related Factors	Other
1. Asmund, 1956	Greatest abundance in Jan.-Feb., under ice. Also in spring.	Eutrophic tendency.	Greatest abundance up to 3, but common to 13.	---	Denmark
2. Asmund and Hilliard, 1961	November-December	Oligotrophic	0.3 - 8.6	6.4 - 7.2	Alaska
3. *Charles and Smol, 1988	---	---	---	AWMpH 6.64	Adirondacks, New York
4. *Christie et al., 1988	---	Oligotrophic	---	---	Also found in brown water. Eastern Finland.
5. Cronberg, 1980	Late Autumn.	Eutrophic	Cold	---	Formed cysts just prior to ice cover.
6. Cronberg and Kristiansen, 1980	---	---	Cold, especially during ice-out.	---	Sweden
7. Dürrschmidt, 1982b	---	---	Considered to be a cold water form.	---	Relatively infrequent in Chile, maybe due to high temperatures of collections.
8. Dürrschmidt, 1984	Mainly in winter.	Oligotrophic to eutrophic.	Cold	---	West Germany
9. Eloranta, 1986	---	Ultra-oligotrophic Oligotrophic.	---	---	Finland
10. Eloranta, 1989	Summer	Variety	---	4.7 - 7.1, mean of 5.8	Finland
11. Hällfors and Hällfors, 1989	Summer	---	---	---	Finland

Table 5.3. continued

Reference	Seasonality	Trophic Status	Temperature (°C)	pH and Related Factors	Other
12. Harris, 1958	All months except September. Very rare in July-August. Most common in winter to early spring.	Variety	Thrives in cold.	Variety	England
13. Ito, 1988	Winter and spring.	---	Cold	---	Lake Biwa, Japan.
14. Kristiansen, 1985; 1988	Maximum abundance in winter and spring.	Eutrophic	0.7 - 10	8.2 - 9.1	Lake Tystrup Sø, Denmark.
15. Péterfi and Momeu, 1976	Early spring.	Mesotrophic peat bog.	---	5.5	Romania
16. Roijackers and Kessels, 1986	---	---	Binodal peaks at 4-10 and 18-20.	Maximum development at 6-7.	Netherlands
17. Siver, 1989a	---	---	---	pH indifferent, although acidophilous tendencies WMpH=6.5 for Connecticut lakes and 6.7 based on literature records.	Common species, found in over 25% of all collections. Connecticut.
18. Siver and Chock, 1986	Winter under the ice.	Oligotrophic	Thrives in cold.	Acidic	Pocono Mountains, Pennsylvania.
19. Takahashi, 1978	Maximal development during winter.	Variety	Eurythermal with maximum near 4.	Acidophilous	Evidence for large cold water form and smaller warm water form. Japan.
20. Wujek, 1984	---	---	Eurythermal	---	Florida

AWMpH = abundance weighted mean pH.
*Indicates a paleolimnological study.

and Smol (1986) reported this taxon to often thrive in acid bog localities (Table 5.2). The majority of samples with a total phosphorus concentration above 20 μg-P/l that contained this organism were from bog localities. Using principal component analysis, Siver and Hamer (1989) found that 74% and 68% of the variance in the temperature and pH data were explained in a one factor model.

Mallomonas akrokomos

Mallomonas akrokomos, a frequently observed species with a worldwide distribution (Takahashi, 1978), has been reported as one of the most common taxa of the genus in England (Harris, 1958), Japan (Takahashi, 1978), Sweden (Cronberg and Kristiansen, 1980), Germany (Dürrschmidt, 1984), Finland (Eloranta, 1989) and from northern boreal regions of North America (Wawrzyniak and Andersen, 1985). This organism has been observed from Alaska (Asmund, 1956; Asmund and Hilliard, 1961) to the tropics (Takahashi, 1978). *Mallomonas akrokomos* was found in 29% and 28% of the collections from Connecticut and the literature survey, respectively.

Although *M. akrokomos* was found during each month of the year, it had a clear preference for colder months and a maximum occurrence under the ice and immediately following ice-out (Figs. 5.5A-B). Populations were found in 37% of the collections from December through April, but in only 9% of those from July and August. Coincident with its preference for colder seasons, *M. akrokomos* was found in 42% of the collections with a temperature less than 6°C, only 20% of the collections above 6°C, and had a weighted mean temperature of 11°C (Fig. 5.5F). The seasonal distribution of this species reported from localities in Connecticut and the Adirondacks is similar to that documented by literature records (Table 5.3). Harris (1958) and Takahashi (1978) both observed *M. akrokomos* to be eurythermal, but with a definite winter-spring maximum and mid-summer minimum. Asmund (1956), Asmund and Hilliard (1961), Siver and Chock (1986), Cronberg and Kristiansen (1980), Dürrschmidt (1984), Ito (1988) and Kristiansen (1985; 1988) also reported a winter-spring maximum for this species and often in waters below 6°C. Roijackers and Kessels (1986) believed that *M. akrokomos* may have two maxima, one between 4°C and 10°C and one between 18°C and 20°C.

Despite its preference for cold water, populations of *M. akrokomos* were found above 25°C. Halläfors and Halläfors (1988) and Eloranta (1989) found this organism to be common in Finnish lakes during the summer months. Takahashi (1978) described winter and summer forms of *M. akrokomos* differing primarily in the shape and size of cells. Winter forms had long, slender and spindle shaped cells, whereas the summer forms had smaller and more rounded cells. There is some evidence that in deeper lakes *M. akrokomos* may migrate vertically in response to changes in water temperature (Bourrelly, 1957).

In this study *M. akrokomos* was found over a pH range from 5.2 to 8.0, however, it had a clear preference for slightly acidic conditions (Fig. 5.5C). The occurrence of this species was maximal between pH 5.5 and 6.0 where it was found in 47% of the collections, and decreased steadily to 17% of the collections between pH 7.5 and 8.0. Its frequency decreased rapidly below pH 5.5 and it was noticeably absent below pH 5.0. Based on its distribution and weighted mean pH of 6.3, *M. akrokomos* is best described as an acidophilous or pH indifferent taxon in Connecticut and Adirondack localities.

Other studies have shown *M. akrokomos* to prefer acidic waters. Takahashi (1978) reported a distribution along a pH gradient similar to that described from Connecticut and the Adirondacks, and listed *M. akrokomos* as an acidophilic species. Roijackers and Kessels (1986) reported the maximum development of this species to be between pH 6 and 7. Eloranta (1989) observed it to have a mean pH of 5.8. In a more encompassing survey of the literature, Siver (1989a) reported *M. akrokomos* to occur over a greater pH range, be distributed more evenly about pH 7 and to have a weighted mean of 6.7. The highest pH reported for this organism was made by Kristiansen (1988) who reported it from a eutrophic lake with a pH of 8.2 to 9.1. A survey of the literature confirms the fact that *M. akrokomos* is rare below pH 5 (Table 5.3).

Mallomonas akrokomos was observed in oligotrophic, eutrophic, clearwater and humic stained waterbodies and over a wide range of total phosphorus and specific conductance values (Figs.

5.5D-E). Weighted mean total phosphorus and specific conductance levels were 19 μg-P/l and 86 μS, respectively. Two maxima were observed along a phosphorus gradient, one below 10 μg-P/l and one above 30 μg-P/l (Fig. 5.5E). The majority of the populations found under high total phosphorus levels were during the warmer summer months, supporting the idea that two distinct ecomorphs exist. A survey of the literature confirms the occurrence of *M. akrokomos* in a wide variety of lake types (Dürrschmidt, 1984) (Table 5.3). This taxon has also been observed in acid peat (pH > 5.0, Péterfi and Momeu, 1976; Harris, 1958), brown water (Christie et al., 1988; Eloranta, 1989), oligotrophic (Siver and Chock, 1986; Eloranta, 1986b; Christie et al., 1988) and eutrophic lakes (Asmund, 1956; Cronberg, 1980; Kristiansen, 1985; 1988).

Mallomonas annulata

Mallomonas annulata is a common and widespread species (Dürrschmidt, 1984) capable of tolerating a wide range of environmental conditions (Harris, 1967). This species was much more common in the literature survey (12%) than in either the Adirondacks or Connecticut (less than 3%). Populations of *M. annulata* existed between October

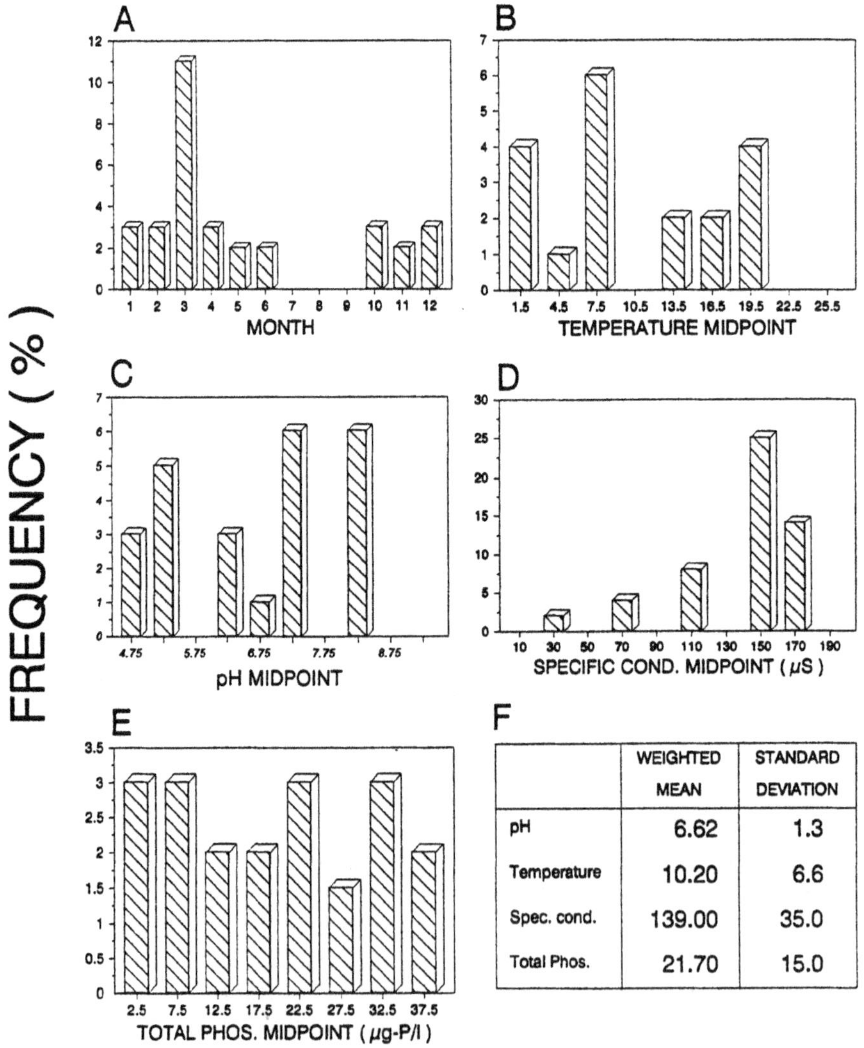

	WEIGHTED MEAN	STANDARD DEVIATION
pH	6.62	1.3
Temperature	10.20	6.6
Spec. cond.	139.00	35.0
Total Phos.	21.70	15.0

Fig. 5.6. The frequencies of occurrence for *Mallomonas annulata* along seasonal (A), temperature (B), pH (C), specific conductance (D) and total phosphorus (E) gradients. Weighted mean and standard deviation values are also given (F).

Table 5.4. Literature records indicating the ecological preferences of *Mallomonas annulata.*

Reference	Seasonality	Trophic Status	Temperature (°C)	pH and Related Factors	Other
1. Dürrschmidt, 1984	April-May	---	---	---	Common and widespread.
2. Dürrschmidt and Croome, 1985	---	---	Cold water and warm tropical waters.	---	Australia and Malaysia
3. Harris, 1967	All months.	Mesotrophic to eutrophic.	---	Acidic	Wide tolerance to many conditions.
4. Ito, 1988	Winter	---	Cold	---	Lake Biwa, Japan
5. Kristiansen, 1985; 1988	Winter-spring and to a lesser extent in fall. Lacking in summer.	Eutrophic	Cold 0.8 to 19.6	8.2 to 9.1	Most abundant at cooler temperatures.
6. Roijackers, 1981	---	---	---	Acidophilous/ acidobiontic group.	Netherlands
7. Roijackers and Kessels, 1986	---	---	Low temperature cluster.	---	Netherlands
8. Siver, 1989a	---	---	---	AWMpH = 6.4 pH indifferent.	Survey of the literature.
9. Takahashi, 1978	---	---	Cold water stenotherm; 0 to 5.	Slightly acidic, pH 6 to 7.	Japan

AWMpH = weighted mean pH.

and June, had a maximum frequency of occurrence just after ice-out in March and were lacking during the warm summer months (Fig. 5.6A). As a result of its distribution during autumn-winter-spring, this species had a low weighted mean temperature of only 10.2°C and was not present above 20°C.

Similar distributional patterns along seasonal and temperature gradients have been documented in the literature (Table 5.4). Although Harris (1967) found *M. annulata* in all months of the year, it has most often been reported during the winter and spring (Dürrschmidt, 1984; Ito, 1988; Kristiansen, 1985; 1988; Roijackers and Kessels, 1986). Based on a survey of the literature, *M. annulata* was not found in collections with a temperature above 21°C and had a weighted mean temperature of 14.5°C. *M. annulata* has also been observed to be a cold water species in Japan (Takahashi, 1978; Ito, 1988), the Netherlands (Roijackers and Kessels, 1986) and Denmark (Kristiansen, 1985; 1988).

Mallomonas annulata was found over a wide pH range of 4.8 to 8.05 and had a weighted mean of 6.6 (Fig. 5.6C). Siver (1989a) reported a similar pH based on literature records and noted that the taxon was as common above pH 7.5 as it was below 5.0. Takahashi (1978) found *M. annulata* primarily between pH 6 and 7, Roijackers (1981) as an acidophilic/acidobiontic species and Kristiansen (1985; 1988) above pH 8, supporting the idea proposed by Siver (1989a) that this species is best described as pH indifferent.

Mallomonas annulata was widely distributed with respect to specific conductance, having been observed between 21 μS and 165 μS (Fig. 5.6D). In addition, populations of this species were found between 1 μg-P/l and 73 μg-P/l (Fig. 5.6E), in oligotrophic and highly eutrophic lakes as well as in clearwater and humic stained localities.

Mallomonas asmundiae

Mallomonas asmundiae was a coldwater form found from January through May at water temperatures of 1°C to 15°C. Although ecological records are few (Table 5.21), Donaldson and Stein (1984) and Dürrschmidt (1984) both reported *M. asmundiae* from spring collections, in agreement with its distribution in Connecticut lakes. Based on its pH range of 5.5 to 7.2, *M. asmundiae* is probably best described as acidophilous, which is in accordance with the findings of Andersen (1978) and Donaldson and Stein (1984).

The most concentrated populations of cells were found in a temporary, humic stained forest pool, similar to that reported by Andersen (1978). *M. asmundiae* was observed in waters of low (31 μS) and high (165 μS) specific conductance and between 4 and 28 μg-P/l.

Mallomonas canina and *Mallomonas hindonii*

Both of these species were very rare, present in 1% or less of the collections from the literature survey (Siver, 1989a), Connecticut and the Adirondacks. The majority of the records of these taxa, especially *M. hindonii*, are from paleolimnological investigations dealing with acidic deposition issues (Tables 5.5 and 5.6). All available data indicate that both taxa may represent the most acidic species of *Mallomonas* and are true acidobionts (Siver, 1989a). Collectively, *M. canina* and *M. hindonii* had a weighted mean pH of only 4.9 and were distributed between pH 4.3 and 5.8. In a principal component analysis Siver and Hamer (1989) found pH and specific conductance to strongly dominate the first principal axis and account for 87% and 77% of the total variance within each variable, respectively, further emphasizing the importance of pH in controlling the distribution of these taxa.

Mallomonas canina, originally described by Kristiansen (1982) from an acidic locality (pH 4.4-4.9), was found in six acidic and slightly humic collections from Connecticut and one acidic and clearwater locality from the Adirondacks (Siver, 1988c). In Finland, Eloranta (1985; 1989) reported *M. canina* to be one of the most acidic species, present in waterbodies with a pH between 4.6 and 6.4 and Tolonen et al. (1986) found the relative percentages of *M. canina* to increase in the surface sediments of several recently acidified lakes. A similar increase in relative abundance was also found in the surface sediments of a small acidic lake in the Bavarian Alps (Hartmann and Steinberg, 1986).

Unlike *M. hindonii* (see below), *M. canina* has been observed in both clearwater acidic (Kristiansen, 1982; Hartmann and Steinberg, 1987) and humic stained acidic (Hartmann and Steinberg, 1987; Eloranta, 1989) waterbodies. In Connecticut and the Adirondacks, *M. canina* was also limited

Table 5.5. Literature records indicating the ecological preferences of *Mallomonas canina*.

Reference	Seasonality	Trophic Status	Temperature (°C)	pH and Related Factors	Other
1. Eloranta, 1985	---	Oligotrophic	---	Acidic lake, 5.6	Forest setting. Finland.
2. Eloranta, 1989	---	Typical of humic stained lakes.	17 - 20	Acidic, 4.6 - 6.4, mean of 5.6.	Finland
3. *Hartmann and Steinberg, 1986	---	---	---	Increase with increase in pH	Alpine Lake, Alps.
4. *Hartmann and Steinberg, 1987	---	---	---	Acidic lakes ranging in pH from 4.1 to 6.3. Clearwater or dystrophic.	Feel they can distinguish from *M. hindonii*.
5. Kristiansen, 1982	July	Oligotrophic	18.4	4.4, clearwater	Forest setting.
6. Nicholls, 1987	---	---	---	---	Evidence for *M. canina* and *M. hindonii* forming one complex.
7. Siver, 1989a	---	---	---	Acidobiontic WMpH = 4.9	---
8. *Tolonen et al., 1986	---	Oligotrophic	---	Acidic, increased concentrations in surface layers.	Found with *M. hindonii*. Finland.

AWMpH = abundance weighted mean pH.
*Indicates a paleolimnological study.

Table 5.6. Literature records indicating the ecological preferences of *Mallomonas hindonii.*

Reference	Seasonality	Trophic Status	Temperature (°C)	pH and Related Factors	Other
1. *Charles et al., 1987	---	---	---	Increase in surface sediments of acidic clearwater lake.	Big Moose Lake, Adirondacks.
2. *Charles and Smol, 1988	---	---	---	AWMpH = 4.93	Adirondacks
3. *Christie et al., 1988	---	---	---	4.9	Finland
4. *Christie and Smol, 1986	---	---	---	Significant increase in surface sediments, pH 4.7 - 5.	Upper Wallface, Adirondacks.
5. *Gibson et al., 1987	---	---	---	Increase in recent sediments of acidic lake. Withstands high Al levels.	Cone Pond, New Hampshire.
6. *Hartmann and Steinberg, 1987	---	---	---	Clearwater localities only.	Lacking in bogs.
7. Siver, 1988c	May	Oligotrophic	17	Acidic clearwater lake, pH 5.4	Lacking in bogs. Adirondacks.
8. *Smol, 1986	---	---	---	Clearwater acidic localities.	Lacking in bogs.
9. *Smol et al., 1984a	---	---	---	4.6 - 7.5 AWMpH = 5.2	Non-bog clearwater sites only. Adirondacks
10. *Smol et al., 1984b	---	---	---	Increase in surface sediments, pH 4.7, clearwater.	Deep Lake, Adirondacks.
11. *Tolonen et al., 1986	---	Oligotrophic	---	Acidic lakes.	Finland

AWMpH = abundance weighted mean pH.
*Indicates a paleolimnological study.

to softwater localities, usually in forest settings, low in specific conductance (11 μS to 50 μS) and total phosphorus. Although only limited data are available, *M. canina* may favor cooler temperatures.

Most of the ecological records for *M. hindonii* are based on paleolimnological surveys from the Adirondacks (Table 5.6). In each study *M. hindonii* was observed in very acidic localities, typical of those harboring true acidobionts. Smol et al. (1984a) noted that this taxon was typical of acidic clearwater localities, but was lacking in humic stained bogs, a characteristic supported by Smol (1986), Siver (1988c) and Hartmann and Steinberg (1987). In addition, the relative importance of *M. hindonii* has increased in the recent sediments of three clear acidic lakes believed to have been adversely affected by acid deposition (Deep Lake, Smol et al., 1984b; Upper Wallface Pond, Christie and Smol, 1986; Big Moose Lake, Charles et al., 1987).

Mallomonas caudata

Like *M. crassisquama*, *M. caudata* is often reported as one of the most common and widespread species of scaled chrysophytes (Asmund and Takahashi, 1969; Takahashi, 1978; Dürrschmidt, 1980; Nicholls, 1982; Kling and Kristiansen, 1983; Wujek and Weis, 1984). Siver (1989a) reported *M. caudata* in 27% of the collections surveyed from the literature. This organism was even more common in Connecticut and Adirondack localities, found in 56% of the collections. In Connecticut lakes, *M. caudata* was easily the most commonly encountered species of *Mallomonas*, found in 20% more of the samples than *M. crassisquama*. Dixit et al. (1988b) and Eloranta (1989) reported *M. caudata* in over 70% of their study lakes from Quebec and Finland, respectively. Cronberg and Kristiansen (1980) reported *M. caudata* to have a distributional pattern in lakes from Sweden similar to that of *M. crassisquama*. It is interesting to also note that, like *M. crassisquama*, *M. caudata* was rare in a study of scaled chrysophytes from the Giessen region of Germany (Dürrschmidt, 1984) and lacking in prelimninary studies from China (Kristiansen, 1989; Kristiansen and Tong, 1989; Siver, unpublished data), Australia and Malaysia (Dürrschmidt and Croome, 1985).

Undoubtedly, the cosmopolitan and widespread occurrence of *M. caudata* is related to its distribution over a wide range of environmental conditions (Figs. 5.7A-F). *M. caudata* was common during each month of the year, found in a minimum of 32% of the collections during January and a maximum of 83% of the collections from August (Fig. 5.7A). Although this taxon was slightly more abundant from August until the initiation of an ice cover in December, it did not exhibit any clear preference for a given season. Other studies support the idea that *M. caudata* does not exhibit a clear seasonal preference (Table 5.7). Asmund (1955) found this species during each month of the year except January, while Donaldson and Stein (1984) and Asmund and Takahashi (1968) observed *M. caudata* during each month of the year except when the lakes were ice covered. Cronberg and Kristiansen (1980) found *M. caudata* in 78% of the lakes surveyed in Sweden during spring and Cronberg (1988) reported whole cells of this species between early spring and autumn.

Because of its widespread seasonal distribution, it is not surprising that *M. caudata* was found over a wide temperature range of 1°C to 28°C (Fig. 5.7B). *M. caudata* had a maximum occurrence between 12°C and 18°C where it was found in over 70% of the collections, however, it was still found in over 47% of the samples with temperatures below 3°C and above 24°C (Fig. 5.7B). It had a weighted mean temperature of 13.8°C. Based on a survey of over 200 collections from the literature, *M. caudata* was found to have a weighted mean temperature of 14°C and a range of 5°C to 21°C. In addition, this organism has been reported over temperature ranges of 0.5°C to 22°C (Asmund, 1955), 5°C to 20°C (Cronberg, 1988) and 2°C to 27°C (Donaldson and Stein, 1984), and to have a maximum occurrence between 12°C and 21°C (Asmund, 1955; Momeu and Péterfi, 1983).

M. caudata was found over a wide range of pH from 4.8 to 9.1, however, it was most abundant between pH 6 and 7.5 (Fig. 5.7C). Its frequency of occurrence decreased significantly as the pH dropped below 5.5 or raised above 8.0. It was absent in 97% of the collections below pH 5.0 and had a weighted mean pH of 6.9. Based on its distribution along a pH gradient, *M. caudata* is best described as pH indifferent.

Numerous studies support the ideas that *M. cau-*

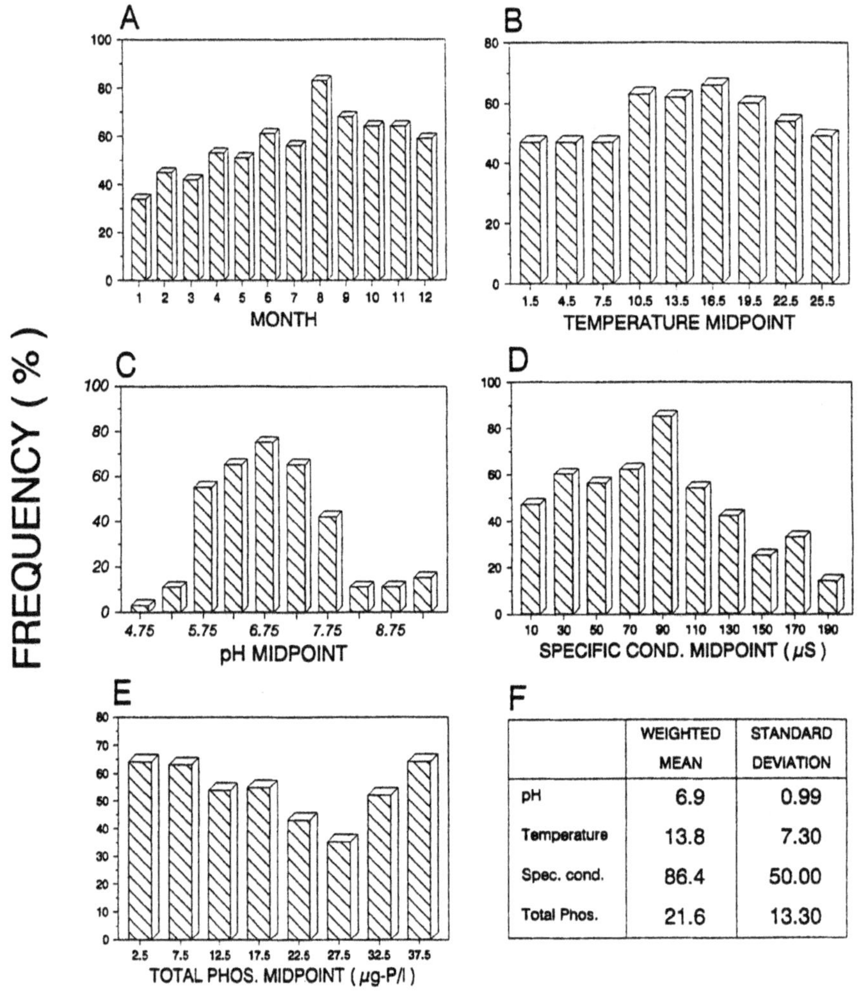

	WEIGHTED MEAN	STANDARD DEVIATION
pH	6.9	0.99
Temperature	13.8	7.30
Spec. cond.	86.4	50.00
Total Phos.	21.6	13.30

Fig. 5.7. The frequencies of occurrence for *Mallomonas caudata* along seasonal (A), temperature (B), pH (C), specific conductance (D) and total phosphorus (E) gradients. Weighted mean and standard deviation values are also given (F).

data is distributed over a wide pH range, has a maximum abundance near pH 7 and is rare below pH 5 (Table 5.7). *M. caudata* has been reported between a pH of 5.4 (Cronberg and Kristiansen, 1980) and 9.1 (Kristiansen, 1988) and to have weighted mean values of 6.6 (Siver, 1989a), 6.9 (Smol et al., 1984a; Dixit et al., 1988a) and 7.0 (Charles and Smol, 1988). This taxon has been observed to prefer slightly acidic (Asmund, 1955; Cronberg, 1988; Eloranta, 1985), slightly acidic to alkaline (Dixit et al., 1988b), circumneutral to alkaline (Smol, 1986; Dixit et al., 1988a) and alkaline (Roijackers and Kessels, 1986; Wee and Gabel, 1989) waterbodies, further supporting the idea that *M. caudata* is best described as a pH indifferent and circumneutral species. Siver (1988c), Smol (1986), Cronberg (1988), Dixit et al. (1988b), Eloranta (1985) and Roijackers and Kessels (1986) also observed this species lacking in waters with a pH less than 5.0. In addition, numerous paleolimnological studies support the idea that *M. caudata* disappears as the lakewater pH declines below the 5.0 to 5.5 pH interval (Smol et al., 1984a; Christie and Smol, 1986; Hartmann and Steinberg, 1986; Christie et al., 1988).

Like *M. crassisquama*, *M. caudata* was found in a wide variety of lake types, ranging from ultra-oligotrophic to eutrophic as well as clearwater to humic stained, although it was lacking from all acidic bog localities. It had a rather high weighted

Table 5.7. Literature records indicating the ecological preferences of *Mallomonas caudata*.

Reference	Seasonality	Trophic Status	Temperature (°C)	pH and Related Factors	Other
1. Asmund, 1955	All months except January.	Oligotrophic dystrophic.	0.5 - 22; max. 15 - 21	Slightly acidic.	Adaptable; Denmark.
2. Arvola, 1983	Summer	Low nutrients, polyhumic.	---	5 - 6	Finland
3. Asmund and Hilliard, 1961	September	Oligotrophic	12 - 13	7.1	Alaska
4. Asmund and Takahashi, 1968	May-September, November.	---	---	---	Worldwide distribution; adaptable; Alaska.
5. *Charles and Smol, 1988	---	---	---	AWMpH = 7.02	Adirondacks
6. *Christie and Smol, 1986	---	---	---	Most abundant when lake was the most alkaline.	Upper Wallface, Adirondacks; occurred with *M. elongata*.
7. *Christie et al., 1988	---	---	---	Found primarily above pH 5.	Finland
8. Cronberg, 1988	Early spring to late autumn.	Oligotrophic to eutrophic: avoids most polluted.	5 - 20	---	Cysts found July-October at 10 - 19°C and pH 5.6 - 6.8.
9. Cronberg and Kristiansen, 1980	78% of spring samples.	Oligotrophic to eutrophic; lacking in most eutrophic.	---	---	Similar distributional pattern as *M. crassisquama*, Sweden.
10. *Dixit et al., 1988a	---	---	---	Circumneutral to alkaline; AWMpH=6.88	Sudbury, Ontario.
11. *Dixit et al., 1988b	---	---	---	Slightly acidic to alkaline	Found in 71% of the study lakes, Quebec.
12. Dürrschmidt, 1980	---	---	---	---	Widespread in temperate and arctic regions; South America.
13. Dürrschmidt, 1984	---	Oligotrophic	---	---	Rare, Giessen region of Germany.

Table 5.7. continued

Reference	Seasonality	Trophic Status	Temperature (°C)	pH and Related Factors	Other
14. Donaldson and Stein, 1984	All months except in winter under the ice.	---	2 - 27	---	British Columbia
15. Eloranta, 1985	---	Humic stained lakes.	---	5.9 - 6.3, acidic	One of the most common species; Finland.
16. Eloranta, 1986b	---	Ultraoligotrophic to oligotrophic.	---	---	Subarctic, Finland.
17. Eloranta, 1989	---	Oligotrohpic humic stained.	---	Acidic, but above 5; cell densities positively correlated with specific conductance.	72% of collection, also found in bog localities. Finland.
18. *Hartmann and Steinberg, 1986	---	---	---	** decreases in surface sediments.	Alpine Lake, Alps.
19. Ito, 1988	Cells in winter.	---	---	---	Lake Biwa, Japan
20. Kling and Kristiansen, 1983	---	---	---	---	Common and widespread; Canada
21. Kristiansen, 1985; 1988	---	Eutrophic	13	8.2 - 9.1	Not common in this Danish lake.
22. Kristiansen, 1989	---	---	---	---	Lacking in all samples, China.
23. Kristiansen and Tong, 1989	---	---	---	---	Lacking in all samples, China.
24. Momeu and Péterfi, 1983	March-April	Eutrophic pools.	12 - 17	7 - 7.5	Romania
25. Nicholls, 1982	---	---	---	---	Common in lakes from Ontario.
26. Roijackers and Kessels, 1986	---	---	4 - 11	6 - 8; lacking at lower pH values.	Netherlands
27. Siver, 1988c	All seasons	Oligotrophic	---	Range: 5.4 - 7.1	53% of lakes and 43% of samples; Adirondacks

Table 5.7. continued

Reference	Seasonality	Trophic Status	Temperature (°C)	pH and Related Factors	Other
28. Siver, 1989a	---	---	---	AWMpH = 6.7 for Connecticut/Adirondack lakes; 6.6 for literature review. Avoids localities below 5.0.	27% of collections surveyed from literature.
29. Siver and Hamer, 1989	---	---	Weighted mean of 15.	First principal component strongly dominated by pH and specific conductance.	Clusters with *M. crassisquama*, *Spiniferomonas trioralis* and *Sp. coronacircumspina*.
30. *Smol, 1980	---	Increased in relative abundance with slightly enriched waters.	---	---	Found Lake and Jake Lake, Algonquin Park, Ontario.
31. Smol, 1986	---	---	---	Circumneutral to alkaline; lacking below 5.	Review article.
32. *Smol et al., 1983	---	Disappeared during very eutrophic period.	---	---	Codominant with *M. crassisquama* and *M. pseudocoronata*; Little Round Lake, Ontario.
33. *Smol et al., 1984a	---	---	---	AWMpH = 6.9 Range: 5.1 - 7.8 alkaline.	Adirondacks
34. *Tolonen et al., 1986	---	Oligotrophic	---	---	Finland
35. Wawrzyniak and Andersen, 1985	---	---	---	---	Common in Northern boreal localities.
36. Wee and Gabel, 1989	---	---	---	Alkaline, high pH	Iowa
37. Wujek and Weis, 1984; Wujek, 1984	---	---	---	---	One of the most widely reported taxa; worldwide.

AWMpH = abundance weighted mean pH.
*Indicates a paleolimnological study.
**Indicates that a decrease in the abundance of scales correlated with a drop in pH below approximately 5.0.

mean total phosphorus concentration of 21.6 μg-P/l, but was found as often in waterbodies with a concentration below 5 μg-P/l as in those with a concentration above 35 μg-P/l (Fig. 5.7E). *M. caudata* was also observed over a wide range of specific conductance values, ranging from 13 μS to 200 μS and to have a weighted mean of 86 μS (Fig. 5.7D). This taxon was found in 58% of the samples with a specific conductance less than 140 μS, but in only 19% of those above 140 μS.

A review of the literature supports the view that *M. caudata* exists in a wide variety of lake types (Table 5.7), ranging from nutrient poor oligotrophic (Krieger, 1930; Eloranta, 1986a; Tolonen et al., 1986; Cronberg, 1988) to eutrophic (Momeu and Péterfi, 1983; Wujek, 1984; Kristiansen, 1985; 1988; Cronberg, 1988) in nature. In addition, *M. caudata* has often been observed in humic colored localities (Nygaard, 1945; Asmund, 1955; Arvola, 1983; Eloranta, 1985) provided that the pH was above 5.0. Eloranta (1989) reported that cell densities of this organism were positively correlated with waters of higher specific conductance. Smol (1980) observed the percentage of scales of *M. caudata* to increase in cores of lakes in the Algonquin Park region of Ontario concurrent with increased human activity within the watershed, presumably the result of a slight increase in the trophic level of the lakes. In other paleolimnological studies, Smol et al. (1983; 1985) found this taxon to disappear under the most eutrophic conditions.

In summary, the ecological requirements of *M. caudata* are very similar to those of *M. crassisquama*. Both taxa are found throughout the year and over wide pH, temperature, specific conductance and total phosphorus levels. *M. caudata* also had a maximum occurrence between pH 6 and 7.5, avoided waters with a pH less than 5.0 or high specific conductance and was slightly less common under the ice.

Mallomonas corymbosa

Mallomonas corymbosa was found in 7% of the collections from Connecticut localities, 3% from the literature survey and lacking from the Adirondacks. Likewise, Smol et al. (1984a) and Charles and Smol (1988) did not report *M. corymbosa* from the Adirondacks. This taxon was observed during every month of the year, except February. Although it did not have a strong preference for any given season, it was most common during September and October where it was found in 18% of the collections (Fig. 5.8A). Because of its occurrence throughout the year, *M. corymbosa* had a wide temperature range from 1°C to 24.5°C and a weighted mean of 14°C (Fig. 5.8B).

Although *M. corymbosa* was found in moderately acidic lakes, it clearly had a maximum occurrence in localities with a pH above 7 (Fig. 5.8C). This species was found in 6% of the collections with a pH between 5.5 and 8.0, but 15% of those above pH 8. It was absent in all collections below a pH of 5.8, had a high weighted mean pH of 8 and is best described as alkaliphilic to alkalibiontic in nature.

Mallomonas corymbosa was found over a wide range of specific conductance from 22 μS to 200 μS, but had a clear preference for waterbodies high in total salt content. This species was found in 32% of the collections from localities with a specific conductance above 160 μS and had a weighted mean of 146 μS. Using principal component analysis, Siver and Hamer (1989) found specific conductance to strongly dominate the first principal axis.

Mallomonas corymbosa was also observed over a wide total phosphorus gradient from 1 μg-P/l to 130 μg-P/l (Fig. 5.8E). Despite the wide range, this species was most common in localities with high total phosphorus levels and had a weighted mean of 29 μg-P/l. Although *M. corymbosa* was recorded in several unproductive lakes, it was most common and had the greatest cell concentrations in moderately productive, eutrophic waterbodies dominated by blue-green algae.

In summary, the preference of *M. corymbosa* for habitats with high pH, high specific conductance and high total phosphorus levels explains its absence from the Adirondacks. Even though ecological records for *M. corymbosa* are few (Table 5.21), they support the ideas that this species favors high pH (Asmund and Hilliard, 1961) and is most abundant during the autumn (Asmund and Hilliard, 1961; Asmund and Takahashi, 1969).

Mallomonas crassisquama

Mallomonas crassisquama is often reported as being

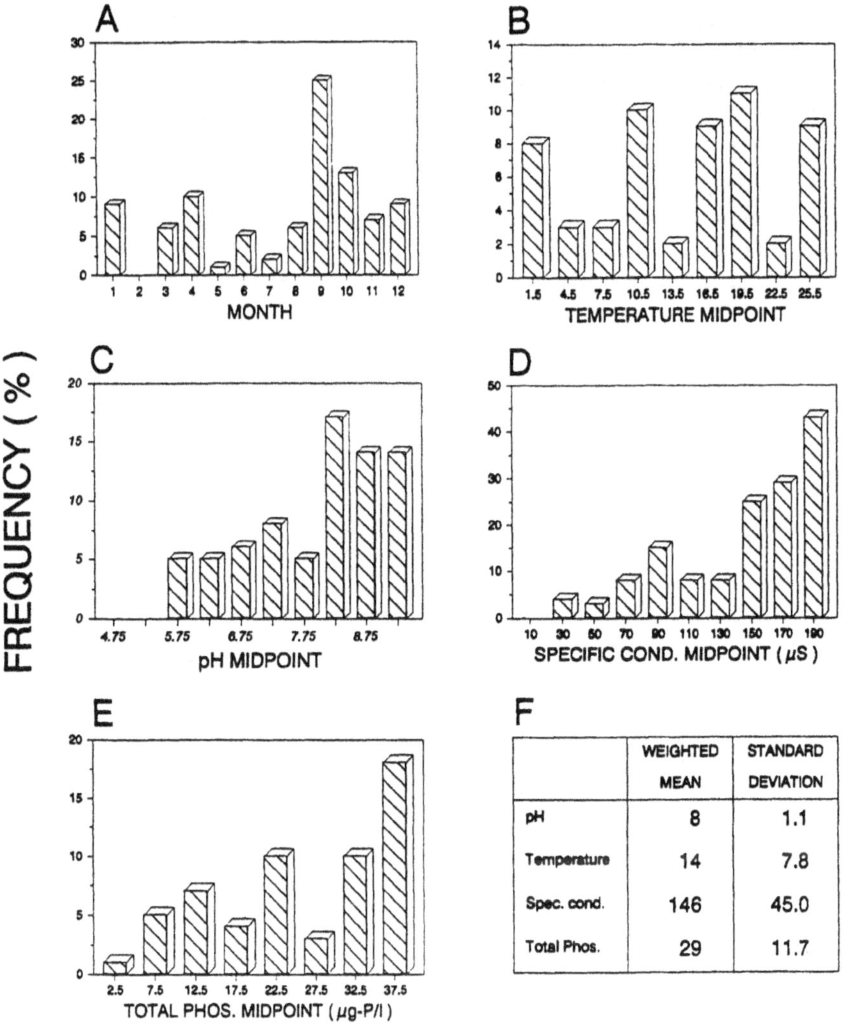

	WEIGHTED MEAN	STANDARD DEVIATION
pH	8	1.1
Temperature	14	7.8
Spec. cond.	146	45.0
Total Phos.	29	11.7

Fig. 5.8. The frequencies of occurrence for *Mallomonas corymbosa* along seasonal (A), temperature (B), pH (C), specific conductance (D) and total phosphorus (E) gradients. Weighted mean and standard deviation values are also given (F).

the most common and widely distributed species of *Mallomonas* (Siver and Skogstad, 1988). It was found in 31 to 36% of collections from Connecticut (Siver and Skogstad, 1988), Norway (Siver and Skogstad, 1988), the Adirondacks (Siver, 1988c), northern boreal regions (Wawrzyniak and Andersen, 1985) and a survey of the literature (Siver, 1989a). Other than *Mallomonas caudata*, *M. crassisquama* was the most common species of *Mallomonas* in this survey. This organism has also been reported as being one of the most common scaled chrysophytes in Japan (Takahashi, 1978), Finland (Eloranta, 1985; 1989), Canada (Kristiansen, 1975; Nicholls, 1982; Kling and Kristiansen, 1983) and Minnesota (Wujek et al., 1981). In addition, *M. crassisquama* has often been found to dominate in the paleolimnological environments of lakes (e.g. Smol et al., 1984b; Christie and Smol, 1986; Charles et al., 1987; Hartmann and Steinberg, 1986; Dixit et al., 1988b), often accounting for over 90% of the total number of scales.

Despite its widespread abundance, *M. crassisquama* was rare in a study of scaled chrysophytes from the Giessen region of Germany (Dürrschmidt, 1984) and lacking in similar studies from South America (Dürrschmidt, 1982b), Australia and Malaysia (Dürrschmidt and Croome, 1985) and in an initial survey of chrysophytes from China (Kristiansen, 1989; Kristiansen and Tong, 1989; Siver, unpublished data). As pointed out in Chapter III, identifications made of *M. crassisquama* in previous studies could have included specimens of *M. duerrschmidtiae*, resulting in an inflated frequency value. However, this would only cause concern in regions where *M. duerrschmidtiae* is found; those with a predominance of clear, soft-water lakes, slightly acidic in nature and low in specific conductance. Thus, a reevaluation of studies from such softwater lakes (e.g. Dixit et al., 1988a,b; Charles et al., 1987) may yield additional information.

The cosmopolitan and widespread distribution of *M. crassisquama* is related to its tolerance to a wide range of environmental conditions (Figs. 5.9A-F). *M. crassisquama* was common during each month of the year and found in over 19% of collections from any given month (Fig. 5.9A). Although it was slightly more abundant during spring and late fall, *M. crassisquama* did not exhibit any clear preference for a given season. Even though *M. crassisquama* had a weighted mean temperature of 13°C, it was common over the entire gradient (Fig. 5.9B), resulting in a rather large standard deviation of 7°C. A review of the literature confirms that *M. crassisquama* is found during all seasons of the year and has a wide temperature tolerance (Asmund, 1959; Asmund and Hilliard, 1961; Takahashi, 1978; Roijackers

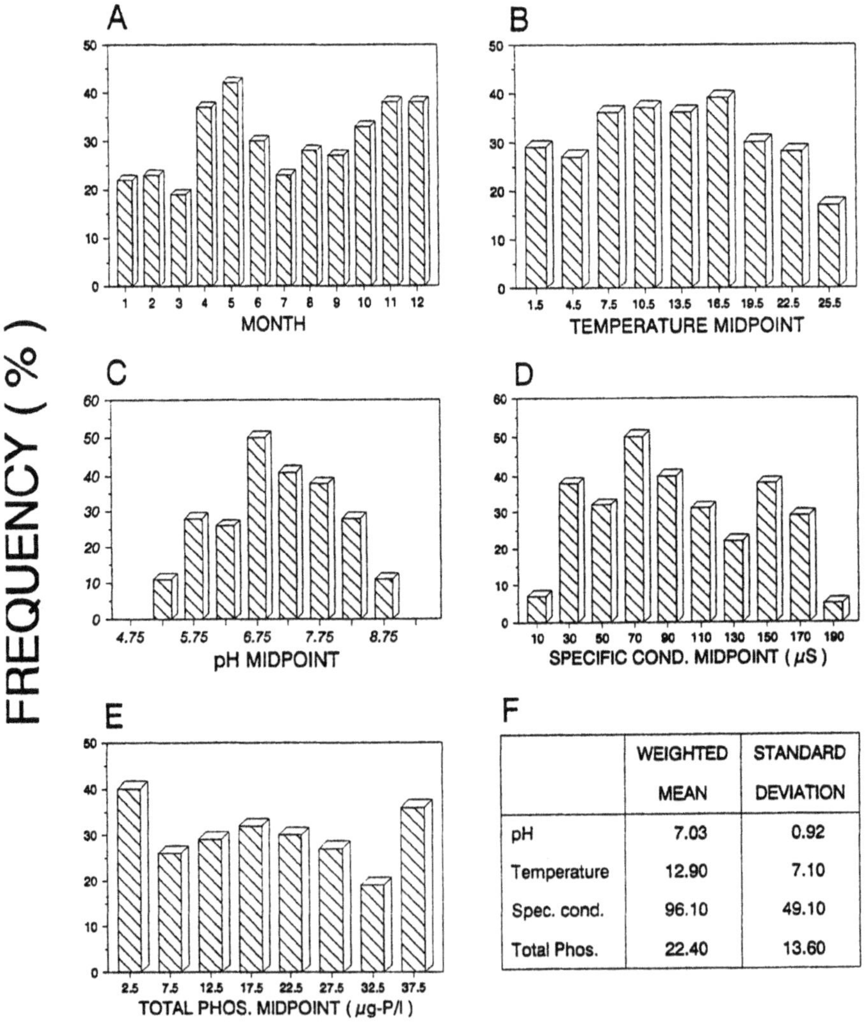

	WEIGHTED MEAN	STANDARD DEVIATION
pH	7.03	0.92
Temperature	12.90	7.10
Spec. cond.	96.10	49.10
Total Phos.	22.40	13.60

Fig. 5.9. The frequencies of occurrence for *Mallomonas crassisquama* along seasonal (A), temperature (B), pH (C), specific conductance (D) and total phosphorus (E) gradients. Weighted mean and standard deviation values are also given (F).

and Kessels, 1986). In addition, Asmund (1959) and Momeu and Péterfi (1983) also found this taxon to be slightly more common in the spring between 10°C and 15°C. Siver and Skogstad (1988) demonstrated that bristle formation in *M. crassisquama* was temperature dependent where cells with only serrated bristles were found below 12°C, and organisms with only helmet bristles in warmer water above 15°C.

Mallomonas crassisquama was found over a wide pH range from 5.2 to 8.7, however, had a maximum occurrence between pH 6.5 and 7.5 (Fig. 5.9C). Eleven percent of the collections between pH 5 and 5.5 had specimens of *M. crassisquama*. The frequency of occurrence increased steadily to 50% in the 6.5 to 7 pH interval and then declined gradually to where it was again found in only 11% of the collections between pH 8.5 and 9 (Fig. 5.9C). This taxon was noticeably absent below pH 5 and above pH 9, had a weighted mean pH of 7 and is best described as pH indifferent.

Similar ranges in pH of 6.3 to 8.6 (Asmund and Hilliard, 1961), 5.5 to 9.0 (Takahashi, 1978), 4.7 to 7.8 (Smol et al., 1984a), 6.2 to 7.5 (Roijackers and Kessels, 1986), 5.4 to 8.0 (Cronberg and Kristiansen, 1980) and 4.8 to 7.2 (Eloranta, 1989) have been reported for *M. crassisquama*. Siver (1989a) calculated a weighted mean pH of 6.6 from literature values, supporting the idea that *M. crassisquama* is best described as pH indifferent (Takahashi, 1978; Roijackers and Kessels, 1986). Asmund (1959) and Roijackers (1981) noted *M. crassisquama* to have a slight preference for alkaline waters, whereas Momeu and Péterfi (1983), Charles and Smol (1988) and Eloranta (1985; 1989) reported a preference for slightly acidic localities (Table 5.8). As observed in this study, reports from the literature consistently confirm the fact that *M. crassisquama* is very rare below a pH of 5.0 (Cronberg and Kristiansen, 1980; Smol, 1986; Siver and Skogstad, 1988; Eloranta, 1989). A large number of paleolimnological studies have correlated a decrease in *M. crassisquama* with a concurrent drop in lakewater pH below 5.0 caused by acidic deposition, confirming the idea that this species disappears at low pH (e.g. Hartmann and Steinberg, 1986; Battarbee et al., 1980; Smol, 1980; Smol et al., 1984b; Christie and Smol, 1986; Dixit et al., 1988b). Preliminary data from several paleolimnological studies suggest that *M. crassisquama* is sensitive to high metal concentrations (Gibson et al., 1987; Dixit et al., 1989).

Dixit et al. (1988a) found *M. crassisquama* below a pH of 5.0 in surface sediments of several lakes in the Sudbury area, contradicting the large body of evidence that supports the idea that this taxon disappears below pH 5.0. However, the scale pictured by Dixit et al. (1988a) to represent *M. crassisquama* is actually of *M. duerrschmidtiae*, indicating that the two taxa were mixed in the counts. This supports the idea that reevaluating collections for taxonomic discrepancies may help to solve apparent contradictions in the literature and eventually yield additional information of ecological importance. It should be noted that *M. duerrschmidtiae* was described after the Dixit et al. (1988a) paper.

Mallomonas crassisquama was found in a wide variety of lake types ranging from ultraoligotrophic to eutrophic as well as clearwater to humic, although it was lacking in all acidic bog localities. It had a weighted mean total phosphorus concentration of 22 μg-P/l, however, it was as abundant above 40 μg-P/l as it was below 5 μg-P/l (Fig. 5.9E). Likewise, populations of *M. crassisquama* were observed over a wide range of specific conductance values, ranging from 18 μS to 219 μS (Fig. 5.9D) and it had a weighted mean of 96 μS. Although *M. crassisquama* was found in an average of 36% of all collections ranging from 20 μS to 180 μS, it occurred in only 7% and 6% of the collections less than 20 μS and greater than 180 μS, respectively.

A review of the literature supports the fact that *M. crassisquama* has a wide ecological tolerance, having been found in ultraoligotrophic (Eloranta, 1986), oligotrophic (Battarbee et al., 1980; Eloranta, 1986; Wawrzyniak and Andersen, 1985), eutrophic (Asmund, 1959; Momeu and Péterfi, 1983), clearwater (Asmund, 1959; Eloranta, 1989) and humic stained (Battarbee et al., 1980; Eloranta, 1985; 1989) lakes (Table 5.8).

In summary, *M. crassisquama* is found throughout the year and over wide ranges of pH, specific conductance, total phosphorus and temperature. It has a slight preference for waters with a pH between 6.5 and 7.5, avoids localities with a pH less than 5 and is less frequent at very low or high specific conductance values.

Table 5.8. Literature records indicating the ecological preferences of *Mallomonas crassisquama*.

Reference	Seasonality	Trophic Status	Temperature (°C)	pH and Related Factors	Other
1. Asmund, 1959	All months.	Eutrophic to highly eutrophic	1 - 23, preference for 10 - 15.	6.8 - 8; alkaline.	Adaptable, widespread, clear water, Denmark.
2. Asmund and Hilliard, 1961	All seasons.	Oligotrophic	0 - 22	6.3 - 8.6	Alaska
3. *Battarbee et al., 1980	---	Oligotrophic	---	---	Finland
4. *Charles et al., 1987	---	---	---	**Decreased with drop in pH near 5.0.	Adirondacks, Big Moose Lake.
5. *Charles and Smol, 1988	---	---	---	AWMpH 6.4	Adirondacks
6. *Christie et al., 1988	---	---	---	Found mainly above 5.0	Finland
7. *Christie and Smol, 1986	---	---	---	**Decrease in surface sediments.	Adirondacks, Upper Wallface Lake.
8. Cronberg and Kristiansen, 1980	---	All but most eutrophic.	---	All but most acidic.	Sweden
9. *Dixit, 1986	---	---	---	pH indifferent	Canada
10. *Dixit et al., 1988a	---	---	---	Found below 5.0	Sudbury, Ontario
11. *Dixit et al., 1988b	---	---	---	**Decrease in surface sediments.	Quebec
12. *Dixit et al., 1989	---	---	---	---	Sensitive to high metal concentrations, Canada.
13. Dürrschmidt, 1982b	---	---	---	---	Lacking in preliminary study of Chile.
14. Dürrschmidt, 1984	---	---	---	---	Rare, Giessen region of Germany.

Table 5.8. continued

Reference	Seasonality	Trophic Status	Temperature (°C)	pH and Related Factors	Other
15. Dürrschmidt and Croome, 1985	---	---	---	---	Not found in study from Australia and Malaysia.
16. Eloranta, 1985	---	Oligotrophic	---	5.5 - 6.3	Humic acid pools, 49-399 mg Pt/l, Finland.
17. Eloranta, 1986a	---	Ultra oligotrophic to oligotrophic.	---	---	Finland
18. Eloranta, 1989	---	Oligotrophic	---	4.8 - 7.2 mean = 5.8, significant drop below 5.0.	Finland, 66% of collections.
19. *Gibson et al., 1987	---	---	---	---	May be sensitive to high Al levels. New Hampshire.
20. *Hartmann and Steinberg, 1986	---	---	---	**Decrease in surface sediments.	Alpine Lake, Alps.
21. Ito, 1988	Spring, winter.	---	---	---	Lake Biwa, Japan
22. Kling and Kristiansen, 1983	---	---	---	---	One of most widespread, Canada.
23. Kristiansen, 1989	---	---	---	---	Lacking in preliminary study of China.
24. Momeu and Péterfi, 1983	March-April.	Mesotrophic to eutrophic.	11 - 12	6 - 6.5	Bogs, Romania
25. Nicholls, 1982	---	---	---	---	Most common and widespread taxon in Ontario.
26. Péterfi and Momeu, 1976	April	Eutrophic	---	5.5 - 6.0	Romania
27. Roijackers, 1981	---	---	---	Alkaline	Netherlands

Table 5.8. continued

Reference	Seasonality	Trophic Status	Temperature (°C)	pH and Related Factors	Other
28. Roijackers and Kessels, 1986	---	---	Scattered	6.2 - 7.5, lacking < 6.2	Netherlands
29. Siver, 1988c	All seasons.	Oligotrophic to mesotrophic.	Wide range.	5.41 - 7.02 lacking in all samples < 5.4	Clearwater to slightly humic, lacking in bogs, Adirondacks.
30. Siver and Skogstad, 1988	All months.	Oligotrophic to eutrophic.	1 - 27	Wide tolerance above 5.5	Avoids high specific conductance, 33% of all samples, Connecticut and Norway. Bristle morphology dependent on water temperature.
31. Smol, 1986	---	---	---	Rarely found below 5.0	Generalist
32. *Smol et al., 1984a	---	Wide	---	4.7 - 7.8 AWMpH 6.6	Adirondacks
33. *Smol et al., 1984b	---	---	---	**Decrease in surface sediments.	Adirondacks, Deep Lake.
34. Takahashi, 1978	All seasons.	---	1 - 26	5.5 - 9.0 pH indifferent with acidophilous tendency.	Japan
35. Wawrzyniak and Andersen, 1985	Spring-summer.	Oligotrophic	---	---	36% of samples from northern boreal regions.

AWMpH = abundance weighted mean pH.
*Indicates a paleolimnological study.
**Indicates that a decrease in the abundance of scales correlated with a drop in pH below approximately 5.0.

Mallomonas doignonii and *Mallomonas dickii*

These closely related species were both restricted to the same three waterbodies, however, they were separated along a seasonal gradient (Figs. 5.10A; 5.11A). Except for one occurrence during May, *M. doignonii* was present only between December and March under the ice, resulting in a low weighted mean temperature of 7.5°C. The few available literature records support the idea that *M. doignonii* is a coldwater organism (Table 5.21). Over a nine year period in a Danish lake, Asmund and Cronberg (1979) reported *M. doignonii* consistently from February through May and to decline shortly after ice-out. Donaldson and Stein (1984) and Wee and Gabel (1989) also reported *M. doignonii* in waters with a temperature below 11°C. In contrast, *M. dickii* was present in the warmwater months of April to October and had a high weighted mean temperature of 21.6°C (Fig. 5.10A-B).

Both taxa had similar distributions along pH and specific conductance gradients. Since *M. doignonii* and *M. dickii* were restricted to a pH range from 5.5 to 6.79 (Figs. 5.10C; 5.11C) and had weighted mean values of 5.8 and 6.3, respectively, they are tentatively classified as acidophilous organisms. The difference in their weighted means

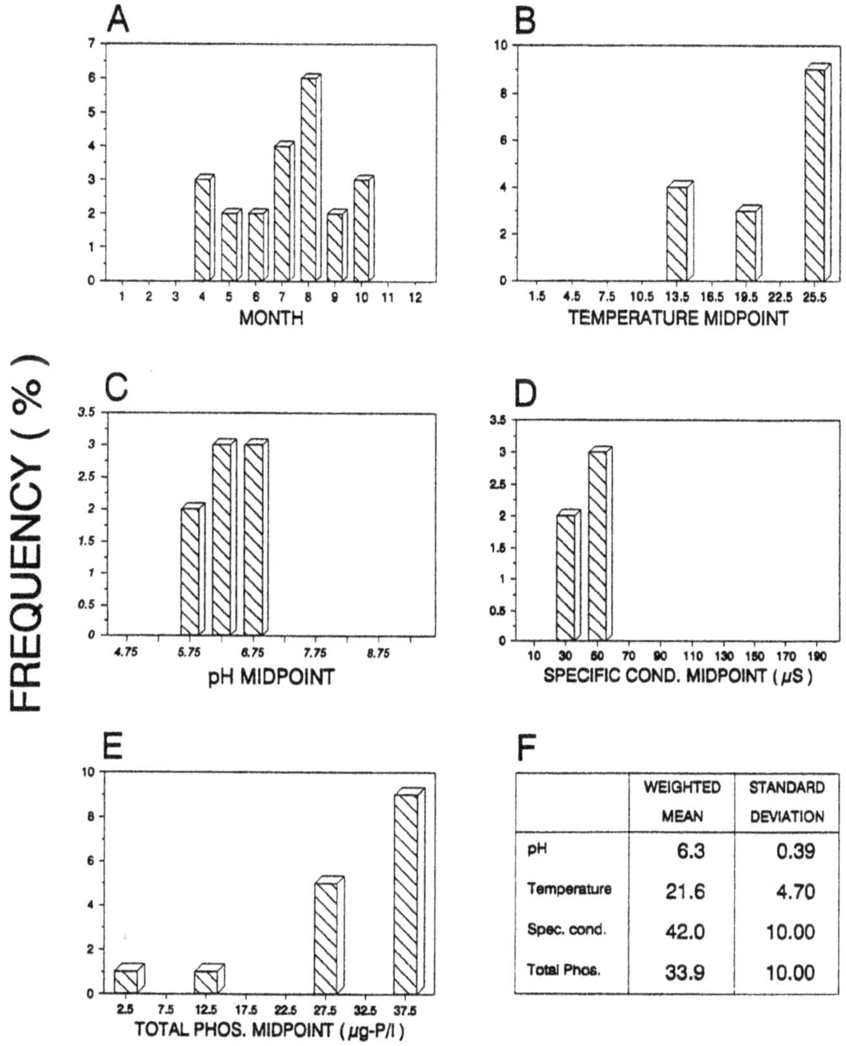

	WEIGHTED MEAN	STANDARD DEVIATION
pH	6.3	0.39
Temperature	21.6	4.70
Spec. cond.	42.0	10.00
Total Phos.	33.9	10.00

Fig. 5.10. The frequencies of occurrence for *Mallomonas dickii* along seasonal (A), temperature (B), pH (C), specific conductance (D) and total phosphorus (E) gradients. Weighted mean and standard deviation values are also given (F).

could reflect seasonal fluctuations in pH. Both taxa were present in collections with specific conductance levels below 53 μS and had weighted means near 40 μS (Figs. 5.10D,F; 5.11D,F). The three softwater localities harboring these two taxa were all slightly humic stained and situated in forest settings.

In a survey of four species within the Torquata group, including *M. doignonii* (as *M. coronifera*), Takahashi (1978) reported a large amount of variation in the sizes of cells and siliceous components. *M. dickii*, described in 1982 by Nicholls, is separated from *M. doignonii* primarily on the basis of differences in the sizes of cells, scales, bristles and posterior spines (see Chapters III and IV). With further research, it will be of interest to see if *M. doignonii* and *M. dickii* are indeed separate species or perhaps ecomorphs of the same species.

Mallomonas duerrschmidtiae

Mallomonas duerrschmidtiae Siver, Hamer and Kling, a newly described species closely related to both *M. crassisquama* and *M. pseudocoronata* (Siver et al., 1990), is found over relatively narrow environmental gradients. In Connecticut and Adirondack waterbodies *M. duerrschmidtiae* was found primarily during the spring and fall with peak frequencies of 17% and 15%, respectively (Fig.

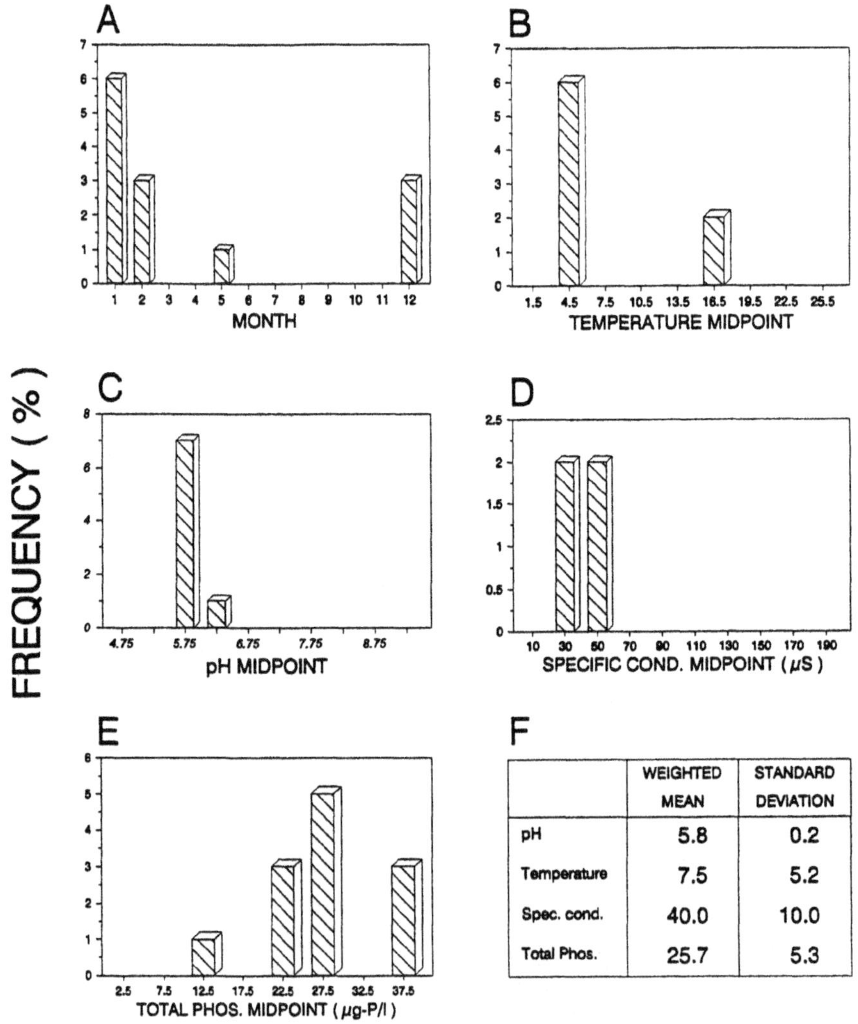

	WEIGHTED MEAN	STANDARD DEVIATION
pH	5.8	0.2
Temperature	7.5	5.2
Spec. cond.	40.0	10.0
Total Phos.	25.7	5.3

Fig. 5.11. The frequencies of occurrence for *Mallomonas doignonii* along seasonal (A), temperature (B), pH (C), specific conductance (D) and total phosphorus (E) gradients. Weighted mean and standard deviation values are also given (F).

5.12A). However, it was present in only 4% of the collections made from under the ice and 3% of those made during the summer. As a result, *M. duerrschmidtiae* has a relatively low weighted mean temperature of 12.4°C, a maximum occurrence between 9°C and 18°C and is much rarer at extreme temperatures (Fig. 5.12B).

Mallomonas duerrschmidtiae is also restricted to acidic softwater lakes low in specific conductance and total phosphorus levels (Figs. 5.12D-E). This species was found in 26% of the collections made between pH 5 and 5.5, and its frequency of occurrence diminished rapidly below and above this interval (Fig. 5.12C). Because it is found almost exclusively below a pH of 7, has a peak in occurrence between pH 5 and 6 and has a weighted mean pH of 5.76, *M. duerrschmidtiae* is best described as a true acidophilous taxon.

Among species of *Mallomonas*, *M. duerrschmidtiae* is one of the most restricted along a specific conductance gradient, having been found only in waterbodies of less than 43 μS (Fig. 5.12D). It had a maximum occurrence below 20 μS and a weighted mean of 18 μS. In addition, *M. duerrschmidtiae* was recorded mostly between 5 and 10 μg-P/l, and was lacking in all collections with a total phosphorus concentration greater than 20 μg-P/l (Fig. 5.12E). Although this taxon was abundant in a few slightly coloured localities, it had a distinct preference for clearwater lakes and was lacking

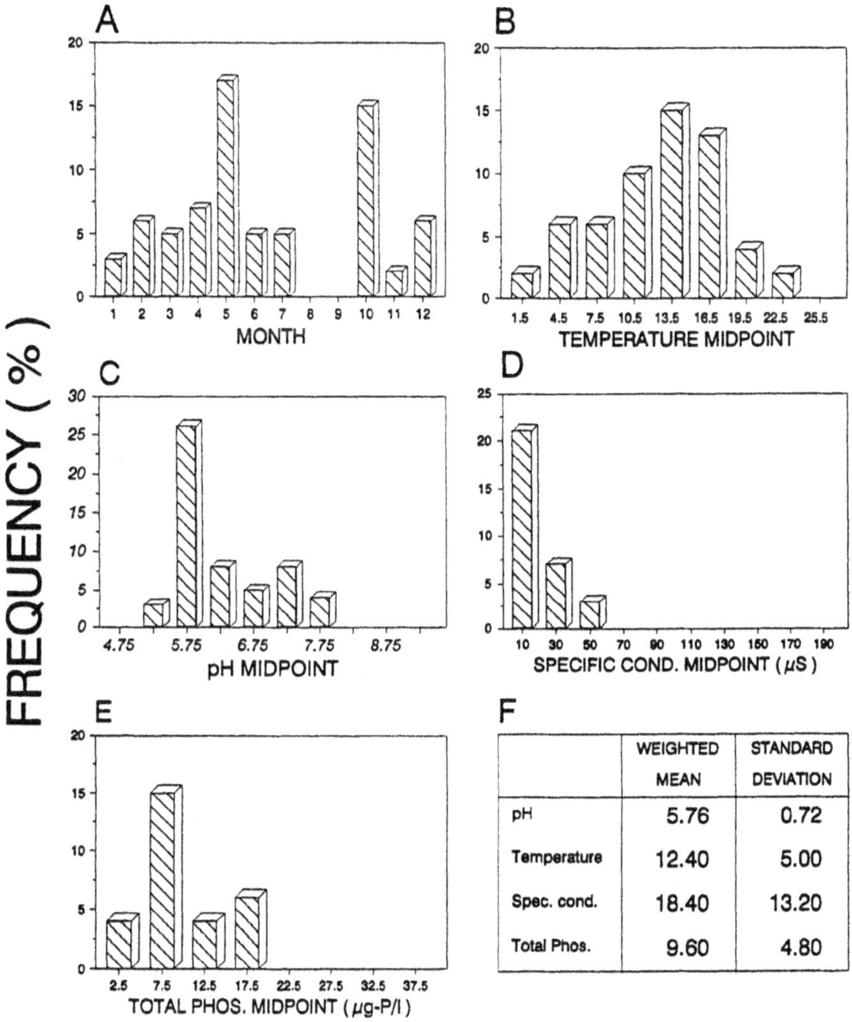

Fig. 5.12. The frequencies of occurrence for *Mallomonas duerrschmidtiae* along seasonal (A), temperature (B), pH (C), specific conductance (D) and total phosphorus (E) gradients. Weighted mean and standard deviation values are also given (F).

from all darkly stained sites.

Mallomonas duerrschmidtiae has inevitably been confused and included with *M. crassisquama* in previous work (Siver et al., 1990), especially in studies from regions with a predominance of soft-water acidic localities low in specific conductance and total phosphorus (e.g. the Adirondacks).

Mallomonas elongata

Although *Mallomonas elongata* has a world-wide distribution (Takahashi, 1978; Smol et al., 1984a; Asmund and Kristiansen, 1986), it is relatively rare both in terms of occurrence and cell concentration. This species was found in 3.5% and 4% of the collections from this survey and the literature review, respectively. Charles and Smol (1988) observed *M. elongata* in the surface sediments of 17% of the lakes they surveyed from the Adirondacks. The greater percent occurrence in the surface sediments, as reported by Charles and Smol (1988), was most likely due to the fact that the surface sediments represented an integration of populations over the yearly cycle.

The seasonal distribution of *M. elongata* was very similar to that of *M. heterospina*. *M. elongata* was observed during the autumn, under the ice and during spring overturn (Fig. 5.13A). This taxon was found in 8% of the collections made between December and May. However, except for one

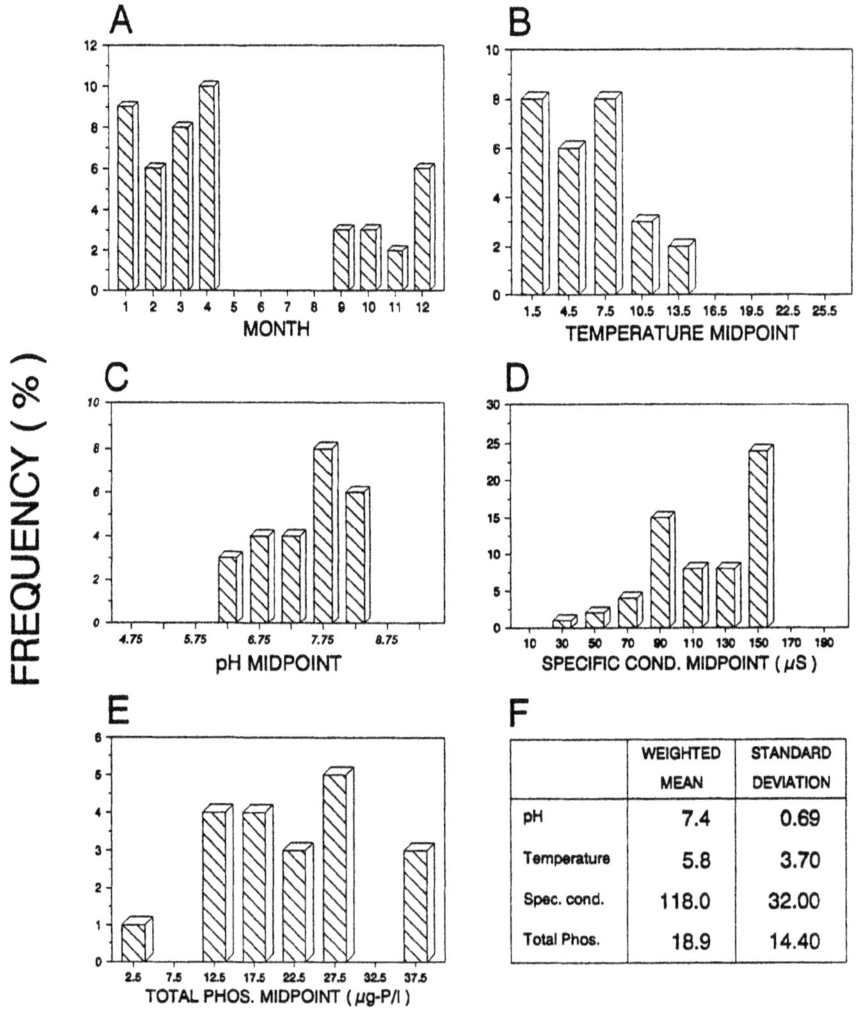

	WEIGHTED MEAN	STANDARD DEVIATION
pH	7.4	0.69
Temperature	5.8	3.70
Spec. cond.	118.0	32.00
Total Phos.	18.9	14.40

Fig. 5.13. The frequencies of occurrence for *Mallomonas elongata* along seasonal (A), temperature (B), pH (C), specific conductance (D) and total phosphorus (E) gradients. Weighted mean and standard deviation values are also given (F).

Table 5.9. Literature records indicating the ecological preferences of *Mallomonas elongata.*

Reference	Seasonality	Trophic Status	Temperature (°C)	pH and Related Factors	Other
1. Asmund, 1959	Winter and spring.	---	9	7.6	Denmark
2. Asmund and Hilliard, 1961	Autumn	Oligotrophic	2 to 12	7.2 to 8.6, slightly alkaline.	Alaska
3. Asmund and Takahashi, 1969	---	Oligotrophic to eutrophic; dystrophic.	2 to 18	---	Alaska
4. *Battarbee, Cronberg and Lowry, 1980	---	Rather more productive habitats.	---	Slightly alkaline.	Sweden
5. *Charles and Smol, 1988	---	---	---	AWMpH = 7.01 neutral to alkaline conditions.	Adirondacks
6. *Christie and Smol, 1986	---	Most abundant when the lake supposedly the most alkaline.	---	---	Upper Wallface Pond. Adirondacks. Co-dominant with *M. caudata*.
7. Ito, 1988	Winter, spring, summer.	---	---	---	Lake Biwa, Japan.
8. Roijackers, 1981	Spring	---	---	Neutral species.	Netherlands
9. Siver, 1988c	November	Slightly productive.	Cold	6.75	Adirondacks
10. Siver, 1989a	---	---	---	Circumneutral to alkaliphilic. WMpH 6.8 for literature survey and 6.9 for Connecticut lakes.	Literature survey. Rather rare species.
11. *Smol et al., 1984	---	Slightly productive habitats.	---	Slightly alkaline, AWMpH = 7.1, range = 5.7 to 7.8	Adirondacks
12. Takahashi, 1978	---	---	Cold stenothermal; 2 to 12; maximum at 8.	Alkalibiontic 7.2 to 8.8	Japan
13. Wujek et al., 1977	---	---	26	8.6	Michigan

AWMpH = abundance weighted mean pH.
*Indicates a paleolimnological study.

sample containing a few isolated scales, *M. elongata* was not found in 202 collections made between May and September. Because of its restricted seasonal distribution, *M. elongata* is a cold water stenotherm with a maximum occurrence below 9°C and a weighted mean of only 5.8°C (Fig. 5.13B,F). Whole cells were observed at temperatures between 1°C and 14.5°C. *M. elongata* was never found in great abundance, perhaps due to it being restricted to cold conditions.

The preference of *M. elongata* for cold water conditions is supported by the literature (Table 5.9). Asmund and Hilliard (1961), Asmund (1959), Takahashi (1978), Cronberg and Kristiansen (1980), Roijackers and Kessels (1981; 1986) and Dürrschmidt (1982b) all observed this organism only in collections with a temperature less than 12°C. In addition, *M. elongata* was reported during the fall in Alaska (Asmund and Hilliard, 1961), spring in the Netherlands (Roijackers and Kessels, 1986) and winter through spring in Denmark (Asmund, 1959) and Japan (Takahashi, 1978). Although Wujek et al. (1977) reported this species from a waterbody in Michigan at 26°C, it is unclear if the observation was of an isolated scale or a whole cell. Based on the body of available evidence, it is clear that *M. elongata* prefers cold conditions.

In Connecticut and Adirondack lakes *M. elongata* was found over a relatively narrow pH range of 6.4 to 8.7 (Fig 5.13C) and had a weighted mean pH of 7.4. This taxon was lacking from over 200 collections with a pH less than 6.4. Except for one record of a pH of 5.5 (Cronberg and Kristiansen, 1980), the pH range of 6.3 to 8.6 found in the literature review was strikingly similar to that found in the Connecticut and Adirondack lakes; the weighted mean was slightly lower at 6.8 (Siver, 1989a). Charles and Smol (1988) reported an abundance weighted mean pH of 7.0 for *M. elongata*. Asmund and Hilliard (1961) and Takahashi (1978) also reported similar pH ranges of 7.2 to 8.6 and 7.2 to 8.8, respectively for *M. elongata*. Battarbee et al. (1980) and Smol et al. (1984a) also cited this taxon as favoring slightly alkaline conditions. Based on all of the available records, *M. elongata* is best described as pH indifferent with alkaliphilic tendencies (Siver, 1989a).

M. elongata was observed mostly in large clear-water lakes that were mesotrophic to eutrophic in nature and associated with summer blooms of blue-green algae. Even though this species had wide and rather scattered distributions over both specific conductance and total phosphorus gradients (Figs. 5.13D-E), it had high weighted mean scores of 118 μS and 19 μg-P/l, respectively. It was absent in 57 collections with a specific conductance less than 20 μS.

In Alaska (Asmund and Hilliard, 1961; Asmund and Takahashi, 1969) *M. elongata* has been reported in waterbodies ranging in trophic status from oligotrophic to eutrophic. Battarbee et al. (1980) believed that an increase in *M. elongata*, along with *Synedra acus*, within the upper sediments of a small Finnish lake signaled an increase in the productivity of the lake. Smol et al. (1983) also reported that increases in *M. elongata* accompanied the early stages of eutrophication. As was observed in Connecticut and Adirondack lakes, Asmund and Takahashi (1969) also found *M. elongata* to be somewhat restricted to larger waterbodies and lacking from the smaller ponds rich in *Mallomonas*. This observation may indicate a requirement for clear, non-humic stained conditions.

In summary, *M. elongata* is a coldwater taxon, found during the fall, winter and spring months in waterbodies that are slightly alkaline and moderately eutrophic in nature.

Mallomonas galeiformis

M. galeiformis, a newly described species (Nicholls, 1988a), has been reported to be restricted along various environmental gradients (Siver, 1988b). This species was found in only 7% of the collections from Connecticut localities, however, it was much more common in Adirondack lakes where it was present in 33% of the samples (Siver, 1988b). In both localities, populations of *M. galeiformis* were primarily observed during the late spring to mid autumn period (Fig. 5.14A). Generally, populations began growth in mid May at water temperatures between 13°C and 17°C (Siver, 1988b), were maintained throughout the summer and disappeared by the end of October. Most observations of this organism made beyond mid November were of isolated scales and bristles and not whole cells. *M. galeiformis* was most frequent between August and October where it was found in 23% of the collections (Fig. 5.14A). However, populations of this taxon were not detected in over 100 collections made between January and mid April.

Because of its predominance during the warmer months, *M. galeiformis* had a relatively high weighted mean temperature of 16.4°C and a maximum occurrence in samples with a temperature above 20°C (Fig. 5.14B). The greatest concentrations of cells were observed in collections above 20°C.

M. galeiformis is predominently found in habitats that are slightly acidic and have low specific conductance (Siver, 1988b). In Connecticut and Adirondack lakes this species was observed between pH 5.6 and 7.1, had a maximum occurrence in the pH 6 to 6.5 interval and had a weighted mean pH of 6.4 (Fig. 5.14C). *M. galeiformis* is clearly lacking from extreme acidic or alkaline habitats (Siver, 1989a). Charles and Smol (1988) reported a similar weighted mean pH of 6.34. In a paleolimnological study of Upper Wallface, a lake with a contemporary pH of less than 5.0, Christie and Smol (1986) observed *M. galeiformis* (as *M. trummensis*) to have appeared along with *M. punctifera* during the initial stages of acidification and later disappear, presumably as the pH continued to drop below 5.5. This observation supports the hypothesis proposed by Siver (1988b) that *M. galeiformis* is a useful paleolimnological indicator for pH. Based on all available data, *M. galeiformis* is best described as an acidoplilous taxon.

M. galeiformis was also found to be restricted to localities with a specific conductance between

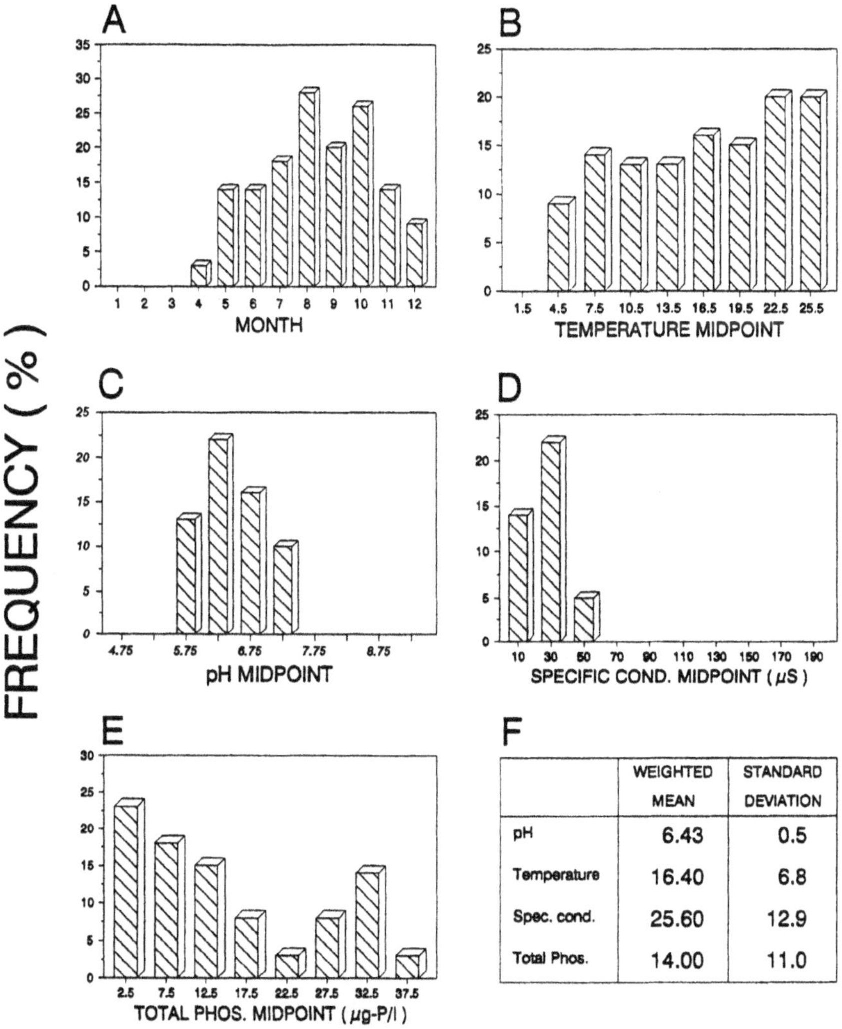

	WEIGHTED MEAN	STANDARD DEVIATION
pH	6.43	0.5
Temperature	16.40	6.8
Spec. cond.	25.60	12.9
Total Phos.	14.00	11.0

Fig. 5.14. The frequencies of occurrence for *Mallomonas galeiformis* along seasonal (A), temperature (B), pH (C), specific conductance (D) and total phosphorus (E) gradients. Weighted mean and standard deviation values are also given (F).

Table 5.10. Literature records indicating the ecological preferences of *Mallomonas galeiformis.*

Reference	Seasonality	Trophic Status	Temperature (°C)	pH and Related Factors	Other
1. *Charles and Smol, 1988	---	---	---	AWMpH = 6.34	Adirondacks
2. *Christie and Smol, 1986	---	---	---	Declines in abundance as pH decreases below 5.	Upper Wallface Pond, Adirondacks. Reported as *M. trummensis.*
3. Siver, 1988b	Summer, warm periods of the year.	Oligotrophic	Primarily between 17 and 25; weighted mean of 20.	Range = 5.8 to 7.1; WMpH = 6.3; Acidophilous.	Requires low specific conductance. Connecticut and the Adirondacks.
4. Siver, 1988c	Summer	---	---	Range = 5.9 to 7.1 absent in very acid localities, including all bogs.	Reported from over 40% of the lakes surveyed in the Adirondacks.
5. Siver, 1989a	---	---	---	WMpH = 6.3. Acidophilous, Range = 5.6 to 7.1.	Literature survey.
6. Siver and Hamer, 1989	Summer	---	Warm water.	---	Specific conductance and water temperature control first principal component.

AWMpH = abundance weighted mean pH.
*Indicates a paleolimnological study.

18 µS and 45 µS (Fig. 5.14D). Siver (1988b) postulated that the lack of populations of *M. galeiformis* in waterbodies with a pH greater than 7.1 was due to its requirement for low specific conductance. In addition, Siver (1988b) pointed out that the greater frequency of occurrence of this taxon in the Adirondacks was most likely the result of a predominance of localities low in specific conductance.

Waterbodies harboring populations of *M. galeiformis* were primarily oligotrophic to mesotrophic, with good light penetration and located in relatively undisturbed woodland settings. *M. galeiformis* was most often encountered in samples with total phosphorus concentrations less than 5 µg-P/l, although populations were also found at higher concentrations (Fig. 5.14E). The weighted mean total phosphorus level was 13.9 µg-P/l.

Mallomonas hamata

Mallomonas hamata is widely distributed in the northern hemisphere, but, to date has not been observed in the southern hemisphere (Cronberg, 1989). In Connecticut and Adirondack waterbodies it was a very common taxon, present in 31% and 50% of the collections, respectively. In previous surveys from the Adirondacks Siver (1988c) and Charles and Smol (1988) found *M. hamata* in 59% and 68% of the lakes, respectively. In a study from

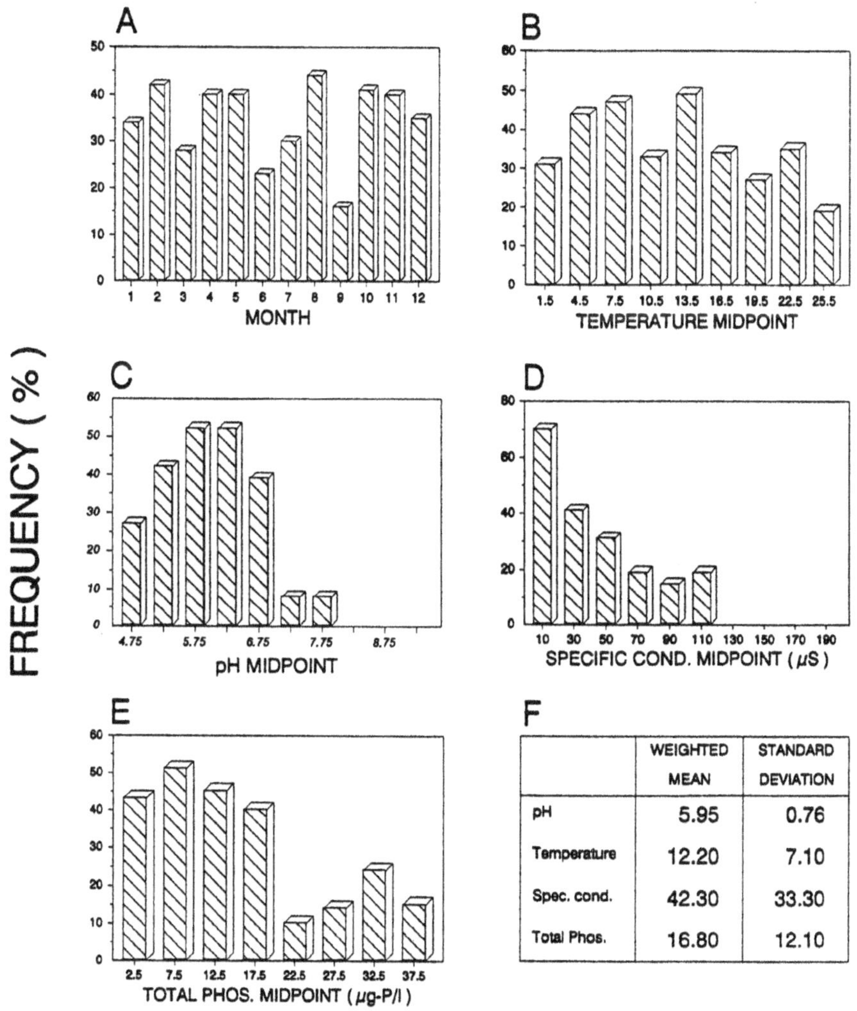

Fig. 5.15. The frequencies of occurrence for *Mallomonas hamata* along seasonal (A), temperature (B), pH (C), specific conductance (D) and total phosphorus (E) gradients. Weighted mean and standard deviation values are also given (F).

Table 5.11. Literature records indicating the ecological preferences of *Mallomonas hamata*.

Reference	Seasonality	Trophic Status	Temperature (°C)	pH and Related Factors	Other
1. Asmund, 1959	---	Mostly oligotrophic, but also eutrophic.	8	6.9	Denmark
2. *Charles et al., 1987	---	---	---	Increase in abundance with increase in acid content.	Big Moose Lake, Adirondacks
3. *Charles and Smol, 1988	---	---	---	AWMpH = 5.4	Adirondacks
4. *Christie and Smol, 1986	---	---	---	Increase in abundance with increase in acid content.	Co-dominant with *Synura echinulata*; Upper Wallface Pond, Adirondacks.
5. Cronberg, 1989	Most frequent in spring, but also found during the summer.	Oligotrophic	Mostly below 10.	5.5 to 6.5	Associated with oligotrophic flora. To date, restricted to northern hemisphere.
6. Cronberg and Kristiansen, 1980	---	Oligotrophic	---	Acidic	Clearwater lakes only, Sweden.
7. *Dixit et al., 1988a	---	---	---	AWMpH = 5.2	Sparse in lakes with high Cu and Ni content. Sudbury.
8. *Dixit et al., 1988b	---	---	---	Acidic taxon	Clearwater habitats; Quebec region.
9. *Dixit et al., 1989	---	---	---	---	Sensitive to high metal concentrations.
10. Donaldson and Stein, 1984	March, May	---	11 and 16	5 and 5.7	Canada
11. Eloranta, 1989	---	---	---	Acidophilous	21% of collections. Finland.
12. Green, 1979; 1980	---	---	---	Acidic tendency	Canada

Table 5.11 continued

Reference	Seasonality	Trophic Status	Temperature (°C)	pH and Related Factors	Other
13. Kling and Kristiansen, 1983	---	Oligotrophic	---	---	Canada
14. Kristiansen, 1979	---	Oligotrophic	---	---	Britain
15. Kristiansen, 1981	---	---	---	Weakly buffered lakes.	Clearwater lakes only, Denmark.
16. Kristiansen, 1986	---	---	---	Weakly buffered lakes	Clearwater lakes. Literature review.
17. Nicholls, 1982	---	---	---	Soft water lakes	Common; Ontario
18. Roijackers and Kessels, 1986	---	Slightly eutrophic	---	Low pH, low alkalinity.	Netherlands
19. Siver, 1988c	---	Oligotrophic	---	4.2 to 7.	Found in 59% of the lakes. Adirondacks. Lacking in bogs.
20. Siver, 1989a	---	---	---	AWMpH = 5.9 for Connecticut and Adirondack lakes and 6.4 for literature review. Lacking in bogs.	Literature review. Much more common in Connecticut and Adirondack lakes than in literature review.
21. Siver and Hamer, 1989	---	---	---	Acidophilous to acidobiontic. AWMpH = 6	Specific conductance and pH controlled the first principal component.
22. *Smol et al., 1984a	---	---	---	Most abundant in acidic lakes; Lacking in bogs.	Adirondacks
23. *Smol et al., 1984b	---	---	---	Indicator of acidic, non-bog habitats.	Adirondacks
24. *Smol, 1986	---	---	---	Most abundant below pH 5	Lacking in bogs.
25. Wujek, 1986	---	Oligotrophic	---	---	Michigan

AWMpH = abundance weighted mean pH.
*Indicates a paleolimnological study.

Finland Eloranta (1989) found *M. hamata* in 21% of the samples. Nicholls (1982) reported this organism to be one of the most abundant species of scaled chrysophytes from softwater localities in Ontario. Despite its abundance in regions with a predominance of softwater lakes, *M. hamata* was found in only 4% of the collections from the literature survey.

Mallomonas hamata was relatively equally distributed throughout the year and exhibited no preference for a given season (Fig. 5.15A). It was most common during August, present in 44% of the collections and least common in September, found in only 16% of the samples. Even though *M. hamata* had a weighted mean temperature of 12°C, it was quite common over the whole gradient (Fig. 5.15B), resulting in a relatively large standard deviation of 7°C. This species exhibited a slight preference for temperatures below 15°C and was less often encountered above 24°C (Fig. 5.15B). Despite the body of information available for *M. hamata*, little has been published concerning its seasonality and temperature preferences (Table 5.11). When reported, it has most often been observed below 15°C (Asmund, 1959; Nygaard, 1978; Skogstad, 1982; Donaldson and Stein, 1984; Green, 1979; 1980; Cronberg and Kristiansen, 1980).

Although *M. hamata* was found over a pH range from 4.7 to 7.9, it was primarily distributed below a pH of 7 and had a low weighted mean of 5.9 (Fig. 5.15C). This organism was found in 51% of the samples with a pH between 5 and 6.5, 39% of those between pH 6.5 and 7 and only 8% of those between pH 7 and 7.9. *Mallomonas hamata* was observed in 27% of the collections below a pH of 5.0.

Unlike temperature and seasonality, a wide body of literature exists describing the distribution of *M. hamata* along a pH gradient (Table 5.11). A review of the literature strongly supports the finding that *M. hamata* is acidic in nature. Asmund (1959), Green (1979; 1980), Cronberg and Kristiansen (1980), Nygaard (1978), Siver (1988c; 1989a), Roijackers and Kessels, (1986), Smol et al. (1984a; 1984b), Charles and Smol (1988), Donaldson and Stein (1984), Dixit et al. (1988a) and Eloranta (1989) have all documented *M. hamata* to prefer acidic localities. Charles and Smol (1988) and Dixit et al. (1988a) recorded abundance weighted mean pH values of 5.4 and 5.2, respectively. Eloranta (1989) reported a pH range of 4.7 to 7.1 and a mean of 6.1 (Table 5.11). It is concluded that *M. hamata* is best described as an acidophilous/acidobiontic taxon.

There is also substantial evidence that *M. hamata* is found primarily in clearwater localities (Kristiansen, 1981; Cronberg and Kristiansen, 1980; Kristiansen, 1986; Dixit et al., 1988b) and lacking from humic stained bog lakes (Siver, 1988c; Smol et al., 1984a; Smol, 1986). If the bog lake samples are removed from the data set, *M. hamata* would have a weighted mean pH of 5.5 and be present in 50% of the collections below pH 5.5.

In paleolimnological surveys of clearwater lakes from the Adirondack region, *M. hamata* has been a valuable bioindicator taxon for the study of the effects of acidic deposition. In Deep Lake (Smol et al., 1984b), Upper Wallface Lake (Christie and Smol, 1986) and Big Moose Lake (Charles et al., 1987) the abundances of *M. hamata* have increased significantly in the surface sediments relative to older layers down core, presumably signaling a decrease in lakewater pH caused by acidic deposition.

Mallomonas hamata also exhibited a greater tolerance for lakes low in specific conductance (Fig. 5.15D). It had a maximum frequency of occurrence of 70% in collections with a specific conductance less than 20 μS, a weighted mean of 42 μS and was not found above 118 μS. In addition, this organism was primarily found in oligotrophic and mesotrophic waterbodies with a total phosphorus level less than 20 μg-P/l. However, it can be found in more eutrophic localities at higher concentrations of phosphorus (Fig. 5.15E). The literature supports the observation that *M. hamata* is more common in oligotrophic softwater lakes (Asmund, 1959; Cronberg and Kristiansen, 1980; Kristiansen, 1979; Nicholls, 1982; Kling and Kristiansen, 1983 and Siver, 1988c). Roijackers and Kessels (1986) and Asmund (1959) reported this taxon from slightly eutrophic sites.

In summary, *M. hamata* is a taxon most commonly found in poorly buffered, dilute waterbodies that are acidic and relatively unproductive in nature. Siver and Hamer (1989) demonstrated that pH and specific conductance both strongly dominated the first principal component, lending further support of the importance of these variables in controlling the distribution of *M. hamata*. There are preliminary data that suggests that this taxon is sensitive to high concentrations of metals, especially Cu and Ni (Dixit et al., 1988a; 1989).

Mallomonas heterospina

Mallomonas heterospina, a geographically widely distributed species (Kling and Kristiansen, 1983; Asmund and Kristiansen, 1986), was found in 9% of the samples from Connecticut localities, but in 26% of those surveyed from the literature (Siver, 1989a). Charles and Smol (1988) reported that *M. heterospina* was relatively rare in sediments from the Adirondacks, found in only 8% of the lakes.

This species begins development during mid-autumn, coincident with a decrease in water temperature, thrives under the ice and disappears soon after ice-out (Fig. 5.16A). *M. heterospina* was found in 18% of the collections made between November and April, however, it was not found in over 175 collections made from May through September (Fig. 5.16A). As a result of its seasonal distribution, *M. heterospina* is a cold water taxon with a maximum occurrence below 9°C (Fig. 5.16B) and a weighted mean temperature of only 7.2°C. Its frequency of occurrence dropped significantly as the temperature increased above 9°C and the taxon was not found above 16°C (Fig. 5.16B).

The preference of *M. heterospina* for winter and

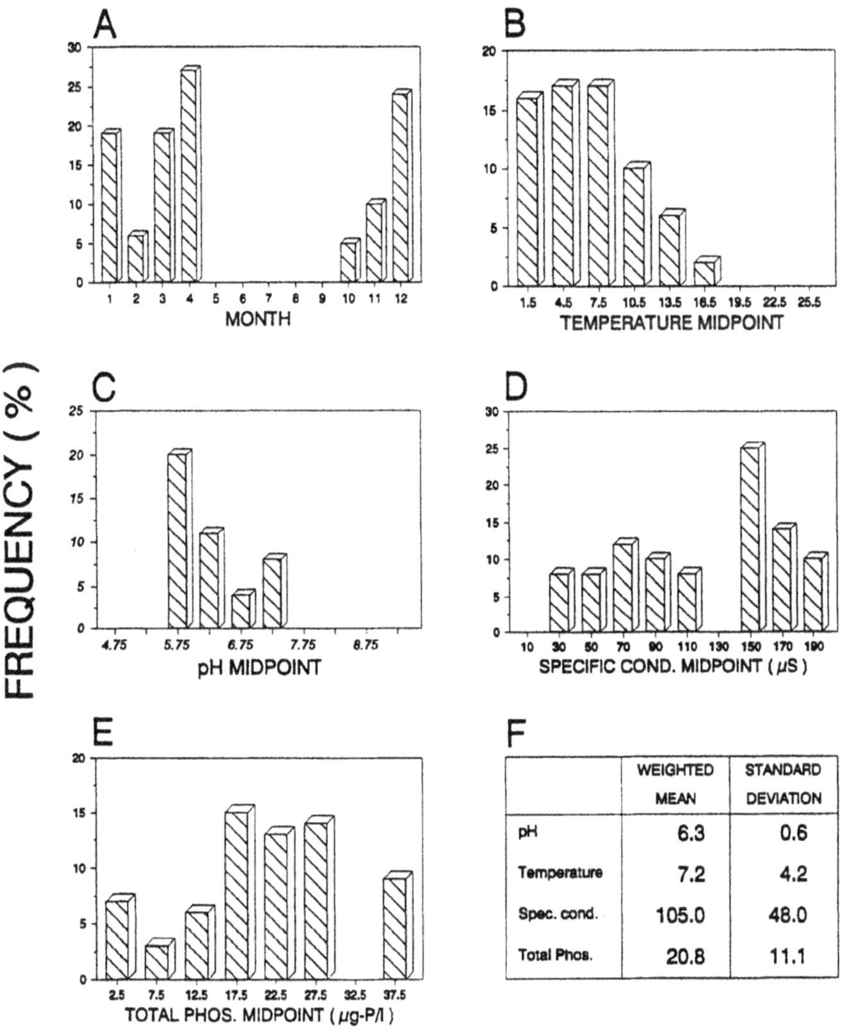

	WEIGHTED MEAN	STANDARD DEVIATION
pH	6.3	0.6
Temperature	7.2	4.2
Spec. cond.	105.0	48.0
Total Phos.	20.8	11.1

Fig. 5.16. The frequencies of occurrence for *Mallomonas heterospina* along seasonal (A), temperature (B), pH (C), specific conductance (D) and total phosphorus (E) gradients. Weighted mean and standard deviation values are also given (F).

Table 5.12. Literature records indicating the ecological preferences of *Mallomonas heterospina*.

Reference	Seasonality	Trophic Status	Temperature (°C)	pH and Related Factors	Other
1. Andersen, 1978	---	Humic stained.	---	Slightly acidic.	Arkansas
2. Asmund, 1956	Most abundant in winter/spring; rare or absent in summer and early autumn.	Most abundant in eutrophic and humic stained ponds.	Maximum development below 10, especially during ice and ice-out periods.	Slightly acidic.	Found in July only in darkly stained localities.
3. Asmund and Takahashi, 1968	Winter and spring.	Humic stained, eutrophic	Cold	Slightly acidic.	Alaska
4. Bradley, 1964	Winter and spring.	---	Cold	Near 7.	England
5. *Charles and Smol, 1988	---	---	---	AWMpH of 6.19	Adirondacks
6. *Cronberg, 1986	---	Cysts increased with increased eutrophication; but disappeared during the most eutrophic phase.	---	---	Lake Växjösjön, Sweden.
7. Cronberg, 1989	Cold water form with maximum development in spring.	Mesotrophic to eutrophic.	Cold temperatures. 1.8 - 7.8	6.3 - 7.7, but tendency for slightly acidic localities.	Found often with eutrophic blue-green and green taxa.
8. Cronberg and Kristiansen, 1980	Spring	Eutrophic	Cold	---	Sweden
9. Dürrschmidt, 1980	October	Oligotrophic	near 15	---	Chile
10. Dürrschmidt, 1984	March-April.	---	Cold	---	Germany
11. Eloranta, 1985	---	Humic lakes.	---	Slightly acidic lakes low in alkalinity. 5.5 - 6.3	Western Finland
12. Eloranta, 1986b	---	Ultra oligotrophic-Oligotrophic sub-artic lakes.	---	---	Finland
13. Eloranta, 1989	Lacking in summer	---	Cold	---	Finland
14. Green, 1979	Cooler months.	---	15 - 19	5.3 - 6.7, slightly acidic.	Lakes on Vancouver Island.
15. Harris and Bradley, 1957	---	Eutrophic-dung contaminated habitats.	---	---	England

Table 5.12. continued

Reference	Seasonality	Trophic Status	Temperature (°C)	pH and Related Factors	Other
16. Ito, 1988	Winter	---	Cold	---	Lake Biwa, Japan
17. Kling and Kristiansen, 1983	---	---	---	---	Widely distributed, Canada.
18. Kristiansen, 1981	---	Eutrophic, cattle watering hole.	---	---	---
19. Kristiansen, 1985; 1988	Winter and spring February to April.	Eutrophic, high total phosphorus.	Cold	Alkaline, 8.2 - 9.1	Lake Tystrup So., Denmark.
20. Kristiansen, 1986	---	One of few scaled chrysophytes to withstand highly eutrophic condition.	---	---	---
21. Nygaard, 1949	Winter	---	Less than 8.	6.8 - 8.0	Denmark
22. Péterfi, 1966	Winter and spring.	Dark-coloured.	Cold, under ice.	Slightly acidic.	Bogs, pools, pits in Romania.
23. Roijackers and Kessels, 1986	Cooler months.	---	Cold water with maximum between 6-10.	---	Netherlands
24. Siver, 1989a	---	---	---	WMpH 6.3 acidophilic.	Connecticut and general literature review.
25. Siver and Hamer, 1989	---	Oligotrophic and eutrophic.	Cold with weighted mean of 6.	Acidophilous	Connecticut
26. Skogstad, 1982	Cooler months.	Varied	4 - 11	5.4 - 8.2	Norway
27. *Smol et al., 1984a	---	Lakes with greater trophic status.	---	Slightly acidic.	Overall, rare in Adirondack lakes.
28. Takahashi, 1978	Winter	---	Cold stenotherm.	Acidophilous	Japan
29. Wujek, Hamilton and Wee, 1975	Fall	---	---	---	Lake Isabella, Michigan.

AWMpH = abundance weighted mean pH.
*Indicates a paleolimnological study.

spring months and cold temperatures is strongly supported by the literature (Table 5.12). *M. heterospina* has been found in Denmark (Asmund, 1956; Kristiansen, 1985; 1988), Alaska (Asmund and Takahashi, 1968), England (Bradley, 1964), Romania (Péterfi, 1966), Sweden (Cronberg and Kristiansen, 1980), Germany (Dürrschmidt, 1984), Japan (Takahashi, 1978), the Netherlands (Roijackers and Kessels, 1986) and Finland (Eloranta, 1989) as being a winter - spring form thriving in cold water. Asmund (1956) and Cronberg (1989) also found *M. heterospina* to be most abundant during the winter to spring period, decrease in importance by early summer and disappear by the warmer summer months. Eloranta (1989) found that although *M. heterospina* was common in Finland, it was lacking in 61 collections taken during the summer. Nygaard (1949), Cronberg and Kristiansen (1980), Skogstad (1982), Kristiansen (1985) and Cronberg (1989) each reported *M. heterospina* to be most abundant in water temperatures less than 10°C. Asmund (1956) found this species during July, but only in a few of the most darkly coloured habitats.

In Connecticut lakes *M. heterospina* was found over a rather narrow pH range, from 5.5 to 7.2, however, it had a clear preference for slightly acidic conditions and a maximum frequency of 23% between pH 5.5 and 6.0 (Fig. 5.16C). Although a slightly wider pH range was evident from the literature survey (Siver, 1989a), the frequency distribution was similar; the maximum occurrence was between pH 5.0 and 6.0 and decreased significantly as the pH increased to 8.5. *M. heterospina* was found in approximately 50% of the collections surveyed from the literature between a pH of 5.0 and 5.5, but was noticeably absent below 5.0. It was also not found in any Connecticut or Adirondack waterbodies below pH 5.0 (Fig. 5.16C). The weighted mean pH from Connecticut localities, the literature survey and in surface sediments from the Adirondacks (Charles and Smol, 1988) were 6.3, 6.3 and 6.2, respectively.

Many studies document the fact that *M. heterospina* favors slightly acidic conditions, is lacking from extreme acid levels and is best classified as acidophilic in nature (Table 5.12). It should be emphasized that even though *M. heterospina* is an acidophilic taxon, it can be found under alkaline conditions (Kristiansen, 1986; 1988).

Mallomonas heterospina was found in both oligotrophic and eutrophic waterbodies and in humic stained localities, provided that the pH was greater than 5.0. As such, it occurred over a wide range of specific conductance and total phosphorus conditions (Figs. 5.16D-E) and had relatively high weighted mean scores of 105 μS and 20.8 μg-P/l, respectively. This organism was not found in lakes with a specific conductance less than 25 μS and was present during the winter in several ponds with large overwintering populations of geese.

The tendencies for *M. heterospina* to exist in both eutrophic (Asmund, 1956; Harris and Bradley, 1957; Cronberg and Kristiansen, 1980; Smol and Boucherle, 1985; Kristiansen, 1981; 1986; 1988) and humic stained (Asmund 1956; Péterfi, 1966; Asmund and Takahashi, 1968; Andersen, 1978; Eloranta, 1985) waterbodies is well documented. In addition, Harris and Bradley (1957) and Kristiansen (1981; 1986) noted *M. heterospina* to be one of the only species of scaled chrysophytes to inhabit dung contaminated localities. Cronberg (1986) found concentrations of cysts from this taxon to increase with increased eutrophication, but then disappear during the most eutrophic period. It was postulated that the lower occurrence of *M. heterospina* in Connecticut (Siver, 1989a) and Adirondack (Smol et al., 1984a; Siver, 1989a; Charles and Smol, 1988) lakes, as compared to the literature review, was due to the smaller percentage of eutrophic waterbodies represented in those studies.

Mallomonas lychenensis

Mallomonas lychenensis and *M. allorgei* are closely related taxa (see Chapter III) that are difficult to separate, especially on the basis of isolated scales (Harris and Bradley, 1960; Nicholls, 1982; Asmund and Kristiansen, 1986). Since the distributional records in the literature of the two organisms are most likely mixed, the ecology of the taxa will be discussed together.

In Connecticut, *M. lychenensis* was predominately found in slightly acidic, softwater, humic stained lakes, although a few cells were also noted from a clearwater locality with a pH above 7. Populations of this taxon were most abundant in the humic stained habitats where they were present throughout most of the year at temperatures between 3°C and 28°C (Fig. 5.17B). The humic stained localities harboring *M. lychenensis* were moderately enriched with phosphorus, primarily as organic phosphorus and the organism had a pH range from 5.6 to 7.3 (Fig. 5.17C).

Both *M. lychenensis* and *M. allorgei* are reported as being rare (Dürrschmidt, 1984; Smol et al., 1984a; Siver, 1989a), however, Eloranta (1989) observed *M. allorgei* in 62% of the collections from southern and central Finland. Waterbodies harboring populations of both taxa are best characterized as being slightly acidic, oligotrophic and humic stained in nature. *Mallomonas allorgei* has been reported to have an abundance weighted mean pH of 6.85 (Charles and Smol, 1988), to be most common between pH 6.5 and 7.5 (Smol et al, 1984a; Dixit et al., 1988b) and to decrease in relative abundance as the pH drops below 6; it is classified as acidophilic (Siver, 1989a). In addition, Eloranta (1989) found *M. allorgei* to be positively correlated with increasing water color and decreasing lake size, while Smol (1980) correlated a decrease in abundance with an increase in nutrient loading.

Mallomonas lychenensis, more often encountered than *M. allorgei*, has been found between pH 4.1 (Cronberg and Kristiansen, 1980) and 7.0 (Skogstad, 1982), in humic localities (Cronberg and Kristiansen, 1980) and has also been classified as acidophilic (Siver, 1989a).

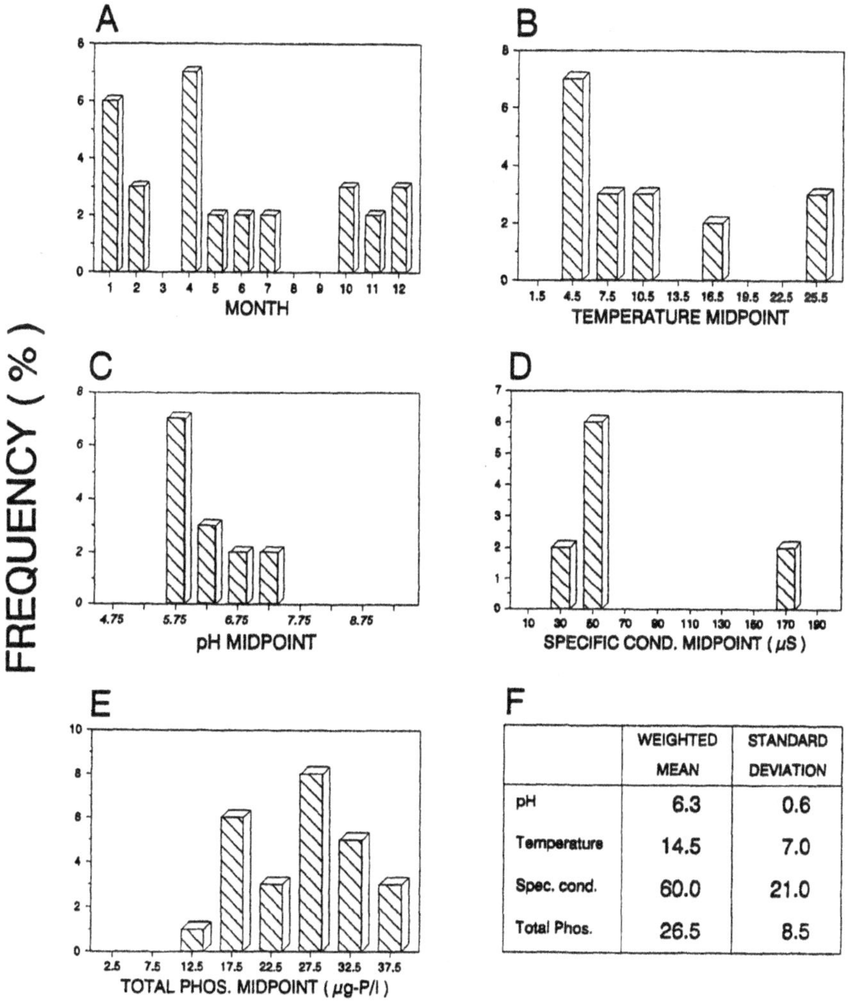

	WEIGHTED MEAN	STANDARD DEVIATION
pH	6.3	0.6
Temperature	14.5	7.0
Spec. cond.	60.0	21.0
Total Phos.	26.5	8.5

Fig. 5.17. The frequencies of occurrence for *Mallomonas lychenensis* along seasonal (A), temperature (B), pH (C), specific conductance (D) and total phosphorus (E) gradients. Weighted mean and standard deviation values are also given (F).

Mallomonas matvienkoae

Mallomonas matvienkoae has a worldwide distribution (Table 5.21) and is found from northern temperate regions to the tropics (Wujek and Timpano, 1984; Dürrschmidt and Croome, 1985). Although rare in both Connecticut and the Adirondacks, present in less than 1% of the collections, it is interesting to note that based on preliminary data *M. matvienkoae* was found in 6 of 10 and 5 of 10 collections from Florida and the People's Republic of China, respectively (unpublished data). Many of the sites in Florida were eutrophic and humic stained in nature, while those in China were severely culturally eutrophied. Based on the few available records this species was present between March and September and had a preference for warm water conditions, although Kristiansen (1985; 1988) and Dürrschmidt (1983b, as f. *litteata*) both reported it under cooler temperatures.

Mallomonas paludosa

Mallomonas paludosa is a rare taxon, found in only 2% of the collections from Connecticut and the Adirondacks, and restricted to very acidic waterbodies (Fig. 5.18C). This species had a maximum occurrence in collections with a pH less than 5.5 and a low weighted mean pH of 5.1 (Fig. 5.18C). *M. paludosa* is well documented in the literature

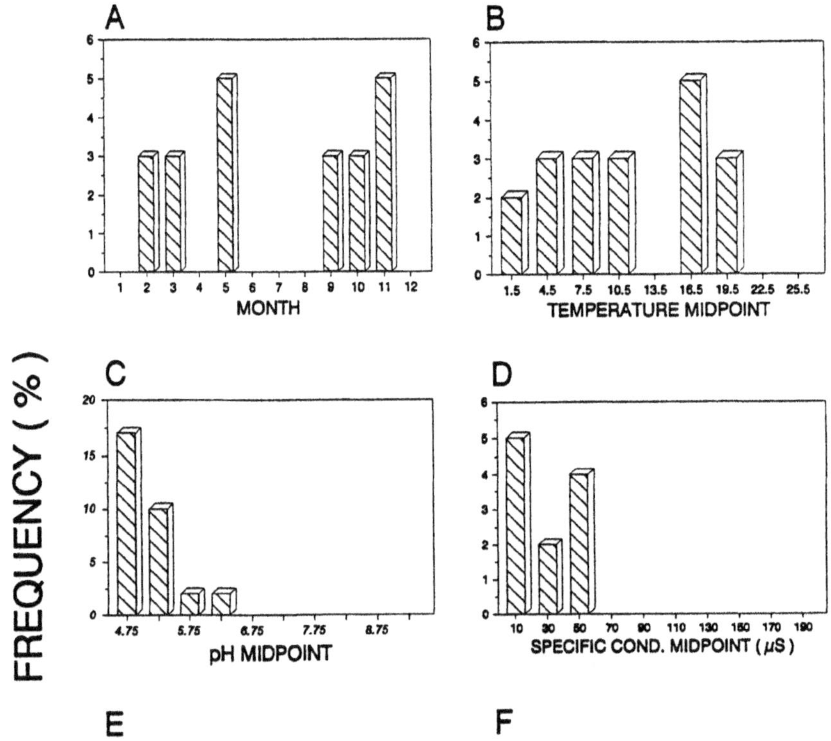

	WEIGHTED MEAN	STANDARD DEVIATION
pH	5.1	0.4
Temperature	13.0	5.0
Spec. cond.	22.0	8.0
Total Phos.	8.9	4.4

Fig. 5.18. The frequencies of occurrence for *Mallomonas paludosa* along seasonal (A), temperature (B), pH (C), specific conductance (D) and total phosphorus (E) gradients. Weighted mean and standard deviation values are also given (F).

Table 5.13. Literature records indicating the ecological preferences of ***Mallomonas paludosa***.

Reference	Seasonality	Trophic Status	Temperature (°C)	pH and Related Factors	Other
1. Cronberg and Kristiansen, 1980	---	Sphagnum bog.	---	Acidic, 4.1	Rare. Sweden
2. Eloranta, 1989	---	Humic localities.	---	Acid humic lakes. Range: 4.7 to 5.6; mean = 5.1	Finland
3. *Hartmann and Steinberg, 1986	---	---	---	Few scales found in the recent sediments of acidified lake.	Alpine Mountain Lake; clear water.
4. Kling and Kristiansen, 1983	---	---	---	Prefers acidic water.	Canada
5. Kristiansen, 1981	---	Humic localities	---	Humic acidic waters.	Rare taxon, found in the same localities as *Synura sphagnicola*.
6. Kristiansen, 1986	---	---	---	Indicator of acid, sometimes humic waters.	Literature review.
7. Péterfi and Momeu, 1976	August	Ombrophilous bog.	---	4.5 to 5.0; acidic bog.	Romania
8. Roijackers and Kessels, 1986	---	---	---	Found in only low pH conditions; 4.3	Netherlands
9. Siver, 1988c	---	---	---	4.77 to 5.42; clear water acidic lakes.	Adirondacks
10. Siver, 1989a	---	---	---	WMpH = 5.1; acidobiontic; clearwater localities.	Literature survey.

*Indicates a paleolimnological study.

as being characteristic of acidic localities (Table 5.13) and is best classified as an acidobiontic species (Siver, 1989a).

Although *M. paludosa* is often reported from humic stained or bog localities (Kristiansen, 1981; 1986; Cronberg and Kristiansen, 1980; Péterfi and Momeu, 1976a; Eloranta, 1989), it has also been found in acidic lakes with very clear water (Hartmann and Steinberg, 1986: Siver, 1988c). In Connecticut and the Adirondacks this species was also limited to habitats low in specific conductance and total phosphorus levels and had weighted mean values of 22 μS and 8.9 μg-P/l, respectively (Figs. 5.18D-E).

Because of limited data the seasonality of *M. paludosa* could not be fully documented, however, a few preliminary remarks are possible. Populations of this species were most abundant in the spring and autumn and lacking during the warm summer months (Fig. 5.18A). *M. paludosa* was found over a temperature range from 3°C to 20°C, but had a relatively low weighted mean of 13.3°C (Fig. 5.18B,F). Although records are scarce, *M. paludosa* has been reported at 6.8°C (Cronberg and Kristiansen, 1980) and 14°C (Roijackers and Kessels, 1986), supporting the idea that it is a cold water organism.

Mallomonas papillosa

Mallomonas papillosa, a widespread species (Asmund and Takahashi, 1968; Kling and Kristi-

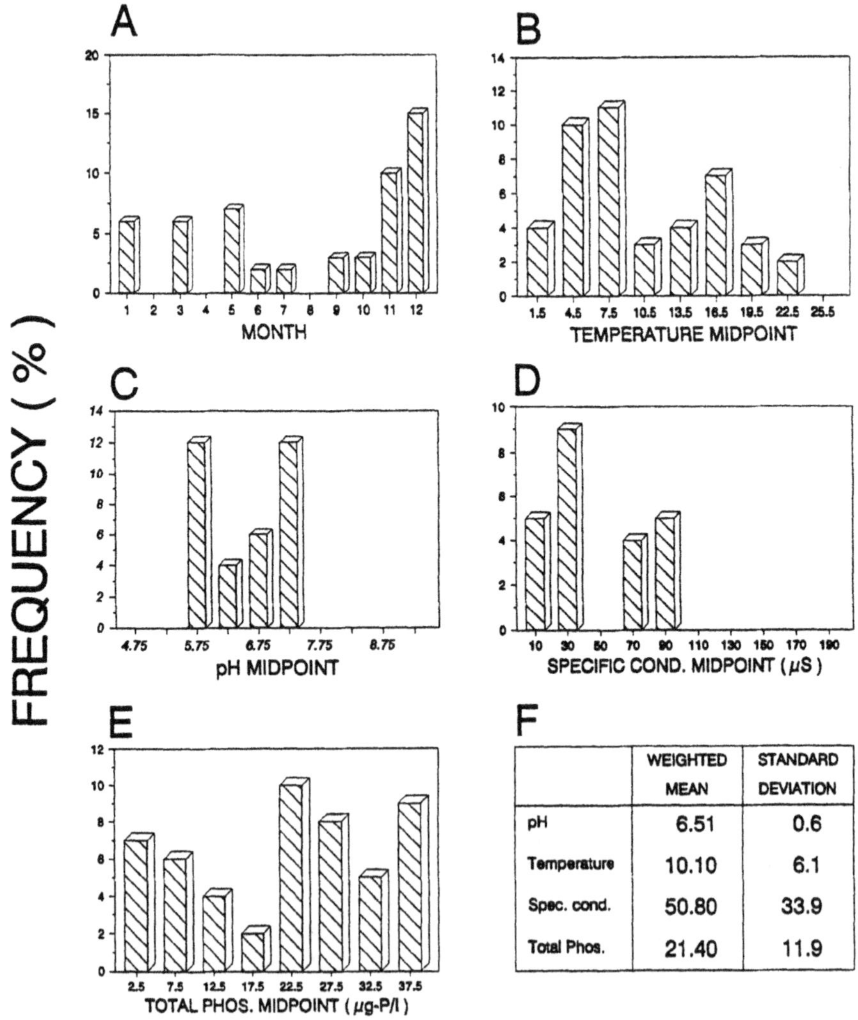

Fig. 5.19. The frequencies of occurrence for *Mallomonas papillosa* along seasonal (A), temperature (B), pH (C), specific conductance (D) and total phosphorus (E) gradients. Weighted mean and standard deviation values are also given (F).

Table 5.14. Literature records indicating the ecological preferences of *Mallomonas papillosa*.

Reference	Seasonality	Trophic Status	Temperature (°C)	pH and Related Factors	Other
1. Asmund and Hilliard, 1961	November-April.	---	Cold water stenotherm.	Acidic	Alaska
2. Asmund and Takahashi, 1969	---	---	---	---	Widespread, adaptable. Alaska
3. Cronberg and Kristiansen, 1980	---	Mostly oligotrophic. Some humic lakes.	---	---	Sweden
4. Donaldson and Stein, 1984	March	---	9	5.1	Canada
5. Dürrschmidt, 1980	---	Low in nutrients.	---	Acidic	Chile
6. Dürrschmidt, 1982b	---	Eutrophic pools.	---	---	As v. *ellipsoida*.
7. Dürrschmidt, 1984	---	All but the most eutrophic.	---	---	Giessen region of West Germany.
8. Dürrschmidt and Croome, 1985	---	---	28°C	---	First record from the tropics. Australia and Malaysia.
9. Eloranta, 1989	---	---	---	4.7 - 7.2, mean 6.0	16% of collections. Finland
10. Harris, 1967	---	---	---	Acid pond	As v. *monilifer*.
11. *Hartmann and Steinberg, 1986	---	---	---	Scales in recent sediments of an acidic lake.	Alpine lake in the Alps. As v. *monilifer*.
12. Kling and Kristiansen, 1983	---	---	---	---	Widespread species.
13. Péterfi and Momeu, 1976	---	Small eutrophic swamp.	---	---	Romania
14. Roijackers and Kessels, 1986	---	---	Cold water.	Acidophilous	Netherlands

Table 5.14. continued

Reference	Seasonality	Trophic Status	Temperature (°C)	pH and Related Factors	Other
15. Siver, 1989a	---	---	---	4.2 - 9.0 pH indifferent.	Literature review.
16. Takahashi, 1978	---	---	Cold water stenotherm, 3 - 15	Acidobiontic 4.5 - 6.5	Japan
17. Wawrzyniak and Andersen, 1985	---	---	---	---	Commonly occurring. Northern boreal regions.

AWMpH = abundance weighted mean pH.
*Indicates a paleolimnological study.

ansen, 1983), was much more common in the literature survey (21%) than in either Connecticut (4%) or Adirondack (9%) localities (Table 5.14). Although *M. papillosa* had a rather scattered seasonal distribution, its maximum occurrence was in November and December and, to a lesser extent, during the early to mid-spring period (Fig. 5.19A). Populations diminished during the warm summer months and periods of maximal ice formation. The seasonal distribution of *M. papillosa* reflects its preference for cooler water temperatures. Even though it was found over a temperature range of 2°C to 24°C, its greatest frequency of occurrence was between 3°C and 9°C and it had a low weighted mean of 10.1°C (Fig. 5.19B,F). *Mallomonas papillosa* has been previously recorded as being a cold water stenotherm in Alaska (Asmund and Hilliard, 1961), the Netherlands (Roijackers and Kessels, 1986) and Japan (Takahashi, 1978). In addition, Dürrschmidt (1982b) and Jacobsen (1985) also observed *M. papillosa* to favor cooler water temperatures. A similar weighted mean temperature of 10.5°C was calculated from over 200 collections surveyed from the literature.

Mallomonas papillosa has been reported as being acidobiontic (Takahashi, 1978), acidophilous (Roijackers and Kessels, 1986) and pH indifferent (Siver, 1989a). A variety, var. *monilifer*, has also been suggested to be acidophilous (Harris, 1967; Hartmann and Steinberg, 1986). In Connecticut and Adirondack localities *M. papillosa* was found over a rather narrow pH range of 5.6 to 7.2 and had a weighted mean pH of 6.5 (Fig. 5.19C). It was missing from all 55 collections with a pH less than 5.6. Since its presence was primarily in waterbodies with a pH less than 7.0 and it was lacking above pH 7.2, *M. papillosa* is probably most appropiately classified as acidophilous in Connecticut and the Adirondacks.

Based on a survey of the literature, Siver (1989a) noted that *M. papillosa* had a much broader distribution with respect to a pH gradient, ranging from 4.2 (Roijackers and Kessels, 1986) to 9.0 (Dürrschmidt, 1984) and a weighted mean pH of 6.6. Siver (1989a) further found this species in over 15% of the collections from the literature with a pH above 7.0 and below 6.0, supporting the classification of *M. papillosa* as pH indifferent. It is very possible that the discrepancy in the classification of this species within Hustedt's pH categories is due to a real difference at the subspecific level.

Mallomonas papillosa was primarily restricted to waters of low specific conductance (Fig. 5.19D), either clear or slightly humic stained, and oligotrophic to mesotrophic in nature. It also had a scattered frequency of occurrence along a total phosphorus gradient (Fig. 5.19E) and a weighted mean of 21.4 μg-P/l. In previous surveys *M. papillosa* was reported in low nutrient, oligotrophic lakes (Cronberg and Kristiansen, 1980; Dürrschmidt, 1980) and slightly eutrophic lakes (Péterfi and Momeu, 1976a; Dürrschmidt, 1982b), but not from the most eutrophied localities (Dürrschmidt, 1984).

Mallomonas portae-ferreae

Mallomonas portae-ferreae, present in 3% of the collections from Connecticut, was found over wide temperature (2°C to 28°C), specific conductance (28 μS to 195 μS) and total phosphorus (1 μg-P/l to 46 μg-P/l) gradients and sporadically throughout the year (Fig. 5.20). Despite its tolerance to a wide specific conductance range, *M. portae-ferreae* was most commonly found in eutrophic waters above 80 μS and had a high weighted mean of 143 μS (Figs. 5.20D-F).

In a previous study, *M. portae-ferreae* was reported to have a weighted mean pH of 7.1 (Siver, 1989a). Based on additional records this species had a pH range of 6.05 to 8.7, a maximum occurrence above pH 8 and a high weighted mean of 7.95 (Fig. 5.20C,F), supporting its classification as an alkaliphile or alkalibiont. The importance of pH and specific conductance in controlling the occurrence of *M. portae-ferreae* was shown by Siver and Hamer (1989) who demonstrated that over 80% of the variability in the pH and specific conductance variables was explained by the first principal component factor. Lastly, it is of interest to note that this organism was observed in the same lakes, often in the same collections, as the closely related taxa *M. corymbosa*, *M. tonsurata*, *M. alpina* and *M. areolata*. Only a few ecological records exist for *M. portae-ferreae* (Table 5.21).

Mallomonas pseudocoronata

Although primarily restricted to North American

localities (Wujek, 1984; Smol, 1988), *M. pseudocoronata* has recently been observed in Lake Biwa, Japan (Ito, 1988), the sediments of a European lake (Smol, 1988) and Panama (Wujek, 1986). Populations of this species, found in 10% of all collections, begin to develop in spring when water temperatures rise above approximately 12°C, are maintained throughout the summer and disappear by mid-autumn (Fig. 5.21A). Smol (1984) observed mass encystment of this species during autumn and believed the process to be temperature related. Whole cells were found in 14% of the samples made from March through October, whereas only isolated scales were found in 2% of the 127 collections made from November through February. As a result, *M. pseudocoronata* has a relatively high weighted mean temperature of 18.1°C. Ito (1988) found active populations only during the warm summer months in Lake Biwa and Wawrzyniak and Andersen (1985) recorded *M. pseudocoronata* in 33% of collections made from northern boreal regions during late spring and summer.

All available evidence suggests that this taxon is alkaliphilic in nature and very rare below pH 6 (Siver, 1989a) (Table 5.15). *M. pseudocoronata* was found in 18% and 2% of the collections above and below pH 6.5, respectively, and was not found below pH 5.8 (Fig. 5.21C). The weighted mean pH of 7.7 was similar to the 7.37 and 7.35 values reported by Charles and Smol (1988) and Dixit

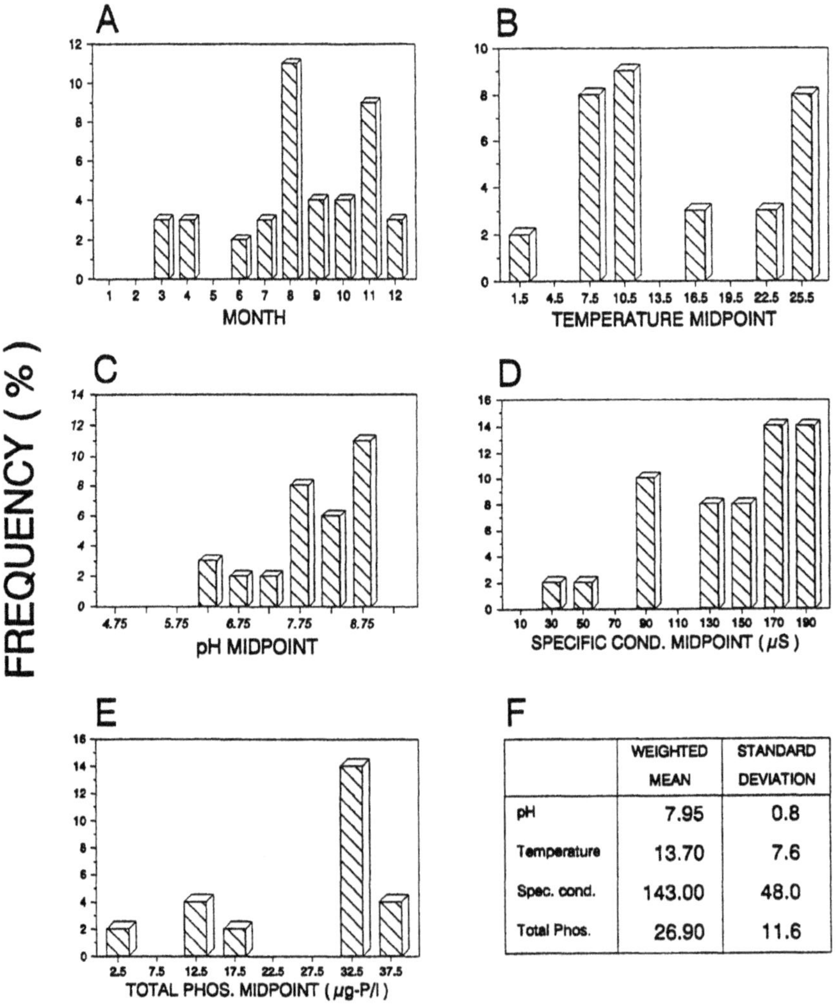

Fig. 5.20. The frequencies of occurrence for *Mallomonas portae-ferreae* along seasonal (A), temperature (B), pH (C), specific conductance (D) and total phosphorus (E) gradients. Weighted mean and standard deviation values are also given (F).

et al (1988a), respectively. In addition, Dixit et al (1988b) reported *M. pseudocoronata* to have a preference for circumneutral to alkaline lakes in Quebec. Wee and Gabel (1989) found this species between a pH of 7.7 and 8.7 in Iowa localities and Asmund and Hilliard (1961) observed it between pH 7.2 and 8.6 in Alaska.

M. pseudocoronata was found over a specific conductance gradient ranging from 21 μS to 170 μS, however, it was significantly less frequent in waterbodies with low specific conductance (Fig. 5.21D). It had a maximum occurrence between 100 μS and 120 μS where it was recorded in 32% of the samples and a weighted mean specific conductance of 112 μS. *M. psuedocoronata* was sporadically distributed over a total phosphorus gradient, being found as often in lakes below 10 μg-P/l as it was in those above 35 μg-P/l (Fig. 5.21E). In general, *M. pseudocoronata* was found in oligotrophic, mesotrophic and eutrophic lakes, avoided the hypereutrophic localities, had a preference for clearwater habitats and was absent in all darkly stained humic samples. The few literature records support the idea that *M. pseudocoronata* exists in a variety of lake types (Table 5.15). Asmund and Hilliard (1961) reported this organism from an oligotophic lake and Wawrzyniak and Andersen (1985) found it to be common in oligotophic boreal

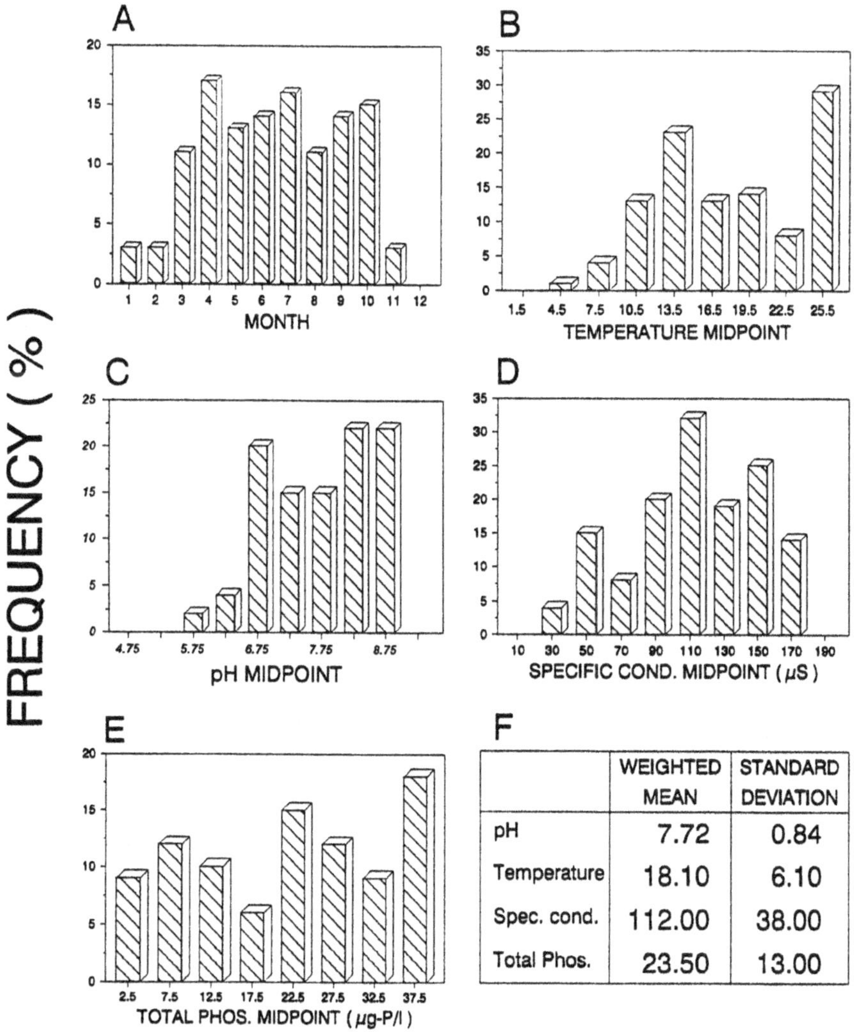

	WEIGHTED MEAN	STANDARD DEVIATION
pH	7.72	0.84
Temperature	18.10	6.10
Spec. cond.	112.00	38.00
Total Phos.	23.50	13.00

Fig. 5.21. The frequencies of occurrence for *Mallomonas pseudocoronata* along seasonal (A), temperature (B), pH (C), specific conductance (D) and total phosphorus (E) gradients. Weighted mean and standard deviation values are also given (F).

Table 5.15. Literature records indicating the ecological preferences of *Mallomonas pseudocoronata*.

Reference	Seasonality	Trophic Status	Temperature (°C)	pH and Related Factors	Other
1. Asmund and Hilliard, 1961	---	Oligotrophic	8 and 11.7	Range: 7.2 to 8.6	Alaska
2. *Charles and Smol, 1988	---	---	---	AWMpH = 7.4; alkaline.	Adirondacks
3. *Dixit et al., 1988a	---	---	---	Greatest abundance in lakes with pH above 7; AWMpH = 7.4	Sudbury region, Canada.
4. *Dixit et al., 1988b	---	---	---	Circumneutral to alkaline preference.	Quebec, Canada.
5. Ito, 1988	Found only in summer.	---	---	---	Lake Biwa, Japan. One of only two records outside of North America.
6. Siver, 1988c	---	---	---	6.94 to 7.1; found in the more alkaline localities.	Found in 18% of the collections. Adirondacks.
7. Siver, 1989a	---	---	---	WMpH = 7.25; found primarily above 6.5, not found below 6.0.	Connecticut and the Adirondacks.
8. Siver and Hamer, 1989	---	---	Temperature dominates second principal component.	Specific conductance and pH control first principal component; WMpH = 7.3; WM specific conductance = 81	Connecticut and Adirondack
9. Smol, 1984	---	---	---	---	Describes characteristics of the stomatocyst.
10. *Smol, 1988	---	---	---	---	Found in sediments of a European lake. One of only two records outside North America.

Table 5.15. continued

Reference	Seasonality	Trophic Status	Temperature (°C)	pH and Related Factors	Other
11. *Smol et al., 1983	---	Lacking in the most eutrophic periods.	---	---	Found with *M. caudata*. Little Round Lake, Ontario, Canada.
12. *Smol et al., 1984a	---	---	---	Lacking below 5.0	Records combined with *M. crassisquama* and *M. duerrschmidtiae*. Adirondacks.
13. Wawrzyniak and Andersen, 1985	Late spring and summer.	Oligotrophic	---	---	Northern boreal regions of North America. Found in 33% of the collections.
14. Wee and Gabel, 1989	---	---	---	Alkaline species; Range: 7.7 to 8.7 mean: 8.3	Iowa
15. Wujek, 1984	---	---	---	---	Restricted primarily to North America.
16. Wujek, 1986	---	---	---	---	Reports this species from Panama and notes that it has not been found in South America.

AWMpH = abundance weighted mean pH.
*Indicates a paleolimnological study.

Table 5.16. Literature records indicating the ecological preferences of *Mallomonas pugio*.

Reference	Seasonality	Trophic Status	Temperature (°C)	pH and Related Factors	Other
1. Bradley, 1964	July (Iceland), but favors winter and spring conditions.	---	Cold	7.0 to 7.2	Found in several stagnant pools in Iceland.
2. *Charles and Smol, 1988	---	---	---	AWMpH = 5.25 acidobiont.	Adirondacks
3. Eloranta, 1989	---	---	---	Acid humic taxon; range: 4.7 to 5.6; mean = 5.2	Finland
4. *Hartmann and Steinberg, 1986	---	---	---	Few scales found in more recent acidified sediments. Acidophilous to acidobiontic.	Alpine mountain lake.
5. Jacobsen, 1985	---	---	---	5.9 to 7.3	Western Greenland
6. Kristiansen, 1979	---	---	---	---	Very restricted distribution.
7. Siver, 1989a	---	---	---	Found in acidic localities.	Connecticut
8. *Smol et al., 1984a	---	---	---	Found in surface sediments from 6 acid lakes. Range 4.9 to 6.6; AWMpH = 5.3	Adirondacks
9. Wee, Wujek and Graebner, 1982	---	---	---	Found in a bog	Michigan

AWMpH = abundance weighted mean pH.
*Indicates a paleolimnological study.

localities. Smol et al. (1983; 1985) found *M. pseudocoronata* to occur during all but the most highly eutrophic periods of the post-glacial history of Little Round Lake, a meromictic lake in Ontario, Canada.

Mallomonas pugio

Like *M. paludosa*, *M. pugio* is a very rare species found primarily in acidic localities and best classified as an acidobiont (Siver, 1989a). It was found over a pH range from 4.3 to 6.5, had a maximum occurrence below pH 6 and a weighted mean pH of 5.0. In a paleolimnological study of the surface sediments from 38 Adirondack lakes, Charles and Smol (1988) found *M. pugio* to have a similar weighted mean pH of 5.25. In another paleolimnological study Hartmann and Steinberg (1986) observed increases of *M. pugio* in the recent sediments from a small acidic lake. In Finnish localities Eloranta (1989) found this taxon between pH 4.7 and 5.6, supporting the idea that this species is a true acidobiont (Siver, 1989a). Jacobsen (1985) observed *M. pugio* in six waterbodies from western Greenland that ranged from pH 5.9 to 7.3 and in Iceland Bradley (1964) reported this species in several stagnant pools near pH 7, suggesting that *M. pugio* can survive at neutral pH.

Even though the amount of data available is somewhat limited several other generalizations can be made (Table 5.16). First, *M. pugio* was found primarily during the cooler parts of the year over a temperature range from 3°C to 19.5°C. Bradley (1964) also noted *M. pugio* to be a cold water organism found during the winter and spring. Second, this taxon was observed in waters of both low specific conductance and total phosphorus concentrations. Third, this species has also been found in humic stained localities (Wee et al., 1982; Eloranta, 1989).

Mallomonas punctifera

Mallomonas punctifera was a relatively common species found in 9% and 12% of all collections from Connecticut and the Adirondacks, respectively. It was reported in a similar percentage of the collections from the literature (Siver, 1989a). This organism has also been observed to be common in Sweden (Cronberg and Kristiansen, 1980), Ontario (Nicholls, 1982), Finland (Eloranta, 1989) and northern boreal habitats of North America (Wawrzyniak and Andersen, 1985). *M. punctifera* was primarily found between late spring and mid-autumn and had a maximum occurrence during August where it was found in 22% of the collections (Fig. 5.22A). Although isolated scales of *M. punctifera* were found in 3 of 76 collections made during November and December, whole cells were lacking. Populations were also lacking in collections taken from under the ice during January and February. Because of its predominance during warmer months, *M. punctifera* had a rather high weighted mean temperature of 18°C. This organism was present in 3%, 12% and 21% of the collections from waters with a temperature less than 9°C, 9°C to 21°C and above 21°C, respectively (Fig. 5.22B).

Little seasonal and temperature data are available from the literature for *M. punctifera* (Table 5.17). Nygaard (1949) reported a similar seasonal distributional pattern for *M. punctifera* and Wawrzyniak and Andersen (1985) and Eloranta (1989) found this species in over 20% of their collections made during the spring and/or autumn. In Malaysian localities it preferred temperatures above 25°C (Dürrschmidt and Croome, 1985). However, Ito (1988) observed *M. punctifera* in Lake Biwa during the winter and spring seasons and Roijackers and Kessels (1986) reported this taxon to prefer temperatures less than 11°C. Thus, there is some evidence to support the idea that both warm water and cold water ecotypes exist.

Mallomonas punctifera was found over a narrow pH range of 5.5 to 6.9, had a weighted mean pH of 6.18 and was noticeably absent from all collections with a pH less than 5.0 (Fig. 5.22C). Based on such a distribution along a pH gradient, *M. punctifera* is best described as an acidophilic taxon (Siver, 1989a). Populations of *M. punctifera* were also quite restricted along a specific conductance gradient, being found primarily in waterbodies less than 40 μS (Fig. 5.22D). Smol et al. (1984a) and Eloranta (1989) reported similar pH ranges of 5.2 to 6.6 and 5.3 to 7.2 and weighted mean pH values of 6.6 and 6.2, respectively, supporting the classification of *M. punctifera* as an acidophilic taxon. Takahashi (1978) also reported this species as acidophilic. Evidence from several paleolimnological studies (e.g. Smol et al., 1984b; Christie and Smol, 1986) supports the idea that *M. punctifera* disappears as the pH drops below 5.0.

Kristiansen (1986) suggested that *M. punctifera* would be expected to be found in habitats with a higher pH and to have a higher weighted mean pH than reported by Smol et al. (1984a) or reported here. Such an argument is supported by results from a survey of the literature where *M. punctifera* was found over a pH range of 6 to 8 and to have a weighted mean pH of 6.8 (Siver, 1989a). In addition, Roijackers and Kessels (1986) reported this species to exist in alkaline habitats from the Netherlands.

Based on preliminary data that *M. punctifera* may be a more acidophilous taxon in northern New England waterbodies than in other regions, Charles and Smol (1988) suggested the possibility that perhaps different ecotypes or undescribed varieties exist. Such an idea is further supported by the fact that populations found in acidic localities are present primarily during warmer parts of the year, while those reported in more alkaline habitats have been found at colder temperatures (e.g. Roijackers and Kessels, 1986). Differences in the sizes of cells and body scales as reported in the literature also support the idea that different ecomorphs may exist. In both Connecticut and

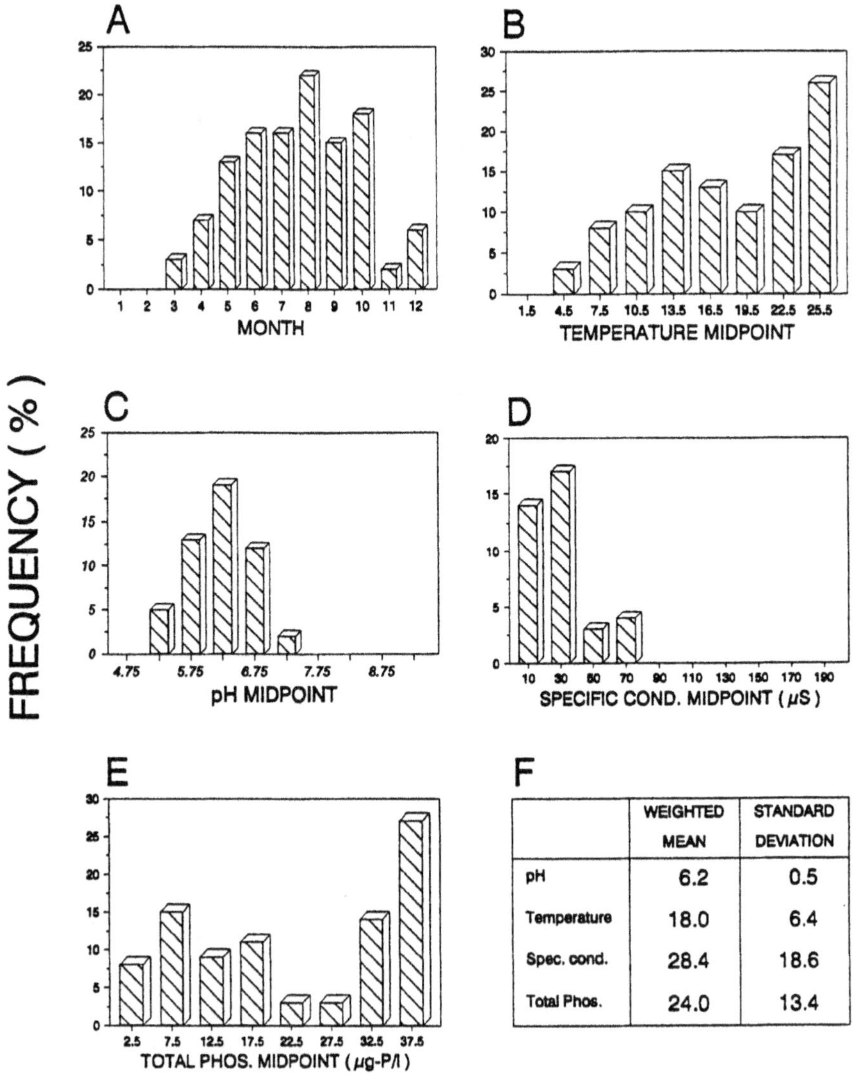

	WEIGHTED MEAN	STANDARD DEVIATION
pH	6.2	0.5
Temperature	18.0	6.4
Spec. cond.	28.4	18.6
Total Phos.	24.0	13.4

Fig. 5.22. The frequencies of occurrence for *Mallomonas punctifera* along seasonal (A), temperature (B), pH (C), specific conductance (D) and total phosphorus (E) gradients. Weighted mean and standard deviation values are also given (F).

Table 5.17. Literature records indicating the ecological preferences of *Mallomonas punctifera*.

Reference	Seasonality	Trophic Status	Temperature (°C)	pH and Related Factors	Other
1. *Charles and Smol, 1988	---	---	---	AWMpH = 6.6	Adirondacks
2. *Christie and Smol, 1986	---	---	---	Decreases in relative abundance as pH drops below about 5 to 5.5.	Upper Wallface, Adirondacks.
3. Cronberg and Kristiansen, 1980	---	---	---	---	Very common in Sweden.
4. Dürrschmidt and Croome, 1985	Wide range in Germany; summer in Malaysia.	---	25 to 30 in Malaysia.	---	---
5. Eloranta, 1989	---	Oligotrophic, forest lakes.	---	Range = 5.3 to 7.2 mean = 6.2	Common in Finland.
6. *Haworth, 1983	---	---	---	---	Increase with *Synura spinosa*, *M. crassisquama* and *Asterionella formosa* and decrease of *Cyclotella* spp. Blelham Tarn, England.
7. Ito, 1988	Winter and Spring.	---	---	---	Lake Biwa, Japan
8. Kristiansen, 1986	---	---	---	Found at neutral pH and above.	Literature review
9. Momeu and Péterfi, 1983	March, June.	---	13, 17	Neutral	Large cells. Romania.
10. Nicholls, 1982	---	---	---	---	Common in Ontario, Canada.
11. Nygaard, 1949	Spring, summer and autumn.	Mesotrophic and Eutrophic.	---	---	Denmark

Table 5.17. continued

Reference	Seasonality	Trophic Status	Temperature (°C)	pH and Related Factors	Other
12. Roijackers and Kessels, 1986	---	---	Primarily in cold water, below 10.	---	Netherlands
13. Siver, 1988c	---	Oligotrophic to mesotrophic.	---	Range = 5.9 to 6.8	Adirondacks
14. Siver, 1989a	---	---	---	AWMpH in Connecticut and Adirondacks = 5.9 and in literature review = 6.8. Absent below pH 5.0. Acidophilous.	Literature review. May be evidence for two different varieties.
15. Siver and Hamer, 1989	---	---	---	pH and specific conductance control first principal component.	Connecticut and the Adirondacks.
16. *Smol, 1980	---	Slight increase with human settlement.	---	---	Canadian shield lakes.
17. *Smol et al., 1984a	---	---	---	Range 4.6 to 7.8 AWMpH = 6.6	Adirondacks
18. *Smol et al., 1984b	---	---	---	Slightly acidic, but decreases as pH drops below 5.	Adirondacks
19. Takahashi, 1978	---	---	Eurythermal, 7 to 23.	Acidophilous, range = 6 to 7.2	Japan
20. Wawrzyniak and Andersen, 1985	Late spring and summer.	---	---	---	Common in northern boreal regions.

AWMpH = abundance weighted mean pH.
*Indicates a paleolimnological study.

Adirondack localities cells and body scales of *M. punctifera* were quite small with a mean size of 12.2 μm · 5.6 μm and 3.2 μm · 2.5 μm, respectively. In contrast, Momeu and Péterfi (1983) reported a range for cells of 22 μm to 67 μm · 9 μm to 17 μm and for body scales of 4.3 μm to 5.9 μm · 3.2 μm to 4.5 μm. Takahashi (1978) also reported the sizes of cells and scales for *M. punctifera* to be significantly larger than those found in either Connecticut or the Adirondacks. The discrepancies in the sizes of cells and scales may reflect different genetic strains or indicate effects from environmental factors.

Mallomonas punctifera was found over a wide range of total phosphorus concentrations, had a rather high weighted mean value of 24.1 μg-P/l (Fig. 5.22E) and was present primarily in either oligotrophic or mesotrophic waterbodies. Eloranta (1989) and Wawrzyniak and Andersen (1985) reported *M. punctifera* from regions dominated with oligotrophic lakes in forest settings, however, Asmund (1969) observed this taxon in small eutrophic ponds (Table 5.17).

Mallomonas striata

Mallomonas striata, a species with a worldwide distribution, has been found under many environmental conditions (Cronberg and Kristiansen, 1980; Dürrschmidt, 1984; Péterfi and Momeu, 1976a). Despite its wide distribution and occurrence in 16% of the collections surveyed from the

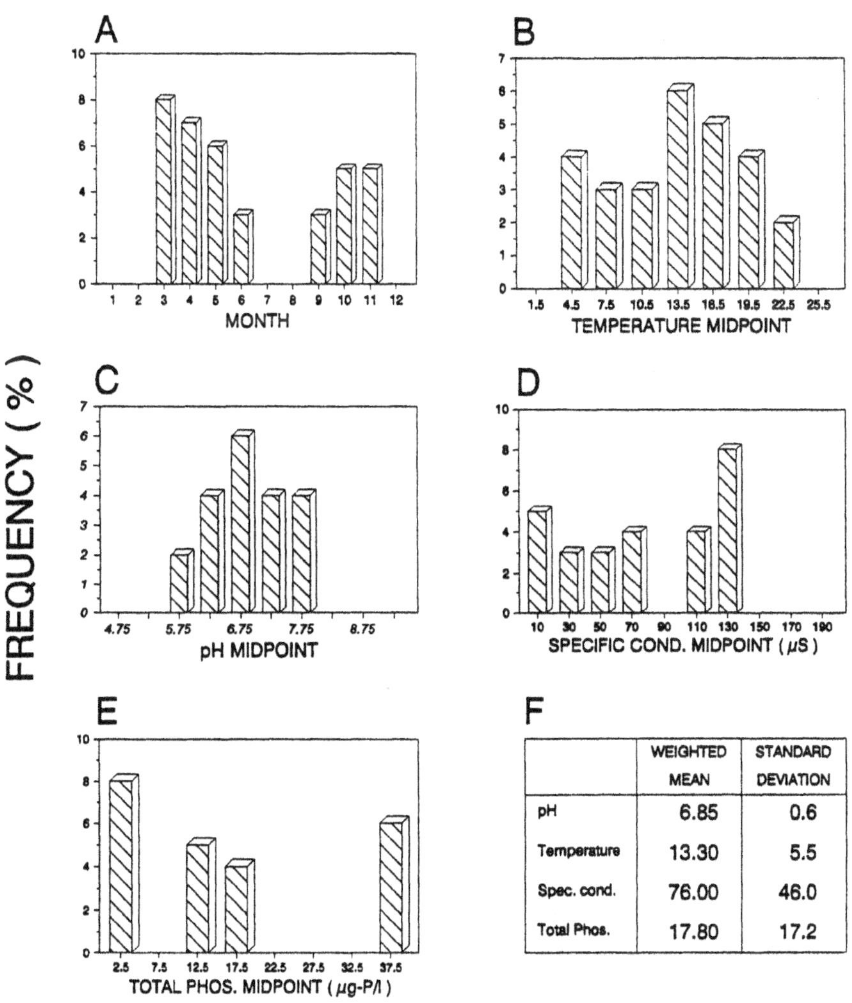

	WEIGHTED MEAN	STANDARD DEVIATION
pH	6.85	0.6
Temperature	13.30	5.5
Spec. cond.	76.00	46.0
Total Phos.	17.80	17.2

Fig. 5.23. The frequencies of occurrence for *Mallomonas striata* along seasonal (A), temperature (B), pH (C), specific conductance (D) and total phosphorus (E) gradients. Weighted mean and standard deviation values are also given (F).

Table 5.18. Literature records indicating the ecological preferences of *Mallomonas striata*.

Reference	Seasonality	Trophic Status	Temperature (°C)	pH and Related Factors	Other
1. Asmund, 1959	Winter/early spring form. Under the ice.	Mesotrophic to eutrophic. Farm pond.	Cold water. 7.5; 8 to 14	Slightly acid to strongly alkaline; 7.25 to 7.9; 6.8 to 8.	Denmark. Found in lakes dominated with blue-greens.
2. Cronberg and Kristiansen, 1980	---	All types, including polluted.	---	---	Widely distributed. Sweden.
3. Dürrschmidt, 1984 (as v. *serrata*)	---	---	---	High pH; alkaline conditions.	Giessen region of Germany. Widespread disribution.
4. Dürrschmidt and Croome, 1985	---	---	---	---	Frequent in several localities in Australia and Malaysia.
5. Ito, 1988 (as v. *striata*)	Spring	---	---	---	Lake Biwa, Japan.
6. Kristiansen, 1985; 1989	Winter/spring.	Highly eutrophic lake.	Cold water. 0.7 to 10	High pH; 8.2 to 9.1	Denmark.
7. Péterfi and Momeu, 1976	Winter and early spring.	All types, including bogs.	---	---	Romania.
8. Roijackers, 1981	---	---	---	Acidophilous to acidobiontic group.	Netherlands.
9. Siver, 1989a	---	---	---	Wide range, but alkaliphilous. WMpH=7.8	Literature review.
10. Smol et al., 1984a	---	Eutrophic	---	Alkaline conditions.	Adirondacks.
11. Takahashi, 1978 (as v. *serrata*)	---	---	Stenotherm; 5 to 17	7 to 8; alkaline	Japan.
12. Wawrzyniak and Andersen, 1985	---	---	---	---	Commonly occurring taxon.

AWMpH = abundance weighted mean pH.

literature (Siver, 1989a), *M. striata* was rare in Connecticut and Adirondack localities, present in only 4% of the samples.

This organism was primarily restricted to the spring and autumn periods, was rare during mid-summer and absent between December and March (Fig. 5.23A). In waterbodies located in Denmark (Asmund, 1959), Japan (Ito, 1988) and Romania (Péterfi and Momeu, 1976a) *M. striata* was observed during late winter and early spring. Because of the spring/autumn nature of this organism, it had a relatively low weighted mean temperature of 13.3°C and was lacking in very cold (<4°C) and very warm (>22°C) water (Fig. 5.23B). Takahashi (1978) also reported *M. striata* var. *serrata* as being a stenothermal taxon with a maximum occurrence between 8°C and 15°C, while Kristiansen (1988) observed *M. striata* between 1°C and 10°C.

Based on a survey of the literature, *M. striata* was found to be widely distributed along a pH gradient from pH 5.2 (Roijackers and Kessels, 1986) to over 9.0 (Kristiansen, 1985), to be most common above pH 7 and to have a weighted mean pH of 7.8. Siver (1989a) classified this taxon as alkaliphilic. In Connecticut and Adirondack localities *M. striata* occurred over a much narrower pH range of 6.0 to 7.6, had a weighted mean pH of 6.85 and would most appropiately be classified as pH indifferent (Fig. 5.23C). In Denmark (Kristiansen, 1988), Japan (Takahashi, 1978) and Germany (Dürrschmidt, 1984) *M. striata* was reported primarily from alkaline waters, however, Roijackers (1981) found it to cluster within the acidophilous/acidobiontic group. The difference may be related, in part, to differences at the variety level and future work should distinguish between subspecific taxa. Despite the differences in the distribution along a pH gradient, it is clear that *M. striata* is often found near or above pH 7 and is lacking in extreme acidic conditions (Fig. 5.23C).

No clear pattern was observed for *M. striata* along either specific conductance or total phosphorus gradients (Figs. 5.23D-E). Populations of this organism were observed mostly in clearwater lakes of an oligotrophic or mesotrophic nature, although several records from eutrophic and/or slightly humic stained waters were also made. This species has been reported from mesotrophic to eutrophic (Asmund, 1959) and more oligotrophic (Warzyniak and Andersen, 1985) habitats, supporting the idea that this species is common in all types of waterbodies (Cronberg and Kristiansen, 1980).

Mallomonas tonsurata

Mallomonas tonsurata was one of the most common species of *Mallomonas* in Connecticut localities, present in 20% of the collections. In a survey of the literature, Siver (1989a) also found *M. tonsurata* to be one of the most common taxa of *Mallomonas*, present in 30% of the collections. Takahashi (1978) reported *M. tonsurata* as the most common species within the genus in Japan. In addition, this organism was common in studies of lakes from Panama (Wujek, 1986), Finland (Eloranta, 1989) and Iowa (Wee and Gabel, 1989). Despite its widespread occurrence Siver (1988c) found *M. tonsurata* in only 6% of the collections from the Adirondacks and Smol et al. (1984a) found it to be rare in a survey of lake sediments from the same region.

Populations of *M. tonsurata* were present during each month of the year, although they were less frequent during the colder winter months (Fig. 5.24A). Generally, populations begin to develop between February and March and persist throughout the summer and into the autumn. *Mallomonas tonsurata* was found in 22% of the collections made from May through November, had a peak in occurrence of 31% during October and decreased to only 6% under the ice. In general, with the onset of colder autumn temperatures populations disappeared. Cyst formation was observed on three occassions during November. Asmund (1959) reported a similar seasonal distribution for *M. tonsurata*, whereby it was present between February and December with early summer and late summer peaks. Kristiansen (1985; 1988) found *M. tonsurata* between April and November whereas Dürrschmidt (1984) observed it during the autumn months (Table 5.19).

Because of its distribution throughout the year, it is not surprising that *M. tonsurata* also had a wide temperature range of 2.5°C to 28°C. Asmund (1959) and Takahashi (1978) also found *M. tonsurata* to be eurythermal with temperature ranges of 0°C to 25°C and 1°C to 31°C, respectively. Roijackers and Kessels (1986) and Kristiansen (1985; 1988) also recorded this organism over a wide temperature gradient (Table 5.19). In this study *Mallomonas tonsurata* had a maximum occurrence at water temperatures above 9°C where it was present in 22% of the collections and a weighted mean of 15.0°C (Fig. 5.24B,F). Asmund (1959) and Kristiansen (1985; 1988) reported this species to be most abundant above 10°C and 19.6°C, respectively.

It is well documented that *M. tonsurata* is an alkaliphilous to alkalibiontic organism (Siver, 1989a). It is common in localities with a pH above 7 and rare in softwater lakes below pH 5.5. This taxon was found in 3%, 18% and 25% of the collections with a pH less than 6, between 6 and 6.5 and above 6.5, respectively (Fig. 5.24C). It had a weighted mean pH of 7.5 and was observed over a pH range from 5.2 to 8.8. In Adirondack localities it was restricted to those waterbodies with higher pH and alkalinity content.

A similar distribution along a pH gradient can be constructed from literature records (Table 5.19). Based on a review of over 200 collections surveyed from the literature, Siver (1989a) reported a weighted mean pH of 8.1 and a pH range of 5.4 to 9.2. Asmund (1959) and Takahashi (1978) reported similar pH ranges for *M. tonsurata* of 6.8 to 8.0 and 5.2 to 9.0, respectively, while Dürrschmidt (1984), Kristiansen (1985; 1988), Roijackers and

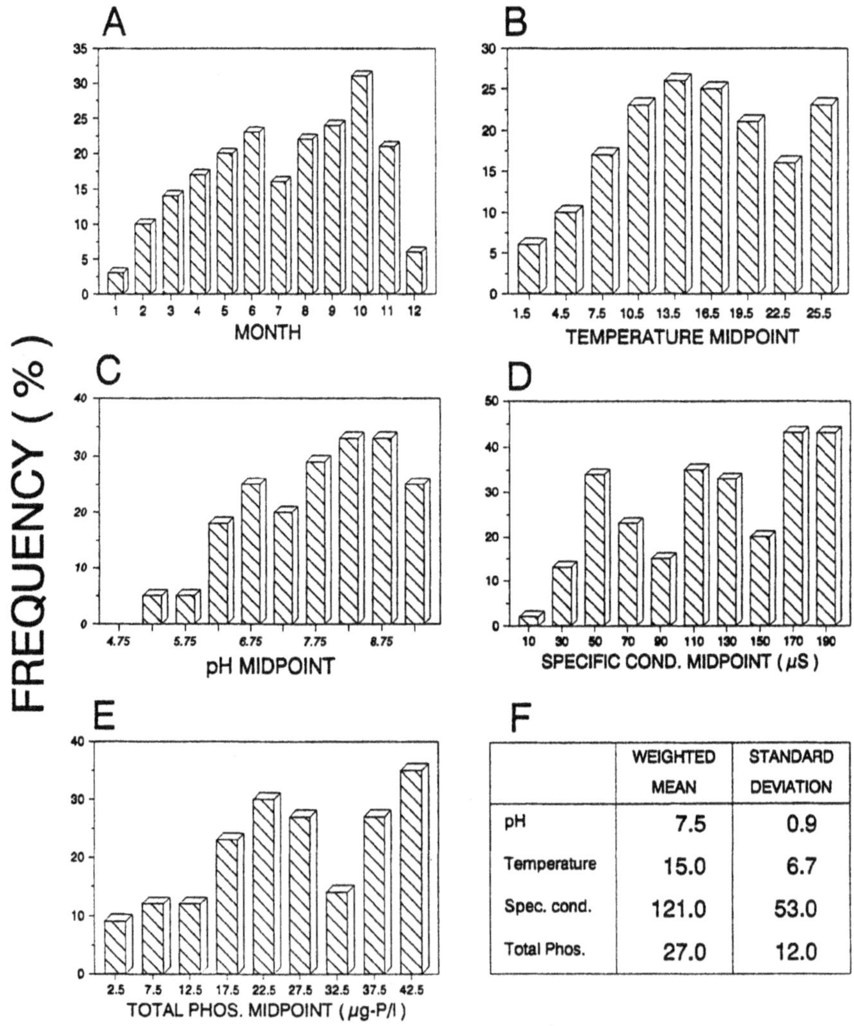

	WEIGHTED MEAN	STANDARD DEVIATION
pH	7.5	0.9
Temperature	15.0	6.7
Spec. cond.	121.0	53.0
Total Phos.	27.0	12.0

Fig. 5.24. The frequencies of occurrence for *Mallomonas tonsurata* along seasonal (A), temperature (B), pH (C), specific conductance (D) and total phosphorus (E) gradients. Weighted mean and standard deviation values are also given (F).

Table 5.19. Literature records indicating the ecological preferences of *Mallomonas tonsurata*.

Reference	Seasonality	Tropic Status	Temperature (°C)	pH and Related Factors	Other
1. Asmund, 1959	February to December with peaks in spring/ early summer and late summer.	Mesotrophic to eutrophic; most abundant in eutrophic ponds.	0 to 25, but most abundant above 10.	Range: 6.8 to 8	Denmark
2. Dürrschmidt, 1984	Autumn	---	---	High pH.	Giessen region of Germany.
3. Dürrschmidt and Croome, 1985	---	---	---	---	Found in Australia and Malaysia.
4. Eloranta, 1989	---	---	---	Range: 5 to 7.2	Common taxon, found in 26% of all collections; Finland.
5. Ito, 1988	Found all year long.	---	---	---	Lake Biwa, Japan
6. Kristiansen, 1985; 1988	April, June, July, October November.	Eutrophic	Wide temperature range; maximum occurrence at 19.6.	Range: 8.2 to 9.1	Denmark
7. Roijackers and Kessels, 1986	---	---	Wide temperature range.	Alkaliphilous with lower end of range near 6.5.	Netherlands
8. Siver, 1988c	---	---	---	Rare, 6.8	Adirondacks
9. Siver, 1989a	---	---	---	Range: 5.5 to 9.1; weighted mean of 8.1; alkaliphilous to alkalibiontic.	Literature records.
10. Siver and Hamer, 1989	---	Total phosphorous and temperature control the second principal component.		pH and specific conductance control the first principal component.	Connecticut and Adirondack lakes.
11. *Smol et al., 1983	---	---	---	Found above 8.0	Little Round Lake, Ontario, Canada.

Table 5.19. continued

Reference	Seasonality	Trophic Status	Temperature (°C)	pH and Related Factors	Other
12. Takahashi, 1959	---	Eutrophic	---	---	Japan
13. Takahashi, 1978	---	---	Wide temperature range from 1 to 31; eurythermal.	Range: 5.2 to 9.0. Acidophilous tendencies, most common between 6 and 7.	Most common species of scaled chrysophyte in Japan.
14. Wee and Gabel, 1989	---	---	---	Range: 8.0 to 8.7; Mean: 8.3	Common, found in 20% of all collections; Iowa.
15. Whitford and Schumacher, 1969	---	Eutrophic	---	---	North Carolina.
16. Wujek, 1986	---	---	---	---	One of the most common taxa; Panama.
17. Wujek and Hamilton, 1973	---	Eutrophic	---	---	Michigan

*Indicates a paleolimnological study.

Kessels (1986) and Wee and Gabel (1989) observed this organism to be even more restricted to alkaline localities (Table 5.19). Thus, even though Takahashi (1978) listed *M. tonsurata* as pH indifferent with acidophilic tendencies, a large body of information supports the classification of this species as alkaliphilous to alkalibiontic in nature. In a principal component analysis explaining the contributions of different environmental variables influencing the occurrence of *M. tonsurata*, Siver and Hamer (1989) found pH and specific conductance to strongly dominate the first principal component factor; 90% and 87% of the variance in the pH and specific conductance variables were accounted for by the first principal component factor.

Mallomonas tonsurata was found over a wide gradient of specific conductance ranging from 20 μS to 255 μS, however, it was significantly less frequent in softwater localities with values below 40 μS (Fig. 5.24D). It also had a high weighted mean value of 121 μS. *Mallomonas tonsurata* was also widely distributed over a total phosphorus gradient, although it was more common in collections above 20 μg-P/l (Fig. 5.24E). This taxon was more often found under eutrophic conditions, but it was also present in a few oligotrophic

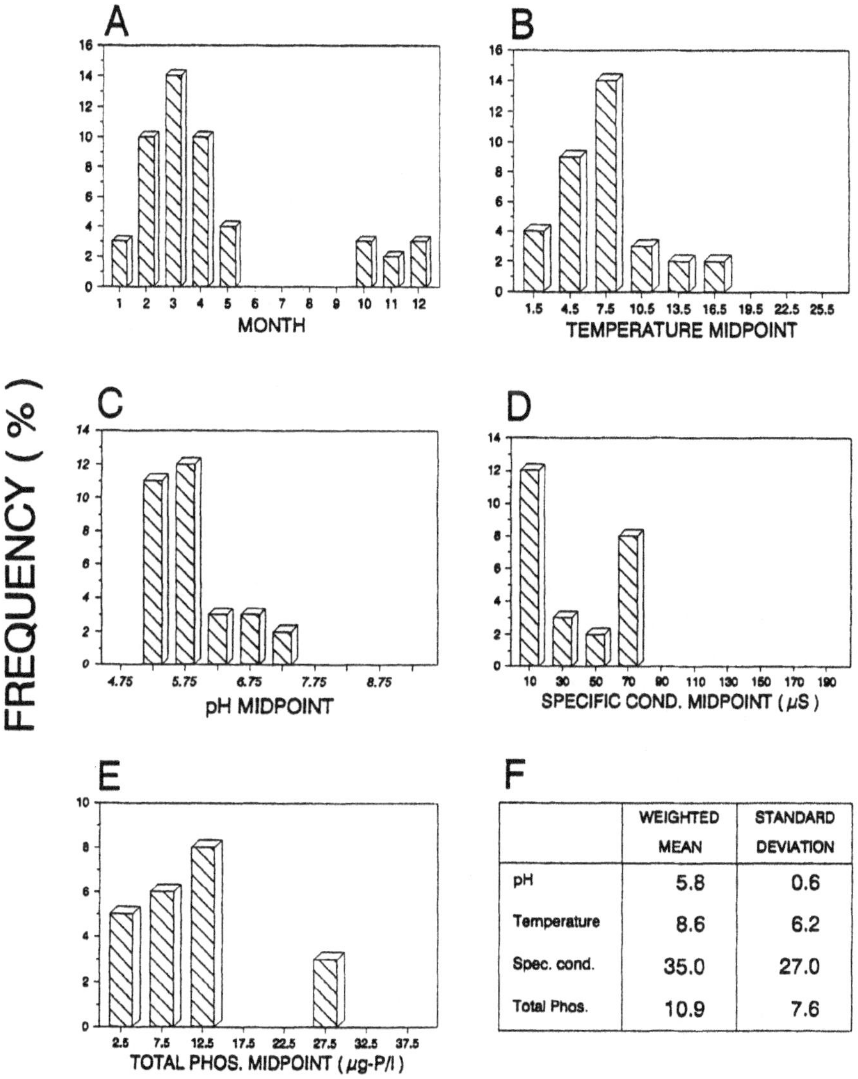

	WEIGHTED MEAN	STANDARD DEVIATION
pH	5.8	0.6
Temperature	8.6	6.2
Spec. cond.	35.0	27.0
Total Phos.	10.9	7.6

Fig. 5.25. The frequencies of occurrence for *Mallomonas torquata* along seasonal (A), temperature (B), pH (C), specific conductance (D) and total phosphorus (E) gradients. Weighted mean and standard deviation values are also given (F).

localities. The eutrophic nature of *M. tonsurata* is supported by the work of Asmund (1959), Whitford and Schumacher (1969), Takahashi (1959) and Wujek and Hamilton (1973). In a paleolimnological study of a small meromictic lake in Ontario, Smol et al. (1983) recorded a peak concentration for *M. tonsurata* coincident with increased eutrophy due to human disturbance.

Mallomonas torquata

Mallomonas torquata is relatively rare, but because it is found over narrow pH, temperature, specific conductance and total phosphorus gradients, it represents another valuable bioindicator species (Fig. 5.25). *M. torquata* is a cold water organism found in acidic, softwater lakes low in specific conductance and total phosphorus (Fig. 5.25). Its distribution and weighted mean values along temperature, pH, specific conductance and seasonal gradients are stikingly similar to those for *M. transsylvanica* and the two species were found in the same habitats.

Mallomonas torquata was found from late October through May, had a maximum occurrence during the late winter to early spring period and was lacking from June through September (Fig. 5.25A). Its initial development during mid-autumn and subsequent decline in spring are coincident with periods of rapid temperature change. As a result of its seasonal distribution, *M. torquata* is a cold water species with a weighted mean temperature of 8.6°C and a maximum occurrence below 9°C (Fig. 5.25B). Its frequency of occurrence dropped significantly as the temperature increased above 9°C and the species was not found above 17.5°C. The distribution of *M. torquata* along temperature and seasonal gradients was also very similar to that for *M. heterospina*. Literature records (Table 5.21) support the idea that *M. torquata* is a cold water organism distributed primarily below 10°C and a common member of the vernal flora (Asmund and Cronberg, 1979; Cronberg and Kristiansen, 1980).

Mallomonas torquata is also restricted to acidic softwater lakes low in specific conductance and total phosphorus levels. Since this taxon was found over a narrow pH range of 5.2 to 7.2, had a maximum frequency of occurrence between 5.2 and 6.0 and a weighted mean pH of 5.8, it is best described as an acidophilous species. In addition, *M. torquata* had a maximum occurrence in waterbodies with a specific conductance less than 20 μS and a total phosphorus concentration below 15 μg-P/l, and had weighted mean values of 35 μS and 10.9 μg-P/l (Figs. 5.25D-E). As reported by Cronberg and Kristiansen (1980) and Kling and Kristiansen (1983), *M. torquata* was most frequently found in oligotrophic lakes situated in an undisturbed forest setting.

Mallomonas transsylvanica

Mallomonas transsylvanica was a relatively rare species found in only 7% and 3% of the collections from Connecticut localities and the literature survey, respectively. Its limited occurrence is due, in part, to its distribution over narrow environmental gradients (Fig. 5.26). In Connecticut lakes *M. transsylvanica* was a true winter-spring form with a maximum occurrence under the ice (Fig. 5.26A). Populations most often developed between November and January and were maintained through spring circulation; active populations were not found in the 141 collections made between mid-June and November (Fig. 5.26A). As a result of its seasonal distribution, *M. transslyvanica* had a maximum occurrence between 0°C and 9°C and a very low weighted mean temperature of 8.4°C (Fig. 5.26B). The few literature records support the idea that *M. transsylvanica* is a cold water taxon (Cronberg and Kristiansen, 1980; Dürrschmidt, 1982b) found primarily during the winter and spring (Dürrschmidt, 1980; Asmund, 1977; Donaldson and Stein, 1984).

Mallomonas transsylvanica is also restricted to acidic softwater lakes low in specific conductance (Figs. 5.26C-D). This species was found between pH 5.4 and 7.05, had a maximum occurrence between pH 5.5 and 6, and a weighted mean pH of 5.9 (Fig. 5.26C). Despite its obvious acidophilic nature, it was noticeably absent in collections below pH 5.5. Based on available literature records *M. transsylvanica* is primarily restricted to a pH range between 5.5 and 7.3 (Donaldson and Stein, 1984; Momeu and Péterfi, 1983; Dürrschmidt, 1982b) and is acidophilous in nature (Cronberg and Kristiansen, 1980; Dürrschmidt, 1982b; Kling and Kristiansen, 1983; Siver, 1989a), supporting the conclusion that *M. transsylvanica* is acidophilic.

The distribution of *M. transsylvanica* along a pH gradient also supports the idea proposed by Siver (1989a) that initial acidification events may trigger an increase in this taxon, however, it would probably not survive extreme acidification episodes (i.e. those where the pH dropped below 5).

The requirement of *M. transsylvanica* for soft-water lakes is reflected in its occurrence within a very narrow range of specific conductance and its low weighted mean value of 29 μS (Fig. 5.26D). *M. transsylvanica* also had a preference for slightly stained humic habitats, a characteristic that is well documented for this species (Dürrschmidt, 1980; 1982b; Cronberg and Kristiansen, 1980; Momeu and Péterfi, 1983; Donaldson and Stein, 1984). Despite its wide distribution along a total phosphorus gradient (Fig. 5.26E), *M. transsylvanica* was most often found in oligotrophic or mesotrophic lakes located within forest settings. The total phosphorus levels of several of the lakes harboring *M. transsylvanica* fluctuated greatly, presumably due to allochthonous inputs of organic phosphorus. The distributions of *M. transsylvanica* along seasonal, temperature, pH and specific conductance gradients were similar to those recorded for *M. torquata*.

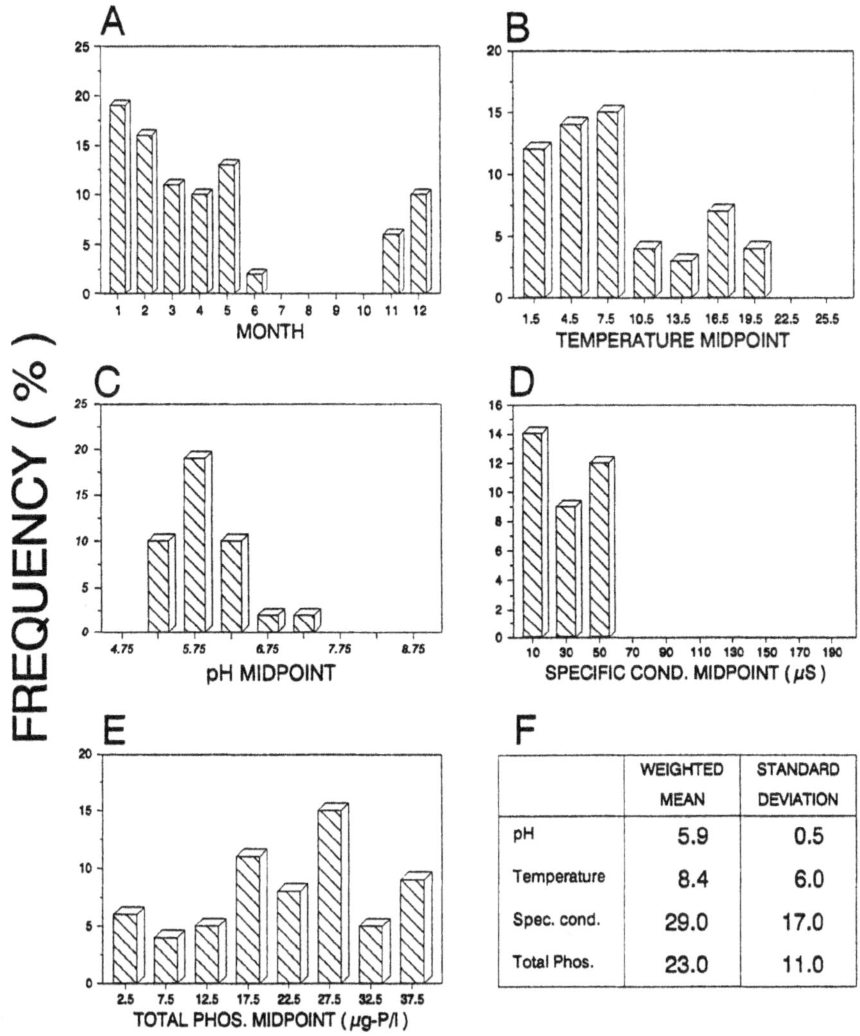

	WEIGHTED MEAN	STANDARD DEVIATION
pH	5.9	0.5
Temperature	8.4	6.0
Spec. cond.	29.0	17.0
Total Phos.	23.0	11.0

Fig. 5.26. The frequencies of occurrence for *Mallomonas transsylvanica* along seasonal (A), temperature (B), pH (C), specific conductance (D) and total phosphorus (E) gradients. Weighted mean and standard deviation values are also given (F).

Table 5.20. Literature records indicating the ecological preferences of *Mallomonas transsylvanica*.

Reference	Seasonality	Trophic Status	Temperature (°C)	pH and Related Factors	Other
1. Asmund, 1977	April	---	Cold, partial ice cover.	Slightly humic stained.	Denmark
2. *Charles and Smol, 1988	---	---	---	AWMpH = 6.6	Adirondacks
3. Cronberg and Kristiansen, 1980	---	---	---	Acidic humic waters.	Sweden
4. Donaldson and Stein, 1984	Spring and autumn.	Lake in a highly urbanized park.	2.8 to 21	5.7 to 6.2, humic stained.	Canada
5. Dürrschmidt, 1980	Spring	Oligotrophic	15	---	Chile
6. Dürrschmidt, 1982b	---	---	---	Acidic humic waters.	Chile
7. *Hartmann and Steinberg, 1986	---	---	---	Few scales in surface sediments of a recently acidified lake.	Alpine mountain lake.
8. Kling and Kristiansen, 1983	---	---	---	Prefers acid waters.	Canada
9. Momeu and Péterfi, 1983	---	Small eutrophic bog.	---	6 to 6.5	Romania
10. Péterfi and Momeu, 1976	---	Mesotrophic bog.	---	Sphagnum bogs.	Romania
11. Siver, 1989a	---	---	---	5.5 to 7.3; Acidic	Connecticut and a survey of the literature.

AWMpH = abundance weighted mean pH.
*Indicates a paleolimnological study.

Other Mallomonas Taxa

Because only a limited amount of data was available for the remaining taxa of *Mallomonas* their distributions along environmental gradients in Connecticut or Adirondack localities could not be fully documented. However, the data were combined with records contained within the literature and used to prepare preliminary remarks concerning the autecology of twelve additional species (Table 5.21).

It is well documented that *M. alata* f. *alata* is primarily found during the cooler months of autumn, winter and spring (Table 5.21). Except for one study (Wujek et al., 1975), this taxon has not been reported above 12°C (Asmund et al., 1982; Dürrschmidt, 1984; Jacobsen, 1985; Kristiansen, 1985; Wee and Gabel, 1989). *Mallomonas alata* f. *alata* has been reported from localities with a pH from 5.9 (Dürrschmidt, 1984) to over 8 (Kristiansen, 1985) and either oligotrophic or eutrophic in nature.

Mallomonas alpina is a widely distributed taxon often reported to occur in the same localities as the closely related species *M. tonsurata* (Asmund, 1959; Dürrschmidt, 1984). In Connecticut, *M. alpina* was also restricted to the same waterbodies that harbored *M. tonsurata* and intermediate forms between the two taxa were often encountered, raising the question as to the validity of both organisms being true species (see Chapter III). In Connecticut, *M. alpina* was observed in the spring and autumn between 5°C and 21°C, and in waterbodies that were eutrophic and alkaline in nature. Such an ecological characterization is well documented in the literature (Table 5.21). *Mallomonas alpina* has most often been observed from autumn to spring at water temperatures below 20°C (Asmund, 1959; Asmund and Takahashi, 1968; Cronberg and Kristiansen, 1980; Kristiansen, 1985; 1988; Roijackers and Kessels, 1986). Based on the literature survey, where it was found in 13% of the collections, *M. alpina* had a weighted mean temperature of 11.9°C.

Similar to *M. tonsurata*, *M. alpina* was repeatedly found in water with a high pH (Table 5.21) and was reported as one of the most alkaline taxa of *Mallomonas* with a weighted mean pH of 8.0 (Siver, 1989a). In Connecticut lakes this species had a similar weighted mean pH of 7.7. Even though a few isolated records for *M. alpina* exist below pH 6.5 (e.g. Cronberg and Kristiansen, 1980; Dürrschmidt, 1984), the vast majority are above pH 7, prompting Siver (1989a) to classify this taxon as a true alkalibiont. As was the case in Connecticut localities, *M. alpina* is most often reported from eutrophic habitats (Table 5.21).

Although rarer than either *M. tonsurata* or *M. alpina*, another closely related species, *M. areolata*, has also been reported primarily during cooler months in habitats that are characterized as eutrophic and alkaline in nature (Table 5.21). Both Asmund (1959) and Dürrschmidt (1980) noted that *M. areolata* was common in eutrophic waterbodies, but lacking from oligotrophic and dystrophic lakes. This species has been observed between pH 6.0 (Momeu and Péterfi, 1983) and 9.1 (Kristiansen, 1985; 1988), but is most common in alkaline waters above pH 7.

Mallomonas calceolus was common in the literature review (12%), but rare in Connecticut (1%) and absent from the Adirondacks. Although this species was found predominately between pH 5 and 7 and had a weighted mean pH of 6.4 (Siver, 1989a), it was reported once below pH 5 and as high as pH 8.8 (Roijackers and Kessels, 1986). Hartmann and Steinberg (1986) reported an increase in scales of *M. calceolus* in the recent sediments of a small alpine lake when the water body first began to acidify; the organism disappeared in the most recent sediments when the inferred pH dropped below 5. Based on the available data *M. calceolus* is classified as acidophilous. Cronberg and Kristiansen (1980) and Eloranta (1989) suggested that *M. calceolus* preferred humic stained lakes.

Based on the literature survey, *M. calceolus* occurred between 5°C and 18°C, had a maximum distribution between 6°C and 10°C and a weighted mean temperature of 9°C, in agreement with the findings of Cronberg and Kristiansen (1980), Dürrschmidt (1982b) and Roijackers and Kessels (1986).

Hartmann and Steinberg (1986) suggested that *Mallomonas clavus* (reported as *Mallomonas allantoides*) was acidobiontic or acidophilic in nature since it increased in relative abundance in the surface sediments of a relatively acidified lake. Further data are needed to support this observation.

Table 5.21. Literature records indicating the ecological preferences of different species of *Mallomonas*.

Reference	Seasonality	Trophic Status	Temperature (°C)	pH and Related Factors	Other
M. alata f. *alata*					
1. Andersen and Meyer, 1977	December	---	---	---	Arkansas
2. Asmund et al., 1982	Autumn to spring.	Oligotrophic, eutrophic.	3 to 12	6.5 to 7.	Chile and Sweden.
3. Balanov, 1978	Spring	---	---	---	U.S.S.R.
4. Bradley, 1966	Winter	---	---	---	Found under the ice. Scotland.
5. Dürrschmidt, 1984	Spring and autumn.	Humic pools.	---	---	Giessen region of West Germany.
6. Kristiansen, 1985	Winter	Eutrophic	0.8 to 6.2	8.2 to 9.1	Denmark
7. Roijackers and Kessels, 1981	March, August.	---	11	7.2	Netherlands
8. Takahashi, 1978	Spring	---	---	---	Japan
9. Wujek et al., 1975	Summer	---	23	6.1	Michigan
M. alpina					
1. Asmund, 1959	Definite spring and autumn form.	Eutrophic	2 to 10 and 15 to 18	high pH.	Found in same waters as *M. tonsurata*. Denmark.
2. Asmund and Takahashi, 1969	Winter	Eutrophic	Cooler waters.	---	Japan; Alaska
3. Cronberg and Kristiansen, 1980	---	---	Cooler waters, below 10.	5.9 to 7.4	Sweden
4. Dürrschmidt, 1984	---	Eutrophic	---	Alkaline	Widely distributed. Found occurring together with *M. tonsurata*. Giessen region of West Germany.

Table 5.21. continued

Reference	Seasonality	Trophic Status	Temperature (°C)	pH and Related Factors	Other
5. Eloranta, 1989	---	---	---	6.1 to 7.2 mean = 6.5 Alkaliphilous to cirumneutral.	Finland
6. Ito, 1988	Spring	---	---	---	Lake Biwa, Japan
7. Kristiansen, 1985; 1988	Spring-winter autumn.	Eutrophic	0.8 to 19.6	8.2 to 9.1	Denmark
8. Roijackers and Kessels, 1986	---	---	Cooler waters.	High	Netherlands
9. Siver, 1989a	---	---	---	WMpH = 8.0 alkalibiontic.	Literature review.
M. areolata					
1. Asmund, 1959	January to May	Mesotrophic to eutrophic. Lacking in oligotrophic and dystrophic ponds.	Cold; 3.5 to 18.5	6.3 to 8.6 7.2 to 8.9 6.8 to 8.0	Denmark
2. Dürrschmidt, 1980	---	Eutrophic. Never found in oligotrophic and dystrophic ponds.	---	---	Chile
3. Dürrschmidt and Croome, 1985	---	---	---	---	First report from tropics.
4. Kristiansen, 1985; 1988	Winter	Eutrophic	2 to 11.5	8.2 to 9.1 alkaline	Denmark
5. Takahashi, 1978	---	---	4 to 18	Near 6	Japan
M. asmundiae					
1. Andersen, 1978	---	---	---	Humic stained, slightly acidic.	Ephemeral pond in Arkansas.
2. Donaldson and Stein, 1984	March, April.	---	9, 16	5.1, 6.25	Canada

Table 5.21. continued

Reference	Seasonality	Trophic Status	Temperature (°C)	pH and Related Factors	Other
3. Dürrschmidt, 1984	March-April.	---	---	---	Rare species. Geissen region of West Germany.
M. calceolus					
1. Bradley, 1964	January under the ice.	Dung contaminated.	Cold	7.0	Ireland
2. Cronberg and Kristiansen, 1980	---	Humic lakes.	5 to 8	---	Sweden
3. Dürrschmidt, 1982b	---	---	---	Acidic	Chile
4. Eloranta, 1989	---	Humic lakes.	---	5.6 to 5.7 mean = 5.7	Rare species. Finland
5. *Hartmann and Steinberg, 1986	---	---	---	Increase in relative abundance in upper sediments presumably due to acid deposition.	Alpine lake in Bavarian Alps.
6. Roijackers, 1981	---	---	---	Acidophilous to acidobiontic.	Netherlands
7. Roijackers and Kessels, 1986	---	---	Low temperature cluster.	---	Netherlands
M. clavus					
1. Bradley, 1964	July	---	---	7.0	Iceland
2. *Hartmann and Steinberg, 1986	---	---	---	Found in upper sediments of a lake believed to be recently acidified.	Alpine lake in Bavarian Alps.
M. corymbosa					
1. Asmund and Hilliard, 1961	September	---	12 to 13	7.1	Alaska

Table 5.21. continued

Reference	Seasonality	Trophic Status	Temperature (°C)	pH and Related Factors	Other
2. Asmund and Takahashi, 1969	May to October, but most abundant in October.	---	Most abundant near 9 to 10	---	Taxon favors larger waterbodies. Alaska
3. Dürrschmidt, 1982b	---	---	---	---	Often found in "lake-sized" waters. Chile.
4. Siver, 1989c	---	---	---	WMpH = 7.7 for Connecticut lakes, 7.1 for literature survey; alkaliphilic.	Literature survey.
5. Siver and Hamer, 1989	---	---	Temperature controls the second principal component.	Specific conductance and pH dominate the first principal component.	Connecticut
6. Takahashi and Hayakawa, 1979	---	---	---	---	Found in the tropics; Bangledesh.
M. cratis					
1. Dürrschmidt, 1982b	---	---	9 to 11	6.3 to 8	Chile
2. Harris, 1967	---	Wooded pool.	---	slightly acidic.	Paper discussed morphological variability.
3. Kristiansen, 1985; 1988	February through May.	Eutrophic	Cold	8.2 to 9.1	Denmark
4. Péterfi and Momeu, 1976	February to April; Winter and spring.	Eutrophic swamp.	Cold	6.0 to 7.5	Romania
5. Wawrzyniak and Andersen, 1985	---	---	---	---	6% of samples from northern boreal localities.
6. Wee and Gabel, 1989	---	---	7 to 18	7.6 to 8.3	Iowa

Table 5.21. continued

Reference	Seasonality	Trophic Status	Temperature (°C)	pH and Related Factors	Other
M. doignonii					
1. Asmund and Cronberg, 1979	Winter/spring.	---	Cold	Close to 6.	Population declined after ice-out. Pattern observed for 9 years. Denmark.
2. Donaldson and Stein, 1984	March	---	9	5.1	As v. *tenuicostis*. Canada.
3. Wee and Gabel, 1989	---	---	2 and 11	6.7 and 7.8	Iowa.
M. eoa					
1. Asmund and Takahashi, 1969	---	---	May be an oligothermol taxon.	---	Alaska
2. Cronberg, 1973; 1980	Under ice with peak in March and cysts produced in April.	Bloom occurred during second year of restoration.	1 to 18, peaks at 6.	Above 7.	Lake Trummen, Sweden.
3. Cronberg, 1982	---	Dominant species in lake after restoration when the lake returned to a more oligotrophic state.	---	---	Believed that 60 year old uncovered cysts seeded the blooms.
4. Cronberg, 1986	---	---	---	---	Maximum cyst formation coincides with maximum pollution levels.
5. Cronberg and Kristiansen, 1980	Vernal period.	Eutrophic and polluted habitats.	Cold 5.8 to 9.8	6.2 to 8.1	Sweden
6. Takahashi, 1978	---	---	Near 0 to above 20.	Acidobiontic with maximum abundance between 6 and 7.	Japan

Table 5.21. continued

Reference	Seasonality	Trophic Status	Temperature (°C)	pH and Related Factors	Other
M. flora					
1. Dürrschmidt, 1982b	---	---	8.7 to 15	6.7 to 7	Chile
2. *Hartmann and Steinberg, 1986	---	---	---	More abundant in recent acidified sediments; acidophilous.	Alpine lake in Bavarian Alps.
3. Jacobsen, 1985	---	---	9, 14	7	Greenland
4. Kristiansen, 1978	---	---	10, 12	5.5	Denmark
5. Roijackers and Kessels, 1986	---	---	Cold water 8 to 10.	5.2 to 6.9	Netherlands
M. mangofera f. *foveota*					
1. Bradley, 1966	Winter	---	Cold	---	Scotland
2. Dürrschmidt, 1983a	Winter	---	7.5	5.7	Chile
3. Takahashi, 1959; 1978	Summer, autumn	---	14.2, 21, 29	6.0, 6.4, 8.3	Japan
4. Takahashi and Hayakawa, 1979	December	---	---	---	Bangladesh
M. mangofera v. *mangofera*					
1. Dürrschmidt, 1983a	Winter	---	8.4	5.5	Low conductance waterbodies. Chile.
2. Harris and Bradley, 1960	Spring	---	Cool water	Acid peat bog.	England
3. Péterfi and Momeu, 1977	Spring	Eutrophic bog.	---	5.5 to 6.	Romania
4. Takahashi, 1978	Summer, winter.	---	15 to 18	mainly 6 to 7.	Japan

Table 5.21. continued

Reference	Seasonality	Trophic Status	Temperature (°C)	pH and Related Factors	Other
M. matvienkoae					
1. Dürrschmidt, 1983b	---	---	Cold, 7.5	*Sphagnum* bog, 5.8	as f. *litteata*. Chile
2. Dürrschmidt and Croome, 1985	---	---	---	---	Worldwide distribution, including Australia and Malaysia.
3. Kristiansen, 1985; 1988	Winter	Eutrophic	Cold, 0.8	8.2 to 9.1	Denmark
4. Péterfi and Momeu, 1976	---	Mesotrophic bog.	---	6	Romania
5. Wujek and Timpano, 1984	---	---	---	---	Northern temperate to the tropics.
M. multiunca					
1. Asmund, 1956	August	Very dark water.	18	Low pH, near 4	Bog lake. Denmark
2. Asmund, 1959	March through June. Partial ice cover.	Eutrophic ponds, oligotrophic ponds, dystrophic ponds.	3 to 25	7.5, 3.7 to 4.6	Denmark
3. Bradley, 1964	Winter to spring and autumn.	---	---	---	Iceland
4. Dürrschmidt, 1984	---	---	---	---	Widespread, but rare. Giessen region of Western Germany.
5. Harris and Bradley, 1960	Autumn for the type; winter/spring for eutrophic form.	Peatlands or dung contaminated.	Cold	Acidic	Indicate that there may be two environmental forms.
6. Kristiansen, 1985	Spring	Eutrophic	Cold	Alkaline 8.2 to 9.1	Denmark
7. Roijackers and Kessels, 1986	---	---	10, 14	4.4, 6.4	Netherlands
8. Takahashi, 1978	---	---	Cold, near 10.	Maximum between 7 and 8.	Japan

Table 5.21. continued

Reference	Seasonality	Trophic Status	Temperature (°C)	pH and Related Factors	Other
9. Wee et al., 1982	---	---	---	Bog	Michigan
10. Wee and Gabel, 1989	---	---	4, 17, 21	6.4, 7.2, 7.4	Iowa
M. oviformis					
1. Asmund, 1959	Winter to spring.	High eutrophic.	3 to 8	Alkaline 7.5, 8.2	Denmark
2. Harris, 1966	November to March.	Ponds on agricultural lands.	Cold	---	England
3. Harris and Bradley, 1960	November to March.	Eutrophic form ponds.	Cold	---	England
4. Kristiansen, 1985; 1988	Winter	Eutrophic	0.8 to 4	Alkaline 8.2 to 9.1	Denmark
5. Momeu and Péterfi, 1983	April	Eutrophic bog.	18	7.5	Romania
6. Wujek and Hamilton, 1973	---	Eutrophic	---	---	Michigan
M. portae - ferreae					
1. Gretz et al., 1985	May and November.	---	10, 17	Alkaliphilous 7.4, 7.7	As v. *reticulata*. Arizona.
2. Ito, 1988	Spring and Autumn.	---	---	---	Lake Biwa, Japan.
3. Kristiansen, 1980; 1981	---	---	---	---	Has a distinct distribution in southern temperate region.
4. Siver, 1989a	---	---	---	WMpH = 7.1	Literature review.
5. Siver and Hamer, 1989	---	Weighted mean total -P of 21 ug-P/l	Weighted mean = 15	WMpH = 7.1	High specific conductance.

Table 5.21. continued

Reference	Seasonality	Trophic Status	Temperature (°C)	pH and Related Factors	Other
M. pumilio v. *munda*					
1. Asmund, 1959	March	Oligohumic; clearwater.	about 3 to 7.5	8.5	Denmark
2. Asmund et al., 1982	March to April.	Oligotrophic to eutrophic.	Less than 10	7.1	Denmark and Sweden
3. Kristiansen, 1978	May	Oligohumic	12	5.5	Denmark
4. Kristiansen, 1985; 1988	Winter	Eutrophic	0.8 to 6.2	8.2 to 9.1	Denmark
5. Péterfi and Momeu, 1976	Spring	Eutrophic	2 to 13	6 to 7	Romania
6. Roijackers and Kessels, 1986	Spring	Eutrophic	8.2 to 10	6.4 to 7	Netherlands
M. pumilio v. *pumilio*					
1. Asmund et al., 1982	Spring	Oligotrophic to eutrophic; humic stained.	5 to 12	6 to 7	Denmark and Sweden.
2. Harris and Bradley, 1957; 1960	November to April.	---	Cold	---	England
3. Roijackers and Kessels, 1981	March to May.	---	10.2	6.4	Netherlands
M. pumilio v. *silvicola*					
1. Asmund et al., 1982	Summer, but cold.	Oligotrophic	4	---	Switzerland
2. Harris and Bradley, 1960	---	---	---	5.5 to 6.5	England
3. Péterfi, 1974	Spring	Eutrophic	0 to 12	6 to 7	Romania

Table 5.21. continued

Reference	Seasonality	Trophic Status	Temperature (°C)	pH and Related Factors	Other
M. torquata					
1. Asmund and Cronberg, 1979	February to May; winter/spring taxon.	---	Cold water species. 0.3, 1, 7.5, 4.1, 20. Often found under an ice cover.	6.4, 8.5, 7.0, 6.4	Alaska and Sweden.
2. Cronberg and Kristiansen, 1980	Vernal flora.	Humic lakes; clean or polluted.	Cooler waters.	Humic stained.	Sweden
3. Kling and Kristiansen, 1983	---	Humic lakes.	---	Acid humic lakes.	Canada
4. Siver, 1988c	---	---	---	Acidic locality; 5.98	Adirondacks

AWMpH = abundance weighted mean pH.
*Indicates a paleolimnological study.

As was observed by Kristiansen (1985; 1988) and Péterfi and Momeu (1976a), *Mallomonas cratis* was restricted to winter and spring periods in Connecticut localities (Table 5.21). Populations were found between 2°C and 17°C in Connecticut and 7°C and 18°C in Adirondack habitats, respectively. Although *M. cratis* had a wide pH range from 6 to over 8, it was most commonly found above 7 and is tentatively classified as pH indifferent. It has been reported from eutrophic (Kristiansen, 1985; 1988; Péterfi and Momeu, 1976a) as well as swampy woodland pools (Harris, 1967; Péterfi and Momeu, 1976a).

Originally classified as acidobiontic by Takahashi (1978), *Mallomonas eoa* has since been observed to be tolerant of a wider pH gradient (Siver, 1989a) (Table 5.21). Much of what is known about the ecology of this taxon stems from a study of Lake Trummen in Sweden (see Cronberg, 1982 and references therein). Lake Trummen was a highly polluted waterbody that was the subject of a large-scale cleanup involving both the diversion of sewage and the suction dredging of nutrient-rich surface sediments. In the post-restoration years Cronberg (1973; 1980) found *M. eoa* to be a primary component of the cold water flora; populations developed under the ice, peaked in March at 6°C and declined thereafter in April when stomatocyst production was maximal. Cronberg (1982) further speculated that 60 year old stomatocysts uncovered by the dredging process provided the seed source for the blooms. The development of a population of *M. eoa* followed a similar course in a Connecticut lake where it was first observed under the ice, peaked at ice-out and declined shortly thereafter.

Based on limited data, *Mallomonas flora* has most commonly been observed between pH 6 and 7 (Dürrschmidt, 1982b; Jacobsen, 1985; Roijackers and Kessels, 1986), has not been reported below pH 5.2 (Kristiansen, 1978; Roijackers and Kessels, 1986) and is tentatively described as acidophilous. The classification of *M. flora* as acidophilous is in agreement with Hartmann and Steinberg (1986) who found scales of this taxon to increase in the surface sediments of a recently acidified lake. *Mallomonas flora* has been found only between 7°C and 15°C, suggesting that it is a cold water species (Table 5.21).

Mallomonas mangofera var. *mangofera* and f. *foveota* are rarely encountered taxa that have been characterized as acidic organisms found primarily in the winter and spring (Table 5.21).

Mallomonas multiunca is a widespread, but rare, taxon (Dürrschmidt, 1984). This species was observed in acidic localities from December to April at water temperatures between 2°C and 10°C. In Connecticut, *M. multiunca* was restricted to a narrow pH range of 5.6 to 6.2 and had a weighted mean of 5.9. Based on literature records, *M. multiunca* had a much wider pH range from below 4 (Asmund, 1959) to over 9 (Dürrschmidt, 1984; Kristiansen, 1985) and a preference for cooler water temperatures. This organism has been characterized from a wide spectrum of lake types, including oligotrophic, eutrophic, dystrophic, bog and dung contaminated localities (Table 5.21), and Harris and Bradley (1960) suggested the existence of both an autumn peatlands form and a winter-spring eutrophic form.

Mallomonas oviformis is best described as a winter to spring form, occurring below 10°C and inhabiting eutrophic to highly eutrophic alkaline ponds (Table 5.21). In Connecticut, this species was found only during the winter in two alkaline ponds kept partially ice-free by a large flock of geese, supporting the idea that *M. oviformis* can withstand highly eutrophic conditions.

Asmund et al. (1982) reviewed the ecological requirements of the three varieties of *Mallomonas pumilio*, a species present in 10% of the collections surveyed from the literature, but only 1% of those from Connecticut and the Adirondacks. Each variety has been shown to occur predominantly in the spring at temperatures less than 13°C (Table 5.21). All three taxa have been found in oligotrophic as well as eutrophic habitats. Although *Mallomonas pumilio* has been observed between pH 5.5 and 9, it is most commonly reported between pH 6 and 7 (Table 5.21). In this study *M. pumilio* was found between 7°C and 21°C and from pH 6 to 7.2. More records would be needed to determine if the three varieties differ in their ecological requirements.

Another rare taxon, *Mallomonas schwemmlei*, was found under the ice in the winter and in the spring between 2°C and 14°C and pH 5.9 to 6.6.

Table 5.22. Ecological characteristics for 28 taxa of *Mallomonas*. The table summarizes the primary conditions under which each taxon is typically found. The numbers under the temperature, specific conductance and total phosphorus sections represent those discussed within the summary section of this chapter. The trophic status is listed under the total phosphorus section.

TAXON	SEASONALITY	TEMPERATURE	pH CONDITION	SPECIFIC CONDUCTANCE	TOTAL PHOSPHORUS/ TROPHIC STATUS
acaroides v. acaroides	late summer to autumn	3, cool	Alkalibiontic	5, high	5, mesotrophic to eutrophic
acaroides v. muskokana	late spring to early autumn	4, warm	Acidobiontic	1, low	3, oligotrophic
akrokomos	All year, but maximum in winter/spring	2, cold to cool	Acidophilous to indifferent	6, but more common below 40 µS	6, oligotrophic
annulata	Autumn to spring	2, cold to cool	Indifferent	4, medium to high	6, oligotrophic to eutrophic
asmundiae	Winter to spring	1, cold	Acidophilous	˙3	˙3
canina hindonii	Winter to spring	1, cold	Acidobiontic	1, low	˙oligotrophic
caudata	All year	5, indifferent	˙˙Indifferent	3, medium	6, oligotrophic to eutrophic
corymbosa	All year, but maximum in autumn	5, indifferent	Alkaliphilous to alkalibiontic	4, medium to high	4, mesotrophic to eutrophic
crassisquama	All year	5, indifferent	˙˙Indifferent	3, almost 6	6, oligotrophic to eutrophic
doignonii	Winter	2, cold to cool	˙˙Acidophilous	1, low	4, oligotrophic to mesotrophic
dickii	Summer	4, warm to cool	˙˙Acidophilous	1, low	4, oligotrophic to eutrophic
duerrschmidtiae	Spring and autumn	3, cool	˙˙Acidophilous	1, low	1, oligotrophic
elongata	Autumn to spring	1, cold	Indifferent to alkaliphilous	4, medium to high	3, mesotrophic to eutrophic
galeiformis	Late spring to autumn	4, warm to cool	˙˙Acidophilous	1, low	2 to 3 oligotrophic to mesotrophic
hamata	All year	5, but cool tendencies	Acidobiontic to acidophilous	2, low to medium	3, oligotrophic to mesotrophic
heterospina	Autumn to spring	1, cold	˙˙Acidophilous	4, medium to high	4, oligotrophic to eutrophic
lychenensis	All year, except summer	3, cool	˙˙Acidophilous	˙	4, mesotrophic to eutrophic
paludosa	Spring and autumn	3, cool	Acidobiontic	1, low	1, oligotrophic

TAXON	SEASONALITY	TEMPERATURE	pH CONDITION	SPECIFIC CONDUCTANCE	TOTAL PHOSPHORUS/ TROPHIC STATUS
papillosa	All year with maximum in early winter	2, cold to cool	··Acidophilous	3, medium	4, oligotrophic to mesotrophic
portae-ferreae	·Scattered	·6	Alkalibiontic to alkaliphilous	4, medium to high	· mesotrophic to eutrophic
pseudocoronata	Spring to autumn	4, warm to cool	Alkaliphilous	4, high	4, oligotrophic to eutrophic
pugio	Winter to spring	2, cold to cool	Acidobiontic	1, low	2, oligotrophic
punctifera	Summer	4, warm to cool	·· Acidophilous	1, low	4 or 6 oligotrophic to mesotrophic
striata	Spring and winter	3, cool	··Indifferent	3, medium	·various
tonsurata	Spring to autumn	4, warm to cool	Alkaliphilous to alkalibiontic	4, high	4, most common in eutrophic
torquata	Late winter to spring	2, cold to cool	··Acidophilous	1, low	2, oligotrophic
transsylvanica	Late autumn to spring	2, cold to cool	··Acidophilous	1, low	6, oligotrophic to mesotrophic

· based on limited data
·· decreases in abundance as the pH drops below the 5 to 5.5 interval

Comparison of the *Mallomonas Floras* from Connecticut, the Adirondacks and the Literature

The distributions of all common taxa of *Mallomonas* along pH, specific conductance, temperature, total phosphorus and seasonal gradients were similar for collections made in Connecticut and the Adirondack localities. For example, if an organism was found in acidic waterbodies during the cooler portions of the year in the Adirondacks, it was found under the same conditions in Connecticut localities. With few exceptions, records from the literature generally supported the ecological niches described from both the Connecticut and Adirondack data sets. Such agreement between data sets suggests that individual taxa are found under similar environmental conditions not only within a given region, but also between geographically separated regions. This, in turn, suggests that individual taxa can serve as indicator organisms not only for lakes within a given area, but also between widespread regions.

There are, however, significant differences in the relative occurrences of species between the Connecticut and Adirondack floras (Fig. 5.27). The differences correspond to differences in the chemical characteristics of lakes from each region. Lakes in Connecticut are higher in pH, specific conductance and total phosphorus than those in the Adirondacks. The watersheds immediately surrounding most of the waterbodies within Connecticut have been developed, whereas many lakes in the Adirondacks remain undisturbed. Consequently, species with higher weighted mean pH, specific conductance and total phosphorus values are more common in Connecticut waterbodies.

This generalization is illustrated in Figure 5.28 where the logarithms of the percent occurrence of the twenty-four most common taxa in the Connecticut lakes are plotted against their correspon-

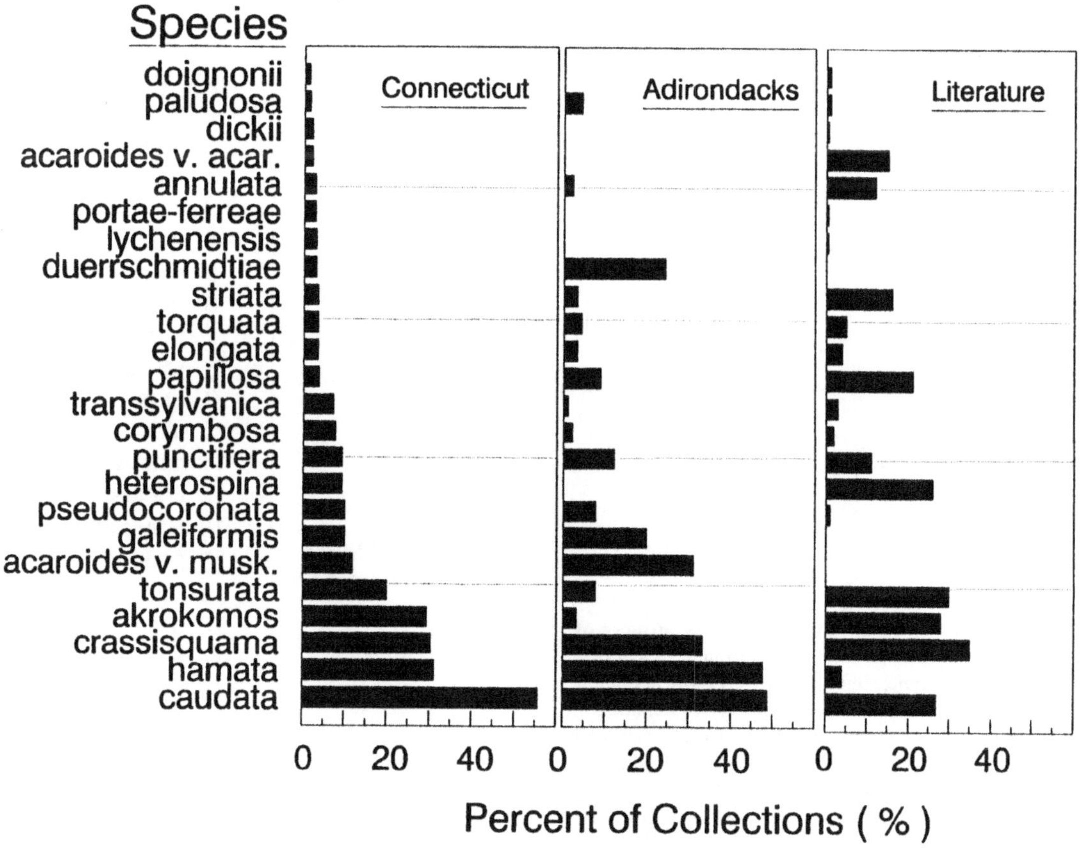

Fig. 5.27. The relative importance of 24 taxa of *Mallomonas* in localities from Connecticut, the Adirondacks and in a survey of the literature. Species are arranged according to their increasing importance in the Connecticut localities.

ding values in the Adirondacks. Values lying close to the 1:1 line depict organisms with a similar occurrence in both regions. Organisms found in a greater number of the collections from Connecticut or the Adirondacks would lie below or above the line, respectively. For example, *M. tonsurata*, *M. corymbosa*, *M. portae-ferreae*, *M. acaroides* var. *acaroides*, *M. pseudocoronata* and *M. elongata*, each with a weighted mean pH above 7, were more common or only found in Connecticut localities (Fig. 5.28). The percentage of collections containing species requiring acidic conditions low in specific conductance, such as *M. duerrschmidtiae*, *M. paludosa*, *M. acaroides* var. *muskokana* , *M. galeiformis*, *M. punctifera* and *M. torquata* were greater in the Adirondacks. Both *M. akrokomos* and *M. heterospina* were much more widespread in Connecticut lakes, presumably because they withstand higher specific conductance and/or nutrient conditions.

Mallomonas caudata, *M. hamata* and *M. crassisquama* were the three most common taxa of *Mallomonas* found in both regions (Fig. 5.27). In Connecticut, *M. caudata* was found in over half of the collections, whereas *M. hamata* and *M. crassisquama* were present in 30%. *M. caudata*, *M. hamata* and *M. crassisquama* were found in 49%, 48% and 33% of the collections from the Adirondacks, respectively. *M. caudata* and *M. crassisquama* are both widely distributed along pH, specific conductance, temperature, total phosphorus and seasonal gradients, accounting for their importance in both localities. *M. hamata* is also found along environmental gradients common to both Connecticut and the Adirondacks.

Mallomonas caudata and *M. crassisquama* are also among the most common taxa of *Mallomonas* reported in the literature, found in 27% and 35% of the collections, respectively (Fig. 5.27). However, significant differences are seen in the relative importance of taxa between the literature and either the Adirondack or Connecticut floras. In general, taxa able to withstand greater dissolved salt, eutrophic or alkaline conditions, are found in a significantly greater percentage of the collections from the literature than from either the Adirondack or Connecticut localities. Species with

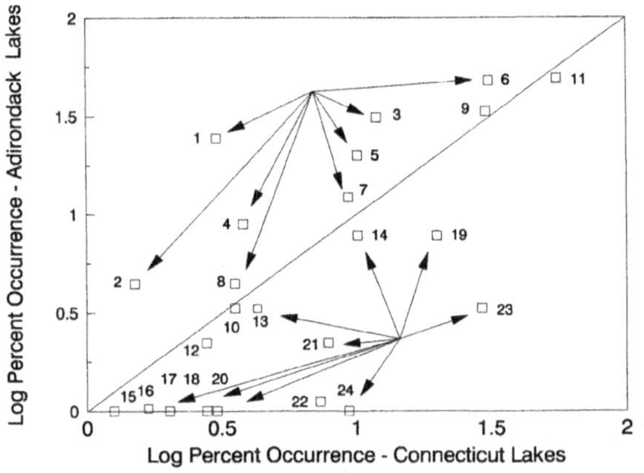

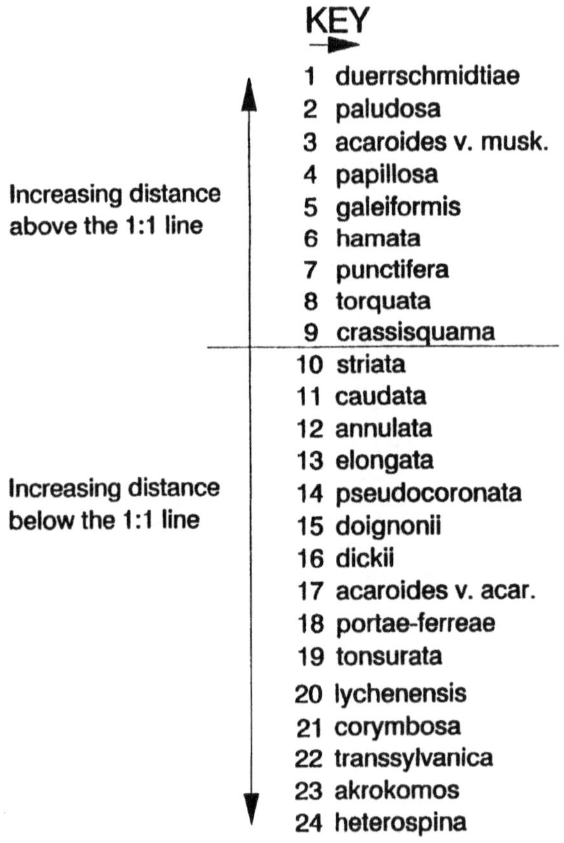

Fig. 5.28. Plot of the logarithms of the percent occurrence of 24 taxa of *Mallomonas* in Connecticut vs. Adirondack localities. Taxa located below the center line are relatively more abundant in Connecticut lakes, while those positioned above the line are more common in Adirondack waterbodies. Species positioned near the line are equally abundant in both regions. The upper set of arrows depicts taxa with low weighted mean pH and specific conductance scores. The lower set of arrows indicate species with relatively high weighted mean pH, specific conductance and/or total phosphorus levels. The numbers correspond to the taxa shown in the key.

tolerances for higher dissolved salt (Fig. 5.30), pH (Fig. 5.31) or eutrophic conditions (Fig. 5.32), such as *M. acaroides* var. *acaroides*, *M. annulata*, *M. heterospina*, *M. tonsurata*, *M. striata* and *M. akrokomos*, are of significantly greater importance in the literature, supporting the idea that the literature records represent collections from more eutrophic localities with higher pH and dissolved salt levels. Each of these taxa were absent or of least importance in the Adirondack flora and most important in the literature.

In comparison, acidic taxa restricted to either low total salt or phosphorus levels were most common in the Adirondack flora. *Mallomonas hamata*, an important member of the flora from both the Adirondacks and Connecticut, was quite rare in the literature (Fig. 5.27). Other taxa, such as *Mallomonas acaroides* var. *muskokana*, *M. duerrschmidtiae* and *M. galeiformis*, were found in 31%, 24% and 20% of the collections from the Adirondacks, but lacking or very rare in the literature. Other acidobiontic species, such as *M. paludosa* and *M. hindonii* are also more important in the Adirondack flora. It should be noted that the lack of importance of *M. galeiformis*, *M. duerrschmidtiae*, and *M. acaroides* var. *muskokana* in the literature may be related, in part, to the fact that they are recently described species. Perhaps with time, these taxa will be found to also be of importance in other regions of the world. In conclusion, many of the differences in the occurrences of individual taxa between geographically separated regions are correlated with major differences in the chemistry of the waterbodies.

Other discrepancies between floras from different regions are the result of the geographical distribution of individual species. For example, *Mallomonas pseudocoronata* and *M. vannigera* are known to be found primarily in North America and Europe, respectively. This would explain the

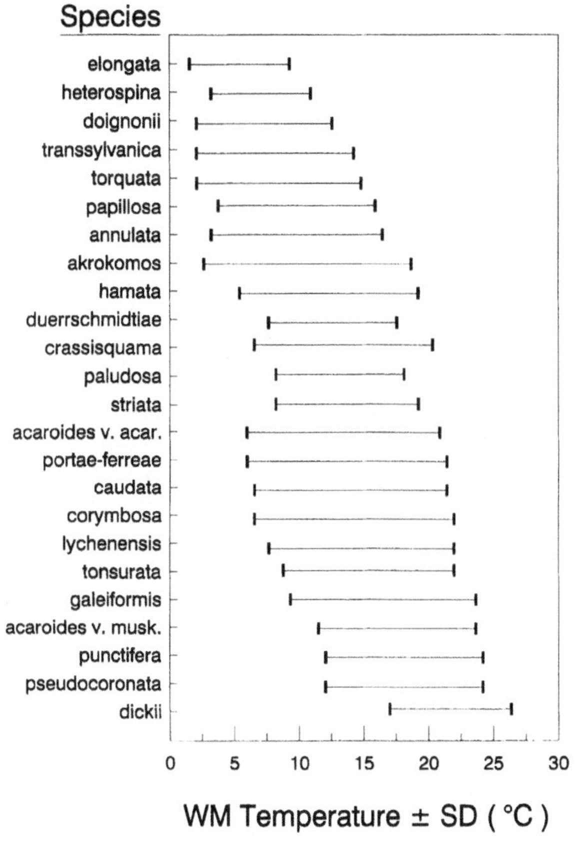

Fig. 5.29. The ranges of weighted mean temperature values ± standard deviations for 24 taxa of *Mallomonas*.

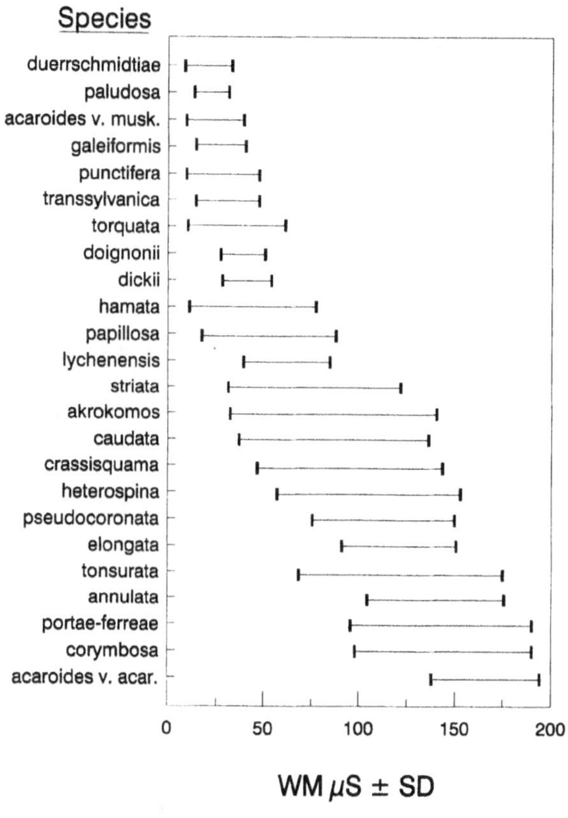

Fig. 5.30. The ranges of weighted mean specific conductance values ± standard deviations for 24 taxa of *Mallomonas*.

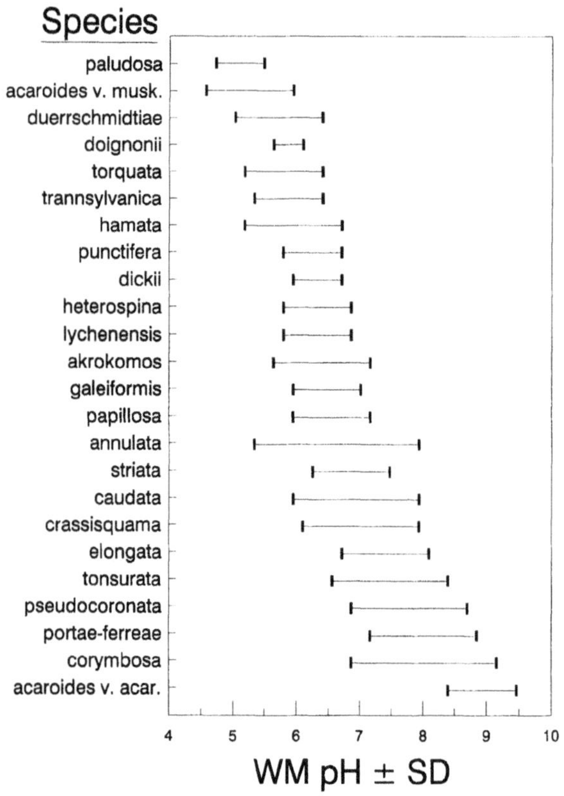

Fig. 5.31. The ranges of weighted mean pH values ± standard deviations for 24 taxa of *Mallomonas*.

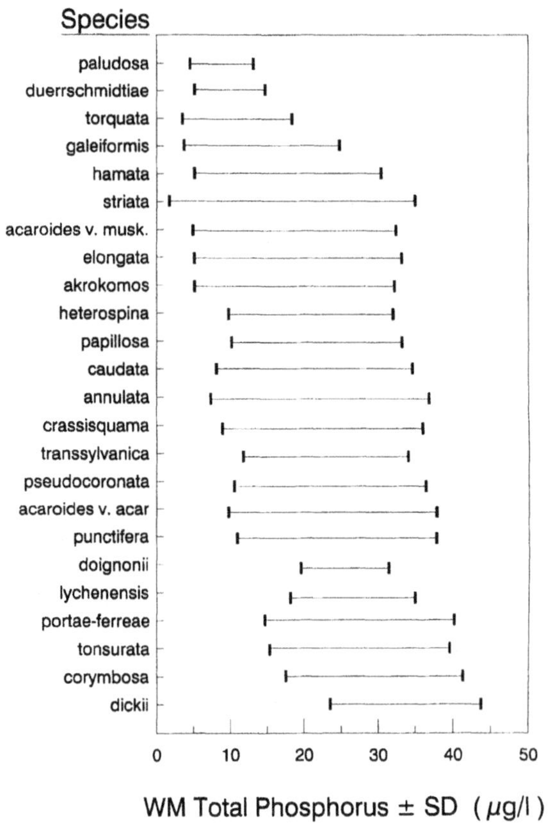

Fig. 5.32. The ranges of weighted mean total phosphorus values ± standard deviations for 24 taxa of *Mallomonas*.

greater importance of *M. pseudocoronata* in both Connecticut and the Adirondacks as opposed to the literature where many of the collections were from European localities. A few species appear to have slightly different niche tolerances, depending on their geographic location. *Mallomonas punctifera*, was limited to acidic waters low in total salt concentrations in Connecticut and the Adirondacks. In the literature survey, however, *M. punctifera* was common at higher pH and conductance levels and had a higher weighted mean pH. Charles and Smol (1988) and Siver (1989a) suggested the possibility that different ecotypes or varieties exist for *M. punctifera*. This idea is supported, in part, by the fact that this taxon is a warm water organism in both the Adirondacks and Connecticut, while in Europe has been found to favor colder conditions (Roijackers and Kessels, 1986). In addition, specimens of *M. punctifera* from the Adirondacks and Connecticut appear to have smaller scales than those from some European localities.

Summary

It is clear that because many taxa of *Mallomonas* are distributed differently along pH, specific conductance, temperature, total phosphorus and seasonal gradients, they represent an important group of indicator organisms. Since each taxon is identified by siliceous cell wall components that remain in lake sediments and become fossils, the genus *Mallomonas* represents an important group of organisms for paleolimnological reconstructions. To date, scaled chrysophytes, including *Mallomonas*, have been utilized in reconstructing historical pH conditions. Highly significant models for pH inference have been developed because of the differential distribution of species along a pH gradient. Since the taxa are also indicative of specific temperature, trophic level and dissolved salt concentrations, they will undoubtedly become important in the reconstruction of other lakewater conditions.

In 1939 Hustedt proposed the use of five pH categories, introduced in the beginning of this chapter, in order to describe the distribution of diatoms along a pH gradient. Although Hustedt's groups are subjective in nature and it can be difficult to categorize an individual species, they serve as an excellent means to summarize a large body of data that include species with similar distributions. In this context and for purposes of discussion, taxa were divided into groups describing their distribution along temperature, specific conductance and total phosphorus gradients (Table 5.22). The groups are defined as follows:

Temperature

1. Cold = Weighted mean temperature below 12°C, rare above 15°C.
2. Cold-Cool = Weighted mean temperature below 12°C, but can be found above 15°C.
3. Cool = Weighted mean between 12°C and 15°C, also found above and below this range.
4. Warm = Weighted mean temperature above 15°C, rare below 12°C.
5. Indifferent = Similar distribution along entire temperature range.

Specific Conductance

1. Low = Weighted mean specific conductance below 40 μS, not found above 90 μS.
2. Low-Medium = Weighted mean specific conductance below 40 μS, can be found above 90 μS.
3. Medium = Weighted mean specific conductance between 40 μS and 90 μS, can be found above and below this range.
4. Medium-High = Weighted mean specific conductance above 90 μS, but can be found below 90 μS.
5. High = Weighted mean specific conductance above 90 μS, not found below 90 μS.
6. Indifferent = Distributed more or less equally along the entire specific conductance gradient.

Total Phosphorus

1. Low = Weighted mean total phosphorus level below 10 μg-P/l, not found above 20 μg-P/l.
2. Low-Medium = Weighted mean total phosphorus level below 10 μg-P/l, but can be found above 20 μg-P/l.
3. Medium = Weighted mean total phosphorus level between 10 μg-P/l and 20 μg-P/l, but can be found above and below this range.
4. Medium-High = Weighted mean total phosphorus level above 20 μg-P/l, can be found below 20 μg-P/l.
5. High = Weighted mean total phosphorus level above 20 μg-P/l, not found below 20 μg-P/l.
6. Indifferent = Distributed more or less equally along the entire total phosphorus gradient.

A group of taxa comprised of *M. acaroides* var. *muskokana*, *M. dickii*, *M. galeiformis*, *M. pseudocoronata*, *M. punctifera* and *M. tonsurata* are warm water organisms distributed along the higher end of a temperature gradient (Fig. 5.29). Populations of each of these taxa begin development in the spring, have a maximum occurrence during the summer months and are no longer found after mid-autumn (Table 5.22). The cold water group of taxa, found only along the lower end of a temperature gradient, include *M. elongata*, *M. heterospina*, *M. asmundiae*, *M. canina* and *M. hindonii* (Table 5.22; Fig. 5.29). The cold water forms first appear in collections made in the late autumn or winter, are often found under the ice and disappear by spring; these taxa are not found during the warmer summer months. Most of the cold -- cool water species, including *M. annulata*, *M. doignonii*, *M. pugio*, *M. torquata* and *M. transsylvanica*, are also primarily found during the late fall, winter and spring months (Table 5.22). Like the cold water group, the cold – cool water forms are distributed along the lower end of the temperature gradient (Fig. 5.29) and are not found during the summer months. Two other cold – cool water species, *M. akrokomos* and *M. papillosa* were also maximal during the colder months, however, they were found during all or most of the year.

Mallomonas acaroides var. *acaroides*, *M. duerrschmidtiae*, *M. paludosa* and *M. striata* are found during the spring and autumn seasons and either rare or lacking during the summer and winter. These four taxa are grouped as cool water forms (Table 5.22) and are distributed primarily along the middle portion of the temperature gradient (Fig. 5.29). The three most common species in both Connecticut and the Adirondacks, *M. caudata*, *M.*

crassisquama and *M. hamata*, were found throughout the year and characterized as temperature indifferent.

Distinct groups of species could also be identified along a specific conductance gradient (Fig. 5.30). A large group of taxa is clearly distributed along the lower end of a specific conductance gradient. This group, referred to as the low specific conductance group, includes *M. acaroides* var. *muskokana*, *M. canina*, *M. hindonii*, *M. dickii*, *M. doignonii*, *M. duerrschmidtiae*, *M. galeiformis*, *M. paludosa*, *M. punctifera*, *M. pugio*, *M. torquata* and *M. transsylvanica*. *Mallomonas striata*, *M. papillosa* and *M. hamata* were found to extend further along the gradient (Fig. 5.30) and grouped within the low-medium or medium categories.

Only one taxon, *M. acaroides* var. *acaroides*, was found within the high specific conductance group (Table 5.22), however, another large number of organisms were distributed primarily along the upper end of the specific conductance gradient (Fig. 5.30). These species, classified within the medium – high specific conductance group, include *M. annulata*, *M. elongata*, *M. heterospina*, *M. portae-ferreae*, *M. pseudocoronata*, *M. tonsurata* and *M. corymbosa*. Three of the more commonly occurring taxa, *M. caudata*, *M. crassisquama* and *M. akrokomos*, were widely distributed along the specific conductance gradient and classified within the Indifferent or medium groups.

A few species of *Mallomonas*, including *M. acaroides* var. *muskokana*, *M. canina*, *M. hindonii*, *M. pugio* and *M. paludosa*, were found to be true acidobionts as defined by Hustedt (1939). These taxa are found exclusively in acidic habitats and are among some of the most acidic organisms known (Fig. 5.31). A large number of the *Mallomonas* taxa found in both Connecticut and Adirondack localities were acidophilous in nature (Table 5.22). The acidophilous species are distri-

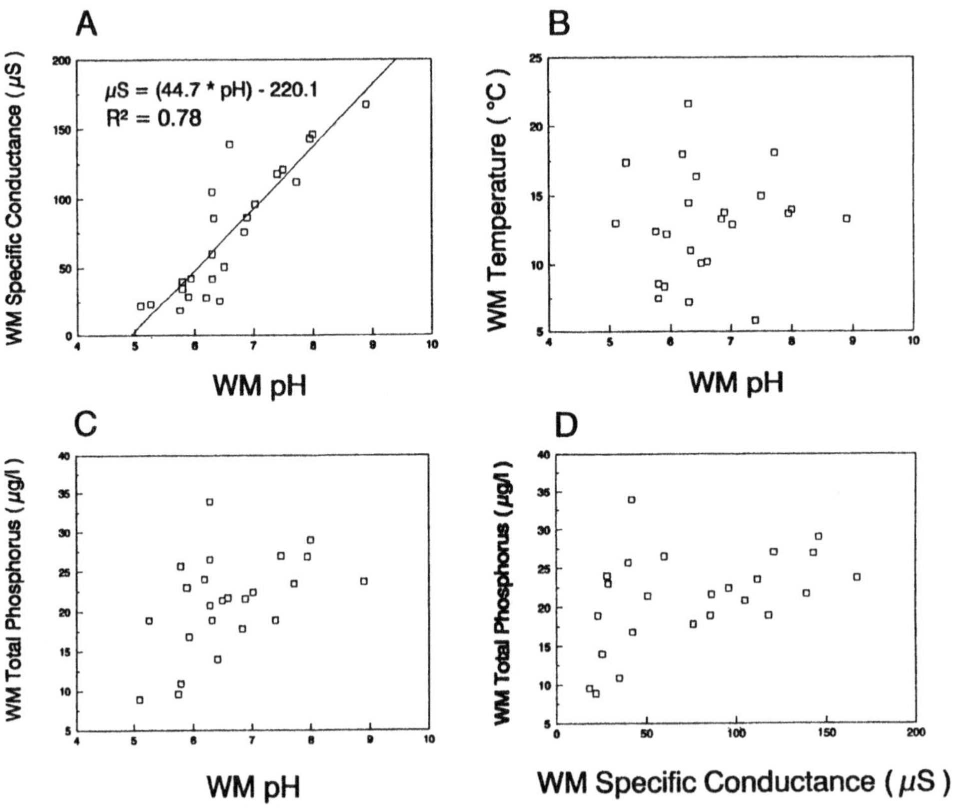

Fig. 5.33. Scatter diagrams of weighted mean values for 24 taxa of *Mallomonas*. A. Weighted mean pH vs. weighted mean specific conductance. B. Weighted mean pH vs. weighted mean temperature. C. Weighted mean pH vs. weighted mean total phosphorus. D. Weighted mean specific conductance vs. weighted mean total phosphorus content.

buted primarily below pH 7.0, but are not as common below pH 5.0 as are the true acidobionts. The most important acidophilous taxa include *M. asmundiae*, *M. dickii*, *M. doignonii*, *M. duerrschmidtiae*, *M. galeiformis*, *M. heterospina*, *M. lychenensis*, *M. punctifera*, *M. torquata* and *M. transsylvanica*. *Mallomonas hamata* is probably best described as an acidobiontic/acidophilous species, whereas both *M. papillosa* and *M. akrokomos* are intermediate between acidophilous and pH indifferent.

At the alkaline end of the pH gradient *M. acaroides* var. *acaroides* is classified as the most alkalibiontic organism. Other species found predominantly above a pH of 7, including *M. elongata*, *M. portae-ferreae*, *M. pseudocoronata*, *M. tonsurata* and *M. corymbosa*, had alkalibiontic and/or alkaliphilous qualities (Table 5.22; Fig. 5.31). *Mallomonas annulata* and *M. striata* were classified as pH indifferent or circumneutral, as were the common taxa *M. caudata* and *M. crassisquama*. Another group of taxa, either classified as acidophilous or pH indifferent, were observed to significantly decrease in occurrence or disappear as the pH dropped below the 5 to 5.5 range (Table 5.22). Species in this group included *M. caudata*, *M. crassisquama*, *M. dickii*, *M. doignonii*, *M. duerrschmidtiae*, *M. galeiformis*, *M. heterospina*, *M. lychenensis*, *M. papillosa*, *M. punctifera*, *M. torquata*, *M. transsylvanica* and *M. striata*.

A group of taxa, including *M. acaroides* var. *muskokana*, *M. asmundiae*, *M. duerrschmidtiae*, *M. galeiformis*, *M. paludosa*, *M. pugio*, *M. torquata* and *M. hamata*, were observed primarily in oligotrophic or mesotrophic waterbodies and were classified in either the low, low - medium or medium total phosphorus groups (Table 5.22). *Mallomonas punctifera*, *M. dickii*, *M. doignonii*, *M. papillosa* and *M. transsylvanica*, were also found mostly in oligotrophic or mesotrophic lakes (Table 5.22), however, they were distributed primarily along the middle or upper portion of the total phosphorus gradient (Fig. 5.32). The collections harboring these taxa were mostly from humic stained localities which tended to have high concentrations of organic matter.

Several species classified within the medium - high or high total phosphorus groups, including *M. tonsurata*, *M. corymbosa*, *M. portae-ferreae*, *M. acaroides* var. *acaroides* and *M. lychenensis*, were reported primarily from mesotrophic or eutrophic habitats. *Mallomonas elongata*, *M. heterospina* and *M. pseudocoronata* tended to favor higher trophic level conditions, but were also found in several collections from oligotrophic sites. *Mallomonas akrokomos*, *M. crassisquama*, *M. caudata* and *M. annulata* were each observed in oligotrophic as well as eutrophic habitats and were classified within the Indifferent total phosphorus group (Table 5.22).

Linear regression and correlation analyses were done for all possible combinations of environmental variables using the weighted mean scores for all species. This exercise was an attempt to identify any significant relationship between environmental variables for species of *Mallomonas*. Four of the scatter plots are shown in Figure 5.33. The only significant relationship was for weighted mean pH and specific conductance scores. Species with higher weighted mean pH scores also had higher weighted mean specific conductance values. Although not statistically significant, species with either high weighted mean pH or specific conductance scores were usually found in collections with higher total phosphorus concentrations.

References

ANDERSEN, R.A. 1978. Scaled Chrysophyceae from Arkansas. II. The genera *Mallomonas*, *Paraphysomonas* and *Spiniferomonas*. Arkansas Acad. Sci. Proc. **32**: 12-17.

ANDERSEN, R.A. and MEYER, R.L. 1977. Scaled Chrysophyceae from Arkansas. Arkansas Acad. Sci. Proc. **31**: 12-16.

ANDERSEN, R.A. and MULKEY, T.J. 1983. The occurrence of chlorophylls c_1 and c_2 in the Chrysophyceae. J. Phycol. **19**: 289-294.

ANDERSEN, R.A. 1985. The flagellar apparatus of the golden alga *Synura uvella*: four absolute orientations. Protoplasma **128**: 94-106.

ANDERSEN, R.A. 1987. Synurophyceae classic nov., a new class of algae. Amer. J. Bot. **74**: 337-353.

ANDERSEN, R.A. 1989. The Synurophyceae and their relationship to other golden algae. Beih. Nova Hedwigia **95**: 1-26.

ARVOLA, L. 1983. Primary production and phytoplankton in two small, polyhumic forest lakes in southern Finland. Hydrobiologia. **101**: 105-110.

ASMUND, B. 1955. Electron microscope observations on *Mallomonas caudata* and some remarks on its occurrence in four Danish ponds. Bot. Tidsskr. **52**: 163-168.

ASMUND, B. 1956. Electron microscope observations on *Mallomonas* species and remarks on their occurrence in some Danish ponds. II. Bot. Tidsskr. **53**: 75-86.

ASMUND, B. 1959. Electron microscope observations on *Mallomonas* species and remarks on their occurrence in some Danish ponds and lakes. III. Dan. Bot. Ark. **18**: 1-50.

ASMUND, B. and HILLIARD, D.K. 1961. Studies on Chrysophyceae from some ponds and lakes in Alaska I – *Mallomonas* species examined with the electron microscope. Hydrobiologia **17**: 237-258.

ASMUND, B. and TAKAHASHI, E. 1969. Studies on Chrysophyceae from some ponds and lakes in Alaska VIII – *Mallomonas* species examined with the electron microscope II. Hydrobiologia **34**: 305-321.

ASMUND, B. 1977. Two new species of *Mallomonas* (Chrysophyceae) in Swedish lakes. Bot. Tidsskr. **71**: 253-258.

ASMUND, B. and CRONBERG, G. 1979. Two new taxa of *Mallomonas* (Chrysophyceae). Bot. Not. **132**: 409-418.

ASMUND, B., CRONBERG, G. and DÜRRSCHMIDT, M. 1982. Revision of the *Mallomonas pumilio* group (Chrysophyceae). Nord. J. Bot. **2**: 383-395.

ASMUND, B. and KRISTIANSEN, J. 1986. The genus *Mallomonas* (Chrysophyceae). Opera Bot. **85**: 1-128.

BALONOV, I.M. 1978. Vidy roda *Mallomonas* Perty (Chrysophyta) v vodajemach bassjejna Volgi. AN SSSR, Inst. biol. vnutrennich vod, Trudy **42:** 1-26.

BATTARBEE, R.W., CRONBERG, G. and LOWRY, S. 1980. Observations on the occurrence of scales and bristles of *Mallomonas* spp. (Chrysophyceae) in the micro-laminated sediments of a small lake in Finnish North Karelia. Hydrobiologia **71**: 225-232.

BEECH, P.L. and WETHERBEE, R. 1990a. Direct observations on flagellar transformation in *Mallomonas splendens* (Synurophyceae). J. Phycol. **26**: 90-95.

BEECH, P.L. and WETHERBEE, R. 1990b. The flagellar apparatus of *Mallomonas splendens* (Synurophyceae) at interphase and its development during the cell cycle. J. Phycol. **26**: 95-111.

BEECH, P.L., WETHERBEE, R. and PICKETT-HEAPS, J.D. 1990. Secretion and deployment of bristles in *Mallomonas splendens* (Synurophyceae). J. Phycol. **26**: 112-122.

BELCHER, J.H. 1969. Some remarks upon *Mallomonas papillosa* Harris and Bradley and *M. calceolus* Bradley. Nova Hedwigia **18**: 257-270.

BOURRELLY, P. 1957. Recherches sur les Chrysophycées. Morphologie, Phylogénie, Systématique. Rev. Algol. Mém. Hors. Sér. 1-412.

BRADLEY, D.E. 1964. A study of the *Mallomonas*, *Synura*, and *Chrysosphaerella* of northern Iceland. J. Gen. Microbiol. **37**: 321-333.

BRADLEY, D.E. 1966. Observations on some chrysomonads from Scotland. J. Protozool. **13**: 143-154.

BRUGEROLLE, G. and BRICHEUX, G. 1984. Actin microfilaments are involved in scale formation of the chrysomonad cell *Synura*. Protoplasma **123**: 203-212.

CARTER, N. 1937. *Pseudomallomonas anglica*: a new British flagellate. New Phytol. **36**: 57-63.

CHARLES, D.F. and SMOL, J.P. 1988. New methods for using diatoms and chrysophytes to infer past pH of low-alkalinity lakes. Limnol. Oceanogr. **33**: 1451-1462.

CHARLES, D.F., WHITEHEAD, D.R., ENGSTROM, D.R., FRY, B.D., HITES, R.A., NORTON, S.A., OWEN, J.S., ROLL, L.A., SCHINDLER, S.C., SMOL, J.P., UUTALA, A.J., WHITE, J.R. and WISE, R.J. 1987. Paleolimnological evidence for recent acidification of Big Moose Lake, Adirondack Mountains, N.Y. (U.S.A.). Biogeochemistry **3**; 267-296.

CHRISTIE, C.E. and SMOL, J.P. 1986. Recent and long-term acidification of Upper Wallface Pond (N.Y.) as indicated by mallomonadacean microfossils. Paleolimnology IV: Proceedings of the Fourth International Symposium on Paleolimnology. Hydrobiologia **143**: 355-360.

CHRISTIE, C.E., SMOL, J.P., HUTTUNEN, P. and MERILÄINEN, J. 1988. Chrysophyte scales recorded in lake sediments from eastern Finland. Hydrobiologia **161**: 237-243.

CONRAD, W. 1927. Essai d'une monographie des genres *Mallomonas* Perty (1852) et *Pseudomallomonas* Chodat (1920). Arch. Protistenk. **59**: 423-505.

CRONBERG, G. 1973. Development of cysts in *Mallomonas eoa* examined by scanning electron microscopy. Hydrobiologia **43**: 29-38.

CRONBERG, G. 1975. *Mallomonas trummensis* nov. spec. (Chrysophyceae) studied by means of scanning and transmission electron microscopy. Bot. Not. **128**: 69-72.

CRONBERG, G. 1980. Cyst development in different species of *Mallomonas* (Chrysophyceae) studied by scanning electron microscopy. Arch. Hydrobiol. Suppl. **56**: 421-434.

CRONBERG, G. 1982. Phytoplankton changes in Lake Trummen induced by restoration. Folia Limnol. Scand. **18**: 1-119.

CRONBERG, G. 1986. Chrysophycean cysts and scales in lake sediments: a review. In Chrysophytes: Aspects and Problems. Edited by J. Kristiansen and R.A. Andersen. Cambridge University Press, Cambridge. pp. 281-315.

CRONBERG, G. 1988. Variability in size and ultrastructure of the statospore of *Mallomonas caudata*. Hydrobiologia **161**: 31-39.

CRONBERG, G. 1989. Stomatocysts of *Mallomonas hamata* and *M. heterospina* (Mallomonadaceae, Synurophyceae) from South Swedish lakes. Nord. J. Bot. **8**: 683-692.

CRONBERG, G. and KRISTIANSEN, J. 1980. Synuraceae and other Chrysophyceae from central Småland, Sweden. Bot. Not. **133**: 595-618.

CRONBERG, G. and HICKEL, B. 1985. *Mallomonas fenestrata* sp. nov. and *M. perforata* sp. nov. (Chrysophyceae, Mallomonadaceae) from tropical lakes. Nord. J. Bot. **5**: 105-110.

CROOME, R.L. and TYLER, P.A. 1983. *Mallomonopsis tasmanica* sp. nov. (Chrysophyceae) and *Mallomonopsis elliptica* Matvienko from Australia. Br. Phycol. J. **18**: 357-365.

CROOME, R.L. and TYLER, P.A. 1986. *Mallomonas sabulosa* (Chrysophyceae), a new species from Australia. Br. Phycol. J. **21**: 93-96.

DIXIT, S.S. 1986. Algal microfossils and geochemical reconstructions of Sudbury lakes: A test of the paleo-indicator potential of diatoms and chrysophytes. Ph.D. Thesis, Queen's University, Kingston, Ontario.

DIXIT, S.S., DIXIT, A.S. and EVANS, R.D. 1988a. Scaled chrysophytes (Chrysophyceae) as indicators of pH in Sudbury, Ontario, lakes. Can. J. Fish. Aquat. Sci. **45**: 1411-1421.

DIXIT, S.S., DIXIT, A.S. and EVANS, R.D. 1988b. Chrysophyte scales in lake sediments provide evidence of recent acidification in two Quebec (Canada) lakes. Water Air Soil Pollut. **38**: 97-104.

DIXIT, S.S., DIXIT, A.S. and EVANS, R.D. 1989. Paleolimnological evidence for trace metal sensitivity in scaled chrysophytes. Environ. Sci. & Technol. **23**: 110-115.

DONALDSON, D.R. and STEIN, J.R. 1984. Identification of planktonic Mallomonadaceae and other Chrysophyceae from selected lakes in the lower Fraser Valley, British Columbia, Canada. Can. J. Bot. **62**: 525-539.

DÜRRSCHMIDT, M. 1980. Studies on the Chrysophyceae from Rio Cruces, Prov. Valdivia, south Chile by scanning and transmission microscopy. Nova Hedwigia **33**: 353-388.

DÜRRSCHMIDT, M. 1981. *Mallomonas cristata* sp. nov. (Chrysophyceae, Synuraceae) from southern Chilean inland waters. Phycologia **20**: 298-302.

DÜRRSCHMIDT, M. 1982a. *Mallomonas parvula* sp. nov. and *Mallomonas retifera* sp. nov. (Chrysophyceae, Synuraceae) from South Chile. Can. J. Bot. **60**: 651-656.

DÜRRSCHMIDT, M. 1982b. Studies on the Chrysophyceae from southern Chilean inland waters by means of scanning and transmission electron microscopy, II. Arch. Hydrobiol. Suppl. **63**: 121-163.

DÜRRSCHMIDT, M. 1983a. A taxonomic study of the *Mallomonas mangofera* group (Mallomonadaceae, Chrysophyceae), including the description of four new taxa. Plant Syst. Evol. **143**: 175-196.

DÜRRSCHMIDT, M. 1983b. Three new species of *Mallomonas* (Chrysophyceae, Mallomonadaceae) from Lake Lanalhue, Chile. Nord. J. Bot. **3**: 423-430.

DÜRRSCHMIDT, M. 1983c. New taxa of the genus *Mallomonas* (Mallomonadaceae, Chrysophyceae) from southern Chile. Nova Hedwigia **38**: 717-726.

DÜRRSCHMIDT, M. 1984. Studies on scale-bearing Chrysophyceae from the Giessen area, Federal Republic of Germany. Nord. J. Bot. **4**: 123-143.

DÜRRSCHMIDT, M. 1986. New species of the genus *Mallomonas* (Mallomonadaceae, Chrysophyceae) from New Zealand. In Chrysophytes: Aspects and Problems. Edited by J. Kristiansen and R.A. Andersen. Cambridge University Press, Cambridge. pp. 87-106.

DÜRRSCHMIDT, M. and CROOME, R. 1985. Mallomonadaceae (Chrysophyceae) from Malaysia and Australia. Nord. J. Bot. **5**: 285-298.

ELORANTA, P. 1985. Notes on the scaled chrysophytes (Synuraceae, Chrysophyceae) in small lakes in and near Salamajarvi National Park, western Finland. Acta Soc. Fauna Flora Fenn. **62**: 41-57.

ELORANTA, P. 1986a. Phytoplankton structure in different lake types in central Finland. Holarctic Ecology **9**: 214-224.

ELORANTA, P. 1986b. The phytoplankton of some subarctic lakes in Finnish Lapland. Memoranda Soc. Fauna Flora Fennica **62**: 41-57.

ELORANTA, P. 1989a. Scaled chrysophytes (Chrysophyceae and Synurophyceae) from national park lakes in southern and central Finland. Nord. J. Bot. **8**: 671-681.

GIBSON, K.N., SMOL, J.P. and FORD, J. 1987. Chrysophycean microfossils provide new insight into the recent history of a naturally acidic lake (Cond Pond, New Hampshire). Can. J. Fish. Aquat. Sci. **44**: 1584-1588.

GLENK, H.O. 1956. *Mallomonas schwemmlei*, eine neue Chrysomonade aus dem Plankton eines fränkischen Teiches. Ber. Deutsch. Bot. Ges. **69**: 189-192.

GREEN, R.B. 1979. Scale-bearing Chrysophyceae from seven lakes on Vancouver Island. Syesis **12**: 156-162.

GREEN, R.B. 1980. Rare scale-bearing Chrysophyceae of Otter Lake, British Columbia. Can. J. Bot. **58**: 599-603.

GRETZ, M.R., SOMMERFELD, M.R. and WUJEK, D.E. 1979. Scaled Chrysophyceae of Arizona: a preliminary survey. J. Ariz. -Nev. Acad. Sci. **14**: 75-80.

GRETZ, M.R., WUJEK, D.E. and SOMMERFELD, M.R. 1983. Scaled Chrysophyceae of Arizona: further additions to the aquatic flora. J. Ariz. -Nev. Acad. Sci. **13**: 17-21.

GRETZ, M.R., SOMMERFELD, M.R. and WUJEK, D.E. 1985. Light and electron microscopical observations of *Mallomonas portae-ferreae* var. *reticulata* var. nov. (Chrysophyceae). Phycologia **24**: 478-481.

HARRIS, K. 1953. A contribution to our knowledge of *Mallomonas*. Bot. J. Linn. Soc. **55**: 88-102.

HARRIS, K. 1958. A study of *Mallomonas insignis* and *Mallomonas akrokomos*. J. Gen. Microbiol. **19**: 55-64.

HARRIS, K. 1966. The genus *Mallomonopsis*. J. Gen. Microbiol. **42**: 175-184.

HARRIS, K. 1967. Variability in *Mallomonas*. J. Gen. Microbiol. **46**: 185-191.

HARRIS, K. 1970. Imperfect forms and the taxonomy of *Mallomonas*. J. Gen. Microbiol. **61**: 73-76.

HARRIS, K. 1970. Species of the Torquata group of *Mallomonas*. J. Gen. Microbiol. **61**: 77-80.

HARRIS, K. and BRADLEY, D.E. 1957. An examination of the scales and bristles of *Mallomonas* in the electron microscope using carbon replicas. Proc. R. Microsc. Soc. **76**: 37-46.

HARRIS, K. and BRADLEY, D.E. 1958. Some unusual Chrysophyceae studied in the electron microscope. J. Gen. Microbiol. **18**: 71-83.

HARRIS, K. and BRADLEY, D.E. 1960. A taxonomic study of *Mallomonas*. J. Gen. Microbiol. **22**: 750-777.

HARTMANN, H. and STEINBERG, C. 1987. On the variability of *Mallomonas canina* Kristiansen (Chrysophyceae). Nova Hedwigia **45**: 169-174.

HARTMANN, H. and STEINBERG, C. 1989. The occurrence of silica-scaled chrysophytes in some central European lakes and their relation to pH. Beih. Nova Hedwigia **95**: 131-158.

HAWORTH, E.Y. 1983. Diatom and chrysophyte relict assemblages in the sediments of Blelham Tarn in the English Lake District. Hydrobiologia **103**: 131-134.

HIBBERD, D.J. 1976. The ultrastructure and taxonomy of the Chrysophyceae and Prymnesiophyceae (Haptophyceae): a survey with some new observations on the ultrastructure of the Chrysophyceae. Bot. J. Linn. Soc. **72**: 55-80.

HIBBERD, D.J. 1977. Ultrastructure of cyst formation in *Ochromonas tuberculata* (Chrysophyceae). J. Phycol. **13**: 309-320.

HIBBERD, D.J. 1978. The fine structure of *Synura sphagnicola* (Korsh.) Korsh. (Chrysophyceae). Br. Phycol. J. **13**: 403-412.

HIBBERD, D.J. 1986. Ultrastructure of the Chrysophyceae – phylogenic implications and taxonomy. In Chrysophytes: Aspects and Problems. Edited by J. Kristiansen and R.A. Andersen. Cambridge University Press, Cambridge. pp. 87-106.

HUSTEDT, F. 1939. Systematische und ökologische Untersuchunger Uber die Diatomeen-Flora von Java, Bali und Sumatra nach dem Material der Deutschen Limnologischen Sunda-Expedition 3. Die Ökologischen Faktoren und ihr Einfluss auf die Diatomeen-flora. Arch. Hydrobiol. Suppl. **16**: 1-394.

HÄLLFORS, G. and HÄLLFORS, S. 1988. Records of chrysophytes with siliceous scales (Mallomonadaceae and Paraphysomonadaceae) from Finnish inland waters. Hydrobiologia **161**: 1-29.

ITO, H. 1988. Scale-bearing chrysophytes in the south basin of Lake Biwa, Japan. Jap. J. Phycol. **36**: 143-153.

JACOBSEN, B.A. 1985. Scale-bearing Chrysophyceae (Mallomonadaceae and Paraphysomonadaceae) from west Greenland. Nord. J. Bot. **5**: 381-398.

KLING, H.J. and KRISTIANSEN, J. 1983. Scale-bearing Chrysophyceae (Mallomonadaceae) from central and northern Canada. Nord. J. Bot. **3**: 269-290.

KRIEGER, W. 1930. Untersuchungen über Plankton-Chrysomonaden. Bot. Arch. **29**: 258-329.

KRISTIANSEN, J. 1975a. Chrysophyceae from Alberta and British Columbia. Syesis **8**: 97-108.

KRISTIANSEN, J. 1975b. On the occurrence of the species of *Synura* (Chrysophyceae). Int. Ver. Theor. Angew. Limnol. Verh. **19**: 2709-2715.

KRISTIANSEN, J. 1976. Studies on the Chrysophyceae of Bornholm. Bot. Tidsskr. **70**: 126-142.

KRISTIANSEN, J. 1978. Studies on the Chrysophyceae of Bornholm. II. Bot. Tidsskr. **73**: 71-85.

KRISTIANSEN, J. 1979. Problems in classification and identification of Synuraceae (Chrysophyceae). Schweiz. Z. Hydrol. **40**: 311-319.

KRISTIANSEN, J. 1981. Distribution problems in the Synuraceae (Chrysophyceae). Int. Ver. Theor. Angew. Limnol. Verh. **21**: 1444-1448.

KRISTIANSEN, J. 1982. *Mallomonas canina* sp. nov. (Chrysophyceae), a new member of sect. Heterospinae. Nord. J. Bot. **2**: 293-296.

KRISTIANSEN, J. 1985. Occurrence of scale-bearing Chrysophyceae in a eutrophic Danish lake. Int. Ver. Theor. Angew. Limnol. Verh. **22**: 2826-2829.

KRISTIANSEN, J. 1986. Silica-scale bearing chrysophytes as environmental indicators. Br. Phycol. J. **21**: 425-436.

KRISTIANSEN, J. 1988. Seasonal occurrence of silica-scaled chrysophytes under eutrophic conditions. Hydrobiologia **161**: 171-184.

KRISTIANSEN, J. 1989. Silica-scaled chrysophytes from China. Nord. J. Bot. **8**: 539-552.

KRISTIANSEN, J. and TONG, D. 1989. Studies on silica-scaled chrysophytes from Wuhan, Hangzhou and Beijing, P.R. China. Nova Hedwigia **49**: 183-202.

LEADBEATER, B.S.C. 1986. Scale-case construction in *Synura petersenii* Korsh. (Chrysophyceae). In Chrysophytes: Aspects and Problems. Edited by J. Kristiansen and R.A. Andersen. Cambridge Univ. Press, Cambridge. pp. 121-131.

LEADBEATER, B.S.C. 1990. Ultrastructure and assembly of the scale case in *Synura* (Synurophyceae Andersen). Br. phycol. J. **25**: 117-132.

MATVIENKO, A.M. 1941. A contribution to the taxonomy of the genus *Mallomonas*. Proc. Inst. Bot. Kharkov **4**: 41-47.

MIGNOT, J.P. and BRUGEROLLE, G. 1982. Scale formation in chrysomonad flagellates. J. Ultrastruct. Res. **81**: 13-26.

MOESTRUP, Ø. 1982. Flagella structure in algae: a review,

with new observations particularly on the Chrysophyceae, Phaeophyceae (Fucophyceae), Euglenophyceae, and *Reckertia*. Phycologia **21**: 427-528.

MOMEU, L. and PÉTERFI, L.S. 1979. Taxonomy of *Mallomonas* based on the fine structure of scales and bristles. Contr. Bot. Cluj-Napoca **1979**: 13-20.

MOMEU, L. and PÉTERFI, L.S. 1983. Electron microscopical study of some *Mallomonas* species. Nov. Hedw. **38**: 369-400.

NICHOLLS, K.H. 1982. *Mallomonas* species (Chrysophyceae) from Ontario, Canada including descriptions of two new species. Nova Hedwigia **34**: 80-124.

NICHOLLS, K.H. 1984. Four new *Mallomonas* species of the *Torquatae* series (Chrysophyceae). Can. J. Bot. **62**: 1583-1591.

NICHOLLS, K.H. 1987a. Form variation in *Mallomonas asmundiae* and a description of *Mallomonas sphagniphila* sp. nov. (series *Corconticae*, Mallomonadaceae). Can. J. Bot. **65**: 627-634.

NICHOLLS, K.H. 1987b. The distinction between *Mallomonas acaroides* v. *acaroides* and *Mallomonas acaroides* v. *muskokana* var. nova (Chrysophyceae). Can. J. Bot. **65**: 1779-1784.

NICHOLLS, K.H. 1987c. Variation in silica scale morphology in *Mallomonas pseudocoronata* (Chrysophyceae). Nord. J. Bot. **7**: 353-357.

NICHOLLS, K.H. 1988a. Descriptions of three new species of *Mallomonas* (Chrysophyceae). *M. hexagonis* sp. nov., *M. liturata* sp. nov., and *M. galeiformis* sp. nov. Br. Phycol. J. **23**: 159-166.

NICHOLLS, K.H. 1988b. Additions to the *Mallomonas* (Chrysophyceae) flora of Ontario, Canada, and a checklist of North American *Mallomonas* species. Can. J. Bot. **66**: 349-360.

NICHOLLS, K.H. 1988c. The identification of some *Mallomonas* species of the *M. doignonii* group (Chrysophyceae). Nord. J. Bot. **8**: 109-116.

NYGAARD, G. 1945. Dansk Planteplankton. Gyldendal, Copenhagen

NYGAARD, G. 1949. Hydrobiological studies on some Danish ponds and lakes II. Kgl. Dan. Vid. Selsk. Biol. Skr. **7**: 1-293.

NYGAARD, G. 1978. Freshwater phytoplankton from the Narssaq area. Bot. Tidsskr. **73**: 191-238.

PASCHER, A. 1913. Chrysomonadinae. In Die Süsswasserflora Deutschlands. Edited by A. Pascher. Österreichs und der Schweiz **2**: 7-95.

PÉTERFI, L.S. 1966. Studies on the Rumanian Chrysophyceae (1). Nova Hedwigia **13**: 117-137.

PÉTERFI, L.S. 1974. Chrysophyceae si Xanthophyceae din mlastinile eutrofe si de trecere de la Salicea (Cluj). Contr. Bot. Cluj-Napoca **1974**: 7-17.

PÉTERFI, L.S. and ASMUND, B. 1972. *Mallomonas portaeferreae* nova species in the light and electron microscopes. Stud. Univ. Babes-Bolyai, Ser. Biologia **17**: 11-18.

PÉTERFI, L.S. and MOMEU, L. 1976a. Romanian *Mallomonas* species studied in light and electron microscopes. Nova Hedwigia **27**: 353-392.

PÉTERFI, L.S. and MOMEU, L. 1976b. *Mallomonas transsylvanica*, spec. nova (Chrysophyceae), light and electron microscopical studies. Plant Syst. Evol. **125**: 47-57.

PÉTERFI, L.S. and MOMEU, L. 1981. *Mallomonas robusta* (Chrysophyceae, Synuraceae): electron microscopical study of silica armour. Plant Syst. Evol. **139**: 137-145.

ROIJACKERS, R.M.M. 1981. Chrysophyceae from freshwater localities near Nijmegen, The Netherlands. Hydrobiologia **76**: 179-189.

ROIJACKERS, R.M.M. and KESSELS, H. 1986. Ecological characteristics of scale-bearing Chrysophyceae from the Netherlands. Nord. J. Bot. **6**: 373-383.

SANDGREN, C.D. 1980. An ultrastructural investigation of resting cyst formation in *Dinobryon cylindricum* Imhof (Chrysophyceae, Chrysophycota). Protistologica **16**: 259-276.

SANDGREN, C.D. 1981. Characteristics of sexual and asexual resting cyst (statospore) formation in *Dinobryon cylindricum* Imhof. J. Phycol. **17**: 199-210.

SANDGREN, C.D. 1983. Morphological variability in populations of Chrysophycean cysts. I. Genetic (interclonal) and encystment temperature effects on morphology. J. Phycol. **19**: 64-70.

SANDGREN, C.D. 1989. SEM investigations of statospore (stomatocyst) development in diverse members of the Chrysophyceae and Synurophyceae. Beih. Nova Hedwigia **95**: 45-69.

SANDGREN, C.D. and CARNEY, H.J. 1983. A flora of fossil Chrysophycean cysts from the recent sediments of Frains Lake, Michigan, U.S.A. Nova Hedwigia **38**: 129-163.

SCHNEPF, E. and DEICHGRABER, G. 1969. Uber die Feinstrucktur von *Synura petersenii* unter besonderen Berucksichtigung der Morphogenese ihrer Kieselschuppen. Protoplasma **68**: 85-106.

SIVER, P.A. 1988a. *Mallomonas retrorsa*, a new species of silica-scaled Chrysophyceae with backwards orientated scales. Nord. J. Bot. **8**: 319-323.

SIVER, P.A. 1988b. Morphology and ecology of *Mallomonas galeiformis* (Chrysophyceae), a potentially useful paleolimnological indicator. Trans. Am. Microsc. Soc. **107**: 152-161.

SIVER, P.A. 1988c. Distribution of scaled chrysophytes in 17 Adirondack (New York) lakes with special reference to pH. Can. J. Bot. **66**: 1391-1403.

SIVER, P.A. 1989a. The distribution of scaled chrysophytes along a pH gradient. Can. J. Bot. **67**: 2120-2130.

SIVER, P.A. 1989b. The separation of *Mallomonas acaroides* v. *acaroides* and v. *muskokana* (Synurophyceae) along a pH gradient. Beih. Nova Hedwigia **95**: 111-117.

SIVER, P.A. 1990. A critical literature review on the usefulness of chrysophytes to detect lake chronic or episodic acidification and/or recovery in the context of a long term monitoring program. U. S. Environmental Protection Agency, Corvallis, OR: In press.

SIVER, P.A. 1991. The stomatocyst of *Mallomonas acaroides* v. *muskokana* (Chrysophyceae). J. Paleolimnology **5**: 11-17.

SIVER, P.A. and CHOCK, J.S. 1986. Phytoplankton dynamics in a chrysophycean lake. In Chrysophytes: Aspects and Problems. Edited by J. Kristiansen and R.A. Andersen. Cambridge Univ. Press, Cambridge. pp. 165-183.

SIVER, P.A. and SKOGSTAD, A. 1988. Morphological variation and ecology of *Mallomonas crassisquama* (Chrysophyceae). Nord. J. Bot. **8**: 99-107.

SIVER, P.A. and HAMER, J.S. 1989. Multivariate statistical analysis of the factors controlling the distribution of scaled chrysophytes. Limnol. Oceanogr. **34**: 368-381.

SIVER, P.A. and HAMER, J.S. 1990. Use of extant populations of scaled chrysophytes for the inference of lakewater pH. Can. J. Fish. Aquat. Sci. **47**: 1339-1347.

SIVER, P.A. and GLEW, J.R. 1990. The arrangement of scales and bristles on *Mallomonas* (Chrysophyceae): a proposed mechanism for the formation of the cell covering. Can. J. Bot. **68**: 374-380.

SIVER, P.A., HAMER, J.S. and KLING, H. 1990. The separation of *Mallomonas duerrschmidtiae* sp. nov. from *M. crassisquama* and *M. pseudocoronata*: implications for paleolimnological research. J. Phycol. **26**: 728-740.

SIVER, P.A. and SMOL, J.P. 1991. The use of scaled chrysophytes in long term monitoring programs for the detection of changes in lakewater acidity. Water, Air and Soil Pollution : In press.

SKOGSTAD, A. 1982. Synuraceae-floraen i 27 lokaliteter i Oslo-området. Hovedfagsoppgave. Ph.D. Thesis, University of Oslo, Norway.

SKOGSTAD, A. 1984. Vegetative cells and cysts of *Mallomonas intermedia* (Mallomonadaceae, Chrysophyceae). Nord. J. Bot. **4**: 275-278.

SKUJA, H. 1950. Körperbau und reproduktion bei *Dinobryon borgei* Lemm. Svensk Bot. Tidskr. **44**: 96-107.

SMOL, J.P. 1980. Fossil synuracean (Chrysophyceae) scales in lake sediments: a new group of paleoindicators. Can. J. Bot. **58**: 458-465.

SMOL, J.P. 1984. The statospore of *Mallomonas pseudocoronata* (Mallomonadaceae, Chrysophyceae). Nord. J. Bot. **4**: 827-831.

SMOL, J.P. 1986. Chrysophycean microfossils as indicators of lakewater pH. In Diatoms and Lake Acidity. Edited by J.P. Smol, R.W. Battarbee, R.B. Davis and J. Meriläinen. Dr. W. Junk, Dordrecht, The Netherlands. pp. 275-287.

SMOL, J.P. 1988. The North American 'endemic' *Mallomonas pseudocoronata* (Mallomonadaceae, Chrysophyta) in an Austrian lake. Phycologia **27**: 427-429.

SMOL, J.P. 1991. Paleolimnology – recent advances and future challenges. In Scientific Perspectives in Theoretical and Applied Limnology. Edited by R. De Bernardi. In press.

SMOL, J.P., BROWN, S.R. and MCNEELY, R.N. 1983. Cultural disturbances and trophic history of a small meromictic lake from central Canada. Hydrobiologia **103**: 125-130.

SMOL, J.P., CHARLES, D.F. and WHITEHEAD, D.R. 1984a. Mallomonadacean (Chrysophyceae) assemblages and their relationships with limnological characteristics in 38 Adirondack (New York) lakes. Can. J. Bot. **62(5)**: 911-923.

SMOL, J.P., CHARLES, D.F. and WHITEHEAD, D.R. 1984b. Mallomonadacean microfossils provide evidence of recent lake acidification. Nature **307**: 628-630.

SMOL, J.P. and M. BOUCHERLE. 1985. Postglacial changes in algal and cladoceran assemblages in Little Round Lake, Ontario. Arch. Hydrobiol. **103**: 25-49.

STARMACH, K. 1980. Chrysophyceae-Zlotowiciowce. Flora Slodkowodna Polski **5**: 1-775.

STARMACH, K. 1985. Chrysophyceae und Haptophyceae. In Süsswasserflora von Mitteleuropa 1. Edited by H. Ettl, J. Gerloff, H. Heynig and D. Mollenhauer. VEB Gustav Fischer, Jena. 515 pages.

TAKAHASHI, E. 1959. Studies on genera *Mallomonas* and *Synura*, and other plankton in fresh-water by electron microscope. I. Bull. Yamagata Univ. (Agri. Sci.) 3, **1**: 117-151.

TAKAHASHI, E. 1978. Electron microscopical studies of the Synuraceae (Chrysophyceae) in Japan, taxonomy and ecology. Tokai University Press, Tokyo.

TAKAHASHI, E. and HAYAKAWA, T. 1979. The Synuraceae (Chrysophyceae) in Bangladesh. Phykos **18**: 129-147.

TAYLOR, W.D., WEE, J.L. and WETZEL, R.G. 1986. A modification of the Utermohl sedimentation technique for improved identification and cell enumeration of diatoms and silica-scaled Chrysophyceae. Trans. Am. Microsc. Soc. **105**: 68-72.

TOLONEN, K., LIUKKONEN, M., HARJULA, R., and PÄTILÄ, A. 1986. Acidification of small lakes in Finland documented by sedimentary diatom and chrysophycean remains. In Diatoms and Lake Acidity. Edited by J.P. Smol, R.W. Battarbee, R.B. Davis and J. Meriläinen. Dr. W. Junk, Dordrecht, The Netherlands. pp. 169-199.

VOLLENWEIDER, R.A., editor. 1969, A manual on methods for measuring primary production in aquatic environments. Int. Biol. Program Handbook 12. Oxford, Blackwell Scientific Publications, 213 pages.

WAWRIK, F. 1979. Eisschluss- und Eisbruchvegetationen in den Teichen des nördlichen Waldviertels 1977/1978. Arch. Protistenk. **122**: 247-266.

WAWRZYNIAK, L.A. and ANDERSEN, R.A. 1985. Silica-scaled Chrysophyceae from North America boreal forest regions in northern Michigan, U.S.A. and Newfoundland, Canada. Nova Hedwigia **5**: 399-401.

WEE, J.L. 1982. Studies on the Synuraceae (Chrysophyceae) of Iowa. Bibl. Phycol. **62**: 1-183.

WEE, J.L., Wujek, D.E. and Graebner, M.P. 1982. Studies on Michigan Chrysophyceae. V. Mich. Bot. **21**: 181-184.

WEE, J.L. and GABEL, M. 1989. Occurrences of silica-scaled chromophyte algae in predominantly alkaline lakes and ponds in Iowa. Am. Midl. Nat. **121**: 32-40.

WEIMANN, R. 1933. Hydrobiologische und hydrographische Untersuchungen an zwei teichartigen Gewässern. Beih. Bot. Centralb. **51**: 397-476.

WETZEL, R.G. 1975. Limnology. W.B. Saunders Co. N.Y. 753 pages.

WHITFORD, L.A. and SCHUMACHER, G.J. 1969. A manual of the fresh-water algae in North Carolina. North Carolina Agric. Exp. Stat., Tech. Bul. No. **188**. 313 pages.

WUJEK, D.E. 1984. Chrysophyceae (Mallomonadaceae) from Florida. Florida Sci. **47**: 161-170.

WUJEK, D.E. 1986. Scale bearing Chrysophyceae from the Panama Canal. Jap. J. Phycol. **34**: 83-86.

WUJEK, D.E. and HAMILTON, R. 1973. Studies on Michigan Chrysophyceae. II. Mich. Bot. **12**: 118-122.

WUJEK, D.E., HAMILTON, R. and WEE, J. 1975. Studies on Michigan Chrysophyceae. III. Mich. Bot. **14**: 91-94.

WUJEK, D.E., GRETZ, M. and WUJEK, M.G. 1977. Studies on Michigan Chrysophyceae. IV. Mich. Bot. **16**: 191-194.

WUJEK, D.E. and KRISTIANSEN, J. 1978. Observations on bristle- and scale-production in *Mallomonas caudata* (Chry-

sophyceae). Arch. Protistenkd. **120**: 213-221.
WUJEK, D.E. and ASMUND, B.C. 1979. *Mallomonas cyathellata* sp. nov. and *Mallomonas cyathellata* var. *kenyana* var. nov. (Chrysophyceae) studied by means of scanning and transmission electron microscopy. Phycologia **18**: 115-119.
WUJEK, D.E., WEIS, M.M. and ANDERSEN, R.A. 1981. Scaled Chrysophyceae from Lake Itasca region. I. *Mallomonas*. J. Minn. Acad. Sci. **47**: 22-24.
WUJEK, D.E. and TIMPANO, P. 1984. The genus *Mallomonopsis* in the United States. Trans. Kans. Acad. Sci. **87**: 73-82.
WUJEK, D.E. and WEIS, M.M. 1984. Mallomonadaceae from Kansas. Trans. Kans. Acad. Sci. **87**: 26-31.

Index

GPSR Compliance

The European Union's (EU) General Product Safety Regulation (GPSR) is a set of rules that requires consumer products to be safe and our obligations to ensure this.

If you have any concerns about our products, you can contact us on ProductSafety@springernature.com

In case Publisher is established outside the EU, the EU authorized representative is:

Springer Nature Customer Service Center GmbH
Europaplatz 3
69115 Heidelberg, Germany

Batch number: 10809836

Printed by Printforce, the Netherlands